AF361900

Pasteur, Toxins/Pathogenicity, Anti-toxins, a Bicentennial Contribution

Pasteur, Toxins/Pathogenicity, Anti-toxins, a Bicentennial Contribution

Guest Editors

Michel R. Popoff
Daniel Ladant

Basel • Beijing • Wuhan • Barcelona • Belgrade • Novi Sad • Cluj • Manchester

Guest Editors

Michel R. Popoff
Unité des Toxines Bactériennes
Institut Pasteur
Université Paris Cité
Paris
France

Daniel Ladant
Unité "Biochimie des Interactions
Macromoléculaires"
Institut Pasteur
Université Paris Cité
Paris
France

Editorial Office
MDPI AG
Grosspeteranlage 5
4052 Basel, Switzerland

This is a reprint of the Special Issue, published open access by the journal *Toxins* (ISSN 2072-6651), freely accessible at: https://www.mdpi.com/journal/toxins/special_issues/Pasteur_Toxins_Pathogenicity_AntiToxins_a_Bicentennial_Contribution.

For citation purposes, cite each article independently as indicated on the article page online and as indicated below:

Lastname, A.A.; Lastname, B.B. Article Title. *Journal Name* **Year**, *Volume Number*, Page Range.

ISBN 978-3-7258-3963-6 (Hbk)
ISBN 978-3-7258-3964-3 (PDF)
https://doi.org/10.3390/books978-3-7258-3964-3

Cover image courtesy of Institut Pasteur/Musée Pasteur - photo Nadar

Contents

Preface

In the 17th century, using an implemented powerful microscope, Dutch Antonie van Leeuwenhoek showed the first observations of the small elements of life, cells and microorganisms, which were called animalcules. Antonie van Leeuwenhoek is considered "the Father of the Microbiology". Two centuries later, L. Pasteur, R. Koch, and other pioneers launched the scientific basis of microbiology and developed two main concepts. First, microorganisms do not proceed from spontaneous generation but by contamination and parental transmission; second, specific microorganisms are responsible for specific diseases in humans and animals. This has led to the development of specific countermeasures through vaccination with attenuated microorganisms. The success of the first human vaccination in preventing rabies, a terrifying fatal disease considered incurable, was the impetus behind the creation of the Institut Pasteur in Paris in 1888, an institution dedicated to disease prevention, research, and training. The following year, Emile Roux created the world's first regular course of microbiology, called the "Cours de Microbie Technique or Cours de Monsieur Roux", at the Institut Pasteur. Then, the question arose as to how bacterial pathogens induce their pathological effects. By comparison with certain plants and animals which produce toxic substances, it was hypothesized that microorganisms could also synthesize toxic compounds that were called toxins. Numerous Pasteurians investigated various bacterial and animal toxins and developed the concept of vaccination with toxoids, non-toxic derivative of toxins, as well as treatment with immune sera. This Special Issue retraces the history of the Pasteurians and the Institut Pasteur's activities related to animal and bacterial toxins (see Editorial).

In addition to the articles listed in the Editorial, a supplemental article is included, related to an investigation about a putative involvement of *Clostridium perfringens* epsilon toxin (ETX) in multiple sclerosis (MS) in humans. In this study, Gougeon et al. found a high prevalence of antibodies against ETX in both MS patients as well as in healthy controls, arguing against ETX as causative agent of MS and instead suggesting that ETX is a mimetic of an autoantigen (contribution 10).

Michel R. Popoff and Daniel Ladant
Guest Editors

Editorial

Toxins, Pathogenicity, Anti-Toxins, a Bicentennial Contribution

Michel R. Popoff [1,*] and Daniel Ladant [2]

[1] Unité des Toxines Bactériennes, Institut Pasteur, Université Paris Cité, CNRS UMR 2001 INSERM U1306, F-75015 Paris, France

[2] Unité de Biochimie des Interactions Macromoléculaires, Institut Pasteur, Université Paris Cité, CNRS UMR 3528, F-75015 Paris, France; daniel.ladant@pasteur.fr

* Correspondence: popoff2m@gmail.com

Citation: Popoff, M.R.; Ladant, D. Toxins, Pathogenicity, Anti-Toxins, a Bicentennial Contribution. *Toxins* **2024**, *16*, 97. https://doi.org/10.3390/toxins16020097

Received: 30 January 2024
Accepted: 8 February 2024
Published: 10 February 2024

The bicentenary of Louis Pasteur's birth raises the opportunity to revisit the activity and influence of L. Pasteur and collaborators in the field of toxins. Microorganisms have been observed since the 17th century, but L. Pasteur clearly demonstrated their main properties, namely that they are living organisms that disseminate by contamination from an infected site to a sterile one and not by spontaneous generation, and that they induce substrate modification such as fermentation. Thus, he defined the basis of a scientific approach of the then nascent scientific domain, microbiology. L. Pasteur and other pioneers in this emerging field, in particular Robert Koch, discovered that certain microorganisms are responsible for specific diseases. The main objective of L. Pasteur was the prevention of infectious diseases, and he developed the concept of prevention with attenuated microorganisms; he called this concept vaccination in honor of Edward Jenner. The success of prevention of rabies pushed the building of the Institut Pasteur in Paris. This institute was devoted to the treatment of rabies, research on infectious diseases, and training on microorganisms. The first microbiology course in the world, called "microbie technique", was taught by Emile Roux (1888), who succeeded to L. Pasteur and Emile Duclaux as Director of the Institut Pasteur. Albeit L. Pasteur knew of the existence of toxic soluble factors produced by putrefying bacteria; he was more convinced that pathogenic microorganisms act by depletion of vital substrates [1,2]. Scientists of the Institut Pasteur characterized various bacterial and animal toxins. This Special Issue retraces the main steps of the involvement of scientists of the Institut Pasteur in bacterial and animal toxins.

The history of the discovery of bacterial toxins with the contribution of Pasteurians is detailed in the article by Jean Marc Cavaillon (Contribution 1). Putrefying bacteria were the first to be recognized as producing potent lethal poisons. This was observed by the Danish Peter Ludvig Panum in the middle of the 19th century. The term toxin was introduced by the German professor of medicine Ludwig Brieger who characterized several toxic compounds from putrefaction. In the end of 19th century, Elie Metchnikoff and other Pasteurians in the Institut Pasteur of Paris investigated the poisons produced by putrefying bacteria. One of the first bacterial toxins produced by a pathogenic bacterium was the diphtheria toxin. Alexandre Yersin and E. Roux showed that culture filtrates of *Corynebacterium diphtheria* are lethal for experimental animals and concluded that this pathogen produces a toxin responsible for the symptoms of disease and death. In addition, they found that sublethal injections of culture filtrates induced protection against virulent *C. diphtheria*, supporting the idea that a soluble substance elaborated by the pathogen, and not the entire microbe as previously used for fowl cholera and anthrax vaccination by L. Pasteur and collaborators, is sufficient to induce a protective immunity. This was called "chemical vaccination". A key step in the preparation of antigens was achieved by Gaston Ramon, a French veterinarian who was in charge of producing horse antisera at the Institut Pasteur. He found that diphtheria toxins treated with formalin were no longer toxic but retained their antigenicity. This was successfully applied to tetanus toxins, and then to other toxins. These toxin antigens were designated anatoxins. G. Ramon also developed the concept of adjuvants of immunity. The contributions of E. Metchnikoff in the discovery of cholera toxin and cholera protection, as well as the deciphering of endotoxins by Richard Pfeifer, Alexandre Besredka, and André Boivin, are introduced in this work. The complementarity between French and German scientists at the dawn of microbiology and immunology closes this historical manuscript.

Investigations into anthrax and its protection were a major milestone in the development of the nascent microbiology and for the notoriety of L. Pasteur. Albeit the causative agent of anthrax, *Bacillus anthracis*, was previously observed by Rayer and Davaine, R. Koch validated the concept that each infectious disease is caused by a specific pathogen. L. Pasteur, in the famous vaccination assay against anthrax at Pouilly Le Fort, acquired a great reputation in promising a fight against infectious diseases. The historical aspects about *B. anthracis* as well as the recent research activities in this field including structural analysis, biochemistry, genetics, bacterial host–cell interactions, in vivo pathogenicity, and therapeutic developments carried out at the Institut Pasteur are reviewed by P. Goossens (Contribution 2).

The article of Camille Locht (Contribution 3) reports the contribution of Pasteurians in the characterization of *Bordetella pertussis* toxins. The agent of whooping cough, *B. pertussis*, was identified and isolated by Jules Bordet and Octave Gengou at the Institut Pasteur of Paris and Bruxelles at the beginning of the 20th century. It appears soon that whooping cough is due to toxic compounds produced by this microorganism. The discovery and characterization of the major toxins, lipo-oligosaccharide, adenylyl cyclase (ACT), and pertussis toxin (PTX) are described. ACT was extensively studied by Pasteurians regarding the regulation of its synthesis, structure-function, and mode of action. ACT binds to calmodulin in a Ca^{++}-dependent manner prior to exerting its toxic and hemolytic activity. Notably, ACT induces apoptosis of macrophages. The unique properties of ACT led to multiple applications such as ACT use as a vehicle to deliver epitopes to immune cells, targeting of dendritic cells, and induction of cytotoxic and humoral immune responses, as well as construction of a two-hybrid system based on the two complementary ACT fragments. PTX is a major virulence factor of *B. pertussis* for which scientists of the Institut Pasteur of Lille largely contributed to deciphering the structure/function and mode of action. PTX retains a complex structure constituted of an enzymatic subunit and a hexamer of four distinct cell receptor binding subunits. Better understanding of the enzymatic mechanism (ADP ribosylation) led to the development of novel vaccines: an acellular vaccine based on genetically inactivated PTX and a live attenuated *B. pertussis* vaccine strain which confers the advantage to induce a mucosal immunity. The live vaccine is under clinical development (Contribution 1).

Some toxigenic bacteria are responsible for respiratory diseases and have been investigated by several Pasteurians such as Lhousseine Touqui, Michel Chignard and collaborators (Contribution 4). A host defense mechanism consists of the secretion of phospholipase A2 (PLA2). Albeit secreted PLA2, notably sPLA2-IIA, have a proinflammatory role, they exert a major bactericidal function, especially against Gram-positive bacteria. Alveolar macrophages are one of the main sources of sPLA2-IIA and the expression of this enzyme can be modulated by various bacterial factors including bacterial toxins. The article of L. Touqui and collaborators reviews the modulation of sPLA2-IIA in airway infection and their pathophysiological consequences (Contribution 4). Lipopolysaccharide from Gram-negative bacteria and exotoxin S of *Pseudomonas* upregulate the expression of sPLA2-IIA, while other bacterial toxins such as *Bacillus anthracis* oedema and lethal toxins are downregulators of this enzyme. The role of other bacterial factors on sPLA2-IIA expression and consequences are described.

Mycobacteria were another important topic investigated by Pasteurians. Indeed, it is noteworthy to remind that Albert Calmette and Camille Guérin at the Institut Pasteur of Lille developed the BCG (Bacille Calmette Guérin) for the prevention of human tuberculosis. Among the mycobacterial virulence factors, a toxin, called mycolactone, has been identified in the pathogenicity of *Mycobacterium ulcerans*, which is responsible for Buruli ulcers, a severe human skin disease. The team of Caroline Demangel further investigates the activities of mycolactone, which are reviewed in (Contribution 5). Mycolactone is a polyketide with cytotoxic and immunosuppressive effects. Notably, mycolactone impairs the expression of cytokines and membrane receptors in multiple immune cells. C. Demangel and collaborators contributed to the deciphering of the mode of action of mycolactone. This toxin blocks the Sec61 translocon in an inactive conformation, thus inhibiting protein secretion. In addition to being a virulence factor for *M. ulcerans*, by preventing the host immune responses, mycolactone is a potent immunosuppressor with promising therapeutic applications such as in the treatment of inflammatory diseases (skin inflammation, rheumatoid arthritis, inflammatory pain) as well as in oncology and virology. A shorter synthetic molecule (mini-mycolactone), which is easier to produce than mycolactone and retains anti-inflammatory properties, is under development.

An important bacterial phylum concerns the cyanobacteria which possess the unique property to perform oxygenic photosynthesis. Cyanobacteria are very widespread on the surface of the Earth and prosper in extremely diverse ecological sites contributing to the oxygenation of the atmosphere. They show a wide diversity regarding their morphology, physiology, and genome. Some species produce toxins which are a threat to animals and humans. The review of Muriel Gugger (Contribution 6) summarizes the historical aspects of the cyanobacteria study at the Institut Pasteur and the current activity of the Collection of Cyanobacteria (Contribution 6). It was initiated by Roger Stanier who was professor of microbiology at the University of Berkeley, California. In 1960–1961, he visited the laboratory of Jacques Monod at the Institut Pasteur of Paris. In 1971, he was appointed Director of the Microbial Physiology Unit at the Institut Pasteur until his death in 1982. His numerous works focused on fundamental aspects of microbiology such as comparative biochemistry and evolution of microorganisms including cyanobacteria for which he started the collection of the Institut Pasteur. Then, the successors of R. Stanier and collaborators enriched, and further documented the cyanobacteria collection, which is a reference collection for these microorganisms. A great effort has been made in strain purification and in maintaining alive these bacteria. Genome sequencing allowed to investigate the diversity and phylogeny of the cyanobacteria phylum, as well as to characterize gene clusters involved in the synthesis of cyanotoxins and other natural products. Indeed, cyanobacteria produce numerous unique natural products which can be useful for pharmacological and/or biotechnological applications and developments.

L. Pasteur discovered life without oxygen, and with Jules Joubert he identified the first anaerobic bacterium responsible for disease, a septicemia in sheep. Subsequently the anaerobic bacteriology was developed at the Institut Pasteur the historical aspects of which are reported in the review by M. Popoff and S. Legout (Contribution 7) of this Special Issue. Adrien Veilon amended the method of anaerobic culture and found that many anaerobic bacteria are saprophytes of the oral cavity leading to the distinction of endogenous and exogenous microflora. Severe diseases such as tetanus and botulism were found to be caused by potent toxins produced by specific anaerobic bacteria. Gas gangrenes, which were widespread harmful diseases, were also recognized to be transmitted by other anaerobic bacteria. The production of therapeutic antisera against diseases due to anaerobic bacteria was one of the main activities at the Institut Pasteur of Paris. They took advantage of the improvement of antigen preparation by G. Ramon. Characterization of anaerobes and their toxins was undertaken by Pasteurians. Among them, André-Romain Prévot had a central role in the development of anaerobic bacteriology, notably by investigating the physiology and taxonomy of anaerobes. In the most recent period, important Pasteurian's contributions, particularly in the team of M. Popoff, have been brought on clostridial toxins including gene sequencing, regulation of synthesis, mode of action, structure/function, and entry into cells.

Among anaerobic bacteria, *Clostridioides difficile* (previously *Clostridium difficile*) has a special place since it is responsible for most nosocomial diarrheas with epidemic evolution in hospitals. Since the emergence of this pathogen in the 1980s, the Institut Pasteur was involved in the identification of *C. difficile* infections and characterization of the causative agent. The team of Bruno Dupuy was interested in the deciphering of the regulation of toxin synthesis which is an important step in the onset of this toxin mediated disease (Contribution 8). The article by B. Dupuy summarizes the major findings in the regulation of toxin synthesis in *C. difficile*. The genes of the major toxins, toxin A (TcdA) and toxin B (TcdB), are located in the chromosomal pathogenicity locus (PaLoc), which also contains two key regulatory genes encoding an alternative sigma factor (TcdR) and an anti-sigma factor (TcdC). Their role was elucidated by B. Dupuy, Linc Sonenshein and collaborators. TcdR is required for the specific transcription of *tcdA* and *tcdB*, while TcdC is a negative regulator. In addition, *C. difficile* toxin synthesis is under the control of a complex regulatory network including general and specific metabolic regulators as well as regulatory loci controlling sporulation, motility, flagellar synthesis, quorum sensing, and SOS response. A comprehensive review of the regulatory genes which have been identified to date, is presented.

Animal toxins are another toxinology field where Pasteurians made a key contribution as reviewed in the article of M. Popoff and colleagues (Contribution 9). The main initial objective was to treat and prevent the envenomations that were a serious threat in certain countries. It was hypothesized that methods used for bacterial toxins could be applied to animal toxins. Cesaire Phisalix, Gabriel Bertrand, and Albert Calmette were pioneers in antivenom serotherapy. They showed that attenuated venoms induce a protective immune response and that hyperimmune sera are efficient therapeutic drugs. Then, Gaston Ramon improved the efficiency of antivenom preparations by using formalin treatment which was successfully developed for bacterial toxins. Thus, the Institut Pasteur of Paris was involved in massive production and distribution of antivenom sera, mainly to Asia and Africa, notably through the international network of the Institut Pasteur. Concomitant characterization of venom composition and mode of action were investigated by Pasteurians. Paul Boquet demonstrated that certain toxins are present in venoms from multiple snake species. His objective was to select the most appropriate antigens to obtain wide polyvalent antisera. P. Boquet investigated the coagulation activity and phospholipases of venoms, and with France Tazieff-Depierre, he analyzed the venom neurotoxic activity. F. Tazieff-Depierre discovered that certain snake toxins bind to acetylcholine receptor and induce paralysis, whereas other toxins from scorpion and sea anemones cause excessive release of acetylcholine and muscle contracture. Cassian Bon who started to work on scorpion venom with F. Tazieff-Depierre, pursued the investigation of neurotoxic activity and phospholipases in snake and scorpion toxins with the perspective to develop therapeutic drugs. More recently, Grazyna Faure continued the exploration of animal phospholipases A2. She notably discovered that a snake toxin modulates certain ion channels opening the way to the development of novel anti-cystic fibrosis agents. In addition to Pasteurians, many French scientists contributed to the exploration of animal toxins.

In the footsteps of L. Pasteur in the deciphering of the origin and transmission of infectious diseases, Pasteurians investigated the mechanisms of pathogenicity. Bacterial and animal toxins were among the first virulence factors that have been identified. Initial concerns were the prevention and treatment of the corresponding diseases and envenomations. Thus, specific toxin-based tools and processes, vaccination and serotherapy, have been developed. As innovative technologies progressed, more and more in-depth characterization of toxins was performed including biochemistry, genetics, structure aspects, enzymatic and cellular activities. Toxins appeared to be not only virulence factors, but also relevant tools to explore physiological cell processes and to develop novel and efficient therapeutics. This Special Issue illustrates the historical aspects of the major Pasteurian contributions to the fascinating story of toxins.

Besides the articles of this Special Issue on Pasteurians and toxins, additional contributions of Pasteurians concern the study of listeriolysins. Indeed, the laboratory of J. Alouf was focused on hemolysins, notably the cholesterol-dependent cytolysin family including listeriolysins. The role of listeriolysins in bacterial invasion was further investigated by the team of Pascal Cossart. The pore-forming toxin listeriolysin O (LLO) is required for *Listeria* escape from phagosome to host cell cytoplasm by disrupting vacuolar membrane [3]. LLO exhibits additional effects contributing to bacterial invasion of cells, notably by remodeling cell surface glycoproteins such as LAMPs (late endosomal membrane protein [4]. Javier Pizarro-Cerda and collaborators demonstrated that the listeriolysin S produced by hypervirulent *Listeria monocytogenes* strains is a membrane associated bacteriocin targeting. Listeriolysin S alters the host gut microbiota and thus facilitates the persistence of *L. monocytogenes* in the intestine and subsequent infection [5,6].

Author Contributions: M.R.P. and D.L. wrote and revised the manuscript. All authors have read and agreed to the published version of the manuscript.

Funding: This editorial received no external funding.

Conflicts of Interest: The authors declare no conflict of interest.

List of Contributions

1. Cavaillon, J.M. From bacterial poisons to toxins: The early works of Pasteurians. *Toxins* **2022**, *14*, 759.
2. Goossens, P.L. *Bacillus anthracis*, "La Maladie du Charbon", Toxins and Institut Pasteur. *Toxins* **2023**, *in press*.
3. Locht, C. Pasteurian Contributions to the Study of Bordetella pertussis Toxins. *Toxins* **2023**, *15*, 176. https://doi.org/110.3390/toxins15030176.
4. Wu, Y.; Pernet, E.; Touqui, L. Modulation of Airway Expression of the Host Bactericidal Enzyme, sPLA2-IIA, by Bacterial Toxins. *Toxins* **2023**, *15*, 440. https://doi.org/410.3390/toxins15070440.
5. Ricci, D.; Demangel, C. From Bacterial Toxin to Therapeutic Agent: The Unexpected Fate of Mycolactone. *Toxins* **2023**, *15*, 369. https://doi.org/310.3390/toxins15060369.
6. Gugger, M.; Boullié, A.; Laurent, T. Cyanotoxins and Other Bioactive Compounds from the Pasteur Cultures of Cyanobacteria (PCC). *Toxins* **2023**, *15*, 388. https://doi.org/310.3390/toxins15060388.
7. Popoff, M.R.; Legout, S. Anaerobes and Toxins, a Tradition of the Institut Pasteur. *Toxins* **2023**, *15*, 43. https://doi.org/10.3390/toxins15010043.
8. Dupuy, B. Regulation of Clostridial Toxin Gene Expression: A Pasteurian Tradition. *Toxins* **2023**, *15*, 413. https://doi.org/410.3390/toxins15070413.
9. Popoff, M.R.; Faure, G.; Legout, S.; Ladant, D. Animal Toxins: A Historical Outlook at the Institut Pasteur of Paris. *Toxins* **2023**, *15*, 462. https://doi.org/410.3390/toxins15070462.

References

1. Cavaillon, J.M. Historical links between toxinology and immunology. *Pathog Dis.* **2018**, *73*, 4923027. [CrossRef] [PubMed]
2. Cavaillon, J.M.; Legout, S. Louis Pasteur: Between Myth and Reality. *Biomolecules* **2022**, *12*, 596. [CrossRef] [PubMed]
3. Hamon, M.A.; Ribet, D.; Stavru, F.; Cossart, P. Listeriolysin O: The Swiss army knife of Listeria. *Trends Microbiol.* **2012**, *20*, 360–368. [CrossRef] [PubMed]
4. Kühbacher, A.; Novy, K.; Quereda, J.J.; Sachse, M.; Moya-Nilges, M.; Wollscheid, B.; Cossart, P.; Pizarro-Cerdá, J. Listeriolysin O-dependent host surfaceome remodeling modulates Listeria monocytogenes invasion. *Pathog. Dis.* **2018**, *76*, fty082. [CrossRef] [PubMed]
5. Quereda, J.J.; Dussurget, O.; Nahori, M.A.; Ghozlane, A.; Volant, S.; Dillies, M.A.; Regnault, B.; Kennedy, S.; Mondot, S.; Villoing, B.; et al. Bacteriocin from epidemic Listeria strains alters the host intestinal microbiota to favor infection. *Proc. Natl. Acad. Sci. USA* **2016**, *113*, 5706–5711. [CrossRef] [PubMed]
6. Quereda, J.J.; Nahori, M.A.; Meza-Torres, J.; Sachse, M.; Titos-Jiménez, P.; Gomez-Laguna, J.; Dussurget, O.; Cossart, P.; Pizarro-Cerdá, J. Listeriolysin S Is a Streptolysin S-Like Virulence Factor That Targets Exclusively Prokaryotic Cells In Vivo. *mBio* **2017**, *8*, e00259-17. [CrossRef] [PubMed]

Review

From Bacterial Poisons to Toxins: The Early Works of Pasteurians

Jean-Marc Cavaillon

National Research Agency (ANR), 75012 Paris, France; jean-marc.cavaillon@pasteur.fr

Abstract: We review some of the precursor works of the Pasteurians in the field of bacterial toxins. The word "toxin" was coined in 1888 by Ludwig Brieger to qualify different types of poison released by bacteria. Pasteur had identified the bacteria as the cause of putrefaction but never used the word toxin. In 1888, Émile Roux and Alexandre Yersin were the first to demonstrate that the bacteria causing diphtheria was releasing a deadly toxin. In 1923, Gaston Ramon treated that toxin with formalin and heat, resulting in the concept of "anatoxin" as a mean of vaccination. A similar approach was performed to obtain the tetanus anatoxin by Pierre Descombey, Christian Zoeller and G. Ramon. On his side, Elie Metchnikoff also studied the tetanus toxin and investigated the cholera toxin. His colleague from Odessa, Nikolaï Gamaleïa who was expected to join Institut Pasteur, wrote the first book on bacterial poisons while other Pasteurians such as Etienne Burnet, Maurice Nicolle, Emile Césari, and Constant Jouan wrote books on toxins. Concerning the endotoxins, Alexandre Besredka obtained the first immune antiserum against lipopolysaccharide, and André Boivin characterized the biochemical nature of the endotoxins in a work initiated with Lydia Mesrobeanu in Bucharest.

Keywords: infection; poison; serotherapy; soluble bacterial chemicals; vaccine

Key Contribution: In 2022 we celebrate the bicentennial anniversary of the birth of Louis Pasteur, and it is an opportunity to recall that Louis Pasteur's close collaborators made key contributions in the field of toxinology, and wrote early books on toxins. Despite their master's Germanophobia, they maintained a close friendship with their German colleagues who also made major discoveries.

Citation: Cavaillon, J.-M. From Bacterial Poisons to Toxins: The Early Works of Pasteurians. *Toxins* **2022**, *14*, 759. https://doi.org/10.3390/toxins14110759

Received: 23 September 2022

Accepted: 31 October 2022

Published: 3 November 2022

Publisher's Note: MDPI stays neutral with regard to jurisdictional claims in published maps and institutional affiliations.

1. From Bacteria of Putrefaction to Putrid Poisons, Bacterial Metabolites and Toxins

Peter Ludvig Panum (1820–1885), a Danish physician and physiopathologist working in Copenhagen was among the very first scientists in 1856 to consider that putrefaction was associated with a putrid poison. It took him a while before he recognized that such poison could derive from the bacteria of putrefaction [1,2]. Efforts were then made to characterize these poisons, and Ernst von Bergmann (1836–1907), a Baltic German surgeon identified what he called sepsin [3]. The word toxin was coined in 1888 by Ludwig Brieger (1849–1919), a professor of medicine at Humboldt University in Berlin [4] who had identified putrescine and cadaverin, organic compounds released by bacteria of putrefaction (1885). In 1878, Francesco Selmi (1817–1881), an Italian chemist from Bologna, coined the word ptomaïnes, from Greek ptōma (corpse), to name the cadaveric alkaloids [5]. Many years later, Elie Metchnikoff (1845–1916) who worked at the Institut Pasteur as soon it had been inaugurated [6,7], also reported that bacteria of putrefaction were producing poisons [8]. He identified bacterial metabolites released by gut bacteria [9]. Two of them (indol and paracresol) were rendered responsible of tissue lesions and vasculature alterations when injected in rabbits and could even induce death when injected into monkeys.

Two other scientists who trained at Institut Pasteur made a significant contribution in the field. Sir Marc Armand Ruffer (1859–1917) who later on became the father of paleopathology [10] worked with Pasteur and Metchnikoff from 1889 to 1890. While in Paris, he also worked with Albert Charrin (1856–1907) at the school of medicine and made a key observation [11]: the filtered supernatants of bacteria (*Pseudomonas aeruginosa*), devoid of any alive or dead bacteria could induce fever once injected into rabbits. Eugenio Centanni, an italian pathologist, named the responsible bacterial poison 'pyrotoxina' [12].

Accordingly, they had identified the effects of endotoxin, three years before Richard Pfeiffer (1858–1945) developed his concept of endotoxin [13]. Most fascinatingly, the authors suspected that fever was the consequence of the activation of macrophages, far before the discoveries of endogenous pyrogens (interleukin-1, IL-1; IL-6, etc.) were made [14]. The second one, Nikolaï Gamaleïa (1859–1949), attended the inaugural ceremony of Institut Pasteur, and was expected to join the institute [7]. In 1886, he was sent by the city of Odessa where he had worked with his mentor, Elie Metchnikoff, to spend four months with Pasteur's team to learn the preparation procedures of the anthrax and rabies vaccines. Pasteur sent him to London to follow the works of the British investigative committee on rabies vaccine. From 1886 to 1892, Gamaleïa shared his life between Odessa and Paris. In 1892, he published his investigations on the cholera poisons [15], in which he described two deleterious substances, one he called «nucléine» present in heated (120 °C) culture media of *Vibrio cholerae*, highly toxic for guinea-pigs, rabbits, pigeons and dogs. The second substance, a nucleo-albumin, obtained after filtration through Chamberland filter, was shown to be heat sensitive and to induce all cholera symptoms, including diarrhea. The same year Gamaleïa further characterized the diphtheria toxin [16]. He showed that both pepsin and trypsin destroyed the diphtheria poison. Still in 1892, Gamaleïa published the first treaty on bacterial poisons [17] (translated in English the following year). He considered that a new science has sprung up: the science of microbial poisons, which is based at once on bacteriology, on biological chemistry, and on general physiology. In his book Gamaleïa addresses all the emerging aspects of this new specialty. After offering a historical background, he argued that infectious diseases were an intoxication by the poisons of the pathogenic microbes, considering that the poisons are intimately linked to the bodies of the bacteria. After addressing the chemical nature of the bacterial poisons, he summarized the knowledges on tetanus, diphtheria, cholera, tuberculosis and anthrax, while referring to the early works on immunity against these toxins, particularly the antisera generated against diphtheria and tetanus toxins.

2. The Soluble Chemical Products Produced by Bacteria and Immunity

Louis Pasteur was aware of the works of Panum and von Bergman, although he mentioned to have failed to reproduce the work of the later [18]. Pasteur did not seem to have been initially fascinated by the concept of bacterial poison. In 1880, he wrote: *"Speak, if you want, of poisoning. Make this hypothesis, I accept it. I do not know the mechanism of death from any disease more than you or anyone else does, no more than we know the mechanism of life. Speak of poison, if you wish, but you will be forced to add that, if a poison causes death, it is the microbe that generates the poison"* [19]. Indeed, he rarely mentioned the word poison and never used the word toxin. However, while working on fowl cholera, in 1880, he acknowledged that a soluble substance was released by the vibrio: *"I have acquired that during the life of the parasite, a narcotic is made, and that it is this narcotic which provokes the morbid symptom so pronounced of sleep in fowl cholera"* [20]. Later, Pasteur realized that these soluble substances could both induce the disease and be used to generate immunity: *"The fine memoir of MM. Roux and Chamberland, contained in the December 1887 issue of the Annales of M. Duclaux, demonstrates with perfect rigor that the life of the septic vibrio develops soluble chemical products which act on it little by little like an antiseptic. Introduced in sufficient quantity into the body of guinea pigs, these products give them immunity to the fatal disease caused by this vibrio. The proof is thus made, that immunity, against a disease so serious and so quickly mortal, can be obtained by the injection of chemical substances which can be measured, and that these substances themselves result from the life of the deadly microbes. [...] The first, among the observers who have occupied themselves with this subject, I had sought to produce immunity in hens by means of the soluble products formed in a broth of culture by the life of the microbe of the fowl cholera. I saw the symptoms of the disease to appear, but not the immunity; which was perhaps, as MM. Roux and Chamberland observed, only a question of the quantity of soluble products used in my experience"* [21]. Despite Pasteur recognized that immunity could be achieved with bacterial soluble substances, as stated by Kendal Smith: *"a careful reading of Pasteur's presentations to the Academy of Sciences reveals that Pasteur was entirely mistaken as to how immunity occurs, in that he reasoned, as a good microbiologist would, that appropriately attenuated microbes would deplete the host of vital trace nutrients absolutely required for their viability and growth, not an active response on the part of the host"* [22]. Indeed, Roux and Chamberland opposed themselves to their master: *"M. Pasteur thought that in the case of fowl cholera, the non-recurrence was due to the disappearance of some substances consumed by the microbe."* When they recognized that *"It is not necessary for the cells of the pathogenic organism to live among the cells of the animal to confer immunity on it"* [23]. On his side, Auguste Chauveau (1827–1917), a French veterinarian, director of the National veterinary school of Alfort disagreed with Pasteur's definition of immunity: *"Also M. Pasteur believed himself more and more authorized to consider the organism as a medium of culture which, by a first attack of the disease, would lose, under the influence of the culture of the parasite, principles that life would not return there or would return there only after a certain time"* [24]. However, the idea of Chauveau was not anymore correct: *"immunity is due to a substance left in the body by the culture of the microbe and which opposes its further development."* To be opposed to the Master was not an easy task: *"This judgment, passed by a master such as M. Pasteur, was of a nature to discourage opposing convictions, even the most robust. I hesitated for a moment to keep mine. Nevertheless, as, meanwhile, new experiments had come to fully confirm my first observations, it was impossible for me not to support the deductions which I had drawn from them. I therefore maintained them with respectful firmness"* [24].

3. Diphtheria Toxin

Known as "the strangling angel of children," diphtheria was a deadly bacterial infection killing thousands of babies and children every year. The germ, *Corynebacterium diphtheria*, was discovered in Germany in 1883 by Edwin Klebs (1834–1913), isolated and cultured the following year by Friedrich Loeffler (1852–1915). From 1888 to 1890, Alexandre Yersin (1863–1943) and Émile Roux (1853–1933) (Figure 1) who was the third director of Institut Pasteur after Louis Pasteur and Émile Duclaux (1840–1904), undertook their investigations on diphtheria [25]. They studied filtrates of 42 days cultures of the bacterium obtained through the porcelain filter set up by Charles Chamberland (1851–1908), one of the closest collaborators of Pasteur. Once injected in guinea pigs, the filtrates ended to the death of the animals, while rabbits and pigeons were more resistant than guinea pigs. Their analyses allowed them to claim that a toxin was elaborated by the germ of diphtheria. Furthermore, they also found the toxin in urines collected from sick children shortly before their deaths. In August 1891, E. Roux had the opportunity to precise his view during the seventh International *Congress of Hygiene and Demography* in London [26]. *"It is natural to conclude that microbes act through their chemical products, true poisons specific to each of them, and which determine the symptoms of the disease in man and in animals [. . .] The infectious disease is therefore a poisoning: the source of the poison is the microbe settled in the tissues; it elaborates his toxin there at the expense of the living being that it is going to kill"*. Furthermore, he addressed *a new way* to induce protective immunity, he called "chemical vaccination" in contrast to the microbial vaccination defined by Pasteur in the case of fowl cholera and by Henry Toussaint (1847–1890) in the case of anthrax: *"It is therefore not necessary for the refractory state to be acquired, that the microbes penetrate into the body, it suffices that the substances prepared in the artificial cultures be introduced into it. If, therefore, by swarming in the organism, microbes give immunity, it is undoubtedly because they produce the same chemicals that we find in* in vitro *cultures"* [26].

Figure 1. Portrait of Émile Roux giving his serotherapy course. Oil on canvas by the Finnish painter Albert Edelfelt (1895). ©Institut Pasteur/Musée Pasteur.

In 1890, the use of immune sera against diphtheria toxin and tetanus toxin was rendered popular by Emil von Behring (1854–1917) and Shibasaburo Kitasato (1853–1931) who offered the basis of serotherapy to treat diphtheria and tetanus [27]. Von Behring and Kitasato prepared immune sera in guinea-pigs, rabbits, sheeps, goats, and horses. Adolf Baginsky (1843–1918) reported the first clinical trial in 220 children with diphtheria who received intraperitoneal injection of immune serum, ending to a healing rate of 77% [28]. Paul Ehrlich (1854–1915), the father of humoral immunity who shared the Nobel prize in 1908 with Metchnikoff, used the anti-toxin responsiveness to demonstrate the transmission of immunity during pregnancy and suckling [29] and developed a standardized method allowing to ensure reproducible titers of high toxin neutralizing activity in the antiserum [30].

Of note, two years before von Behring, two French physicians and scientists, friends since secondary school, Jules Héricourt (1850–1938) and Charles Richet (1850–1935) reported that the peritoneal transfusion of whole blood of dogs previously inoculated with *Staphylococcus pyosepticus* into rabbits was able to transfer immunity in these rabbits once challenged with the same bacteria, in contrast to the absence of protection provided by the blood from uninfected dogs [31]. Their failure to provide protection with a similar approach in the case of tuberculosis and cancer led to the oblivion of their pioneer works [32]. Later, Richet was awarded the Nobel prize 1913 for his discovery of anaphylaxis, and Emil von Behring was awarded with the very first Nobel prize in medicine or physiology in 1901. In his Nobel lecture, von Behring paid tribute to his precursors: *"Without the preliminary works by Loeffler and Roux, there would be no serum treatment for diphtheria."*

In addition to have discovered the toxin, to prepare the immune sera, Émile Roux developed the use of horses on a large scale with the help of Edmond Nocard (1850–1903). Nocard had joined Pasteur's laboratory in 1880 and was Director of the veterinary school of Alfort. Horses were first hosted at the veterinary school before a large number of horses could join the stables of the annex of Institut Pasteur in Marnes-La-Coquette (Figure 2). With an initial number of a dozen in September 1894, 136 horses were hosted by the beginning of 1895. Because the Parisian production remained insufficient to cover the need of the whole country, institutions were created in various cities (including Institut Pasteur de Lille, and production centers in Lyon, Le Havre, Grenoble, Bordeaux, Marseille, Rouen). In Nancy, the center was created thanks to a donation of Osiris (1825–1907), a generous donator who had regularly supported the Institut Pasteur [33].

Figure 2. Inoculation of diphtheria toxin to horses producing antidiphtheria serum, at the stables of Marne la Coquette. Engraving after a drawing by Alexis Lemaistre (1896). ©Institut Pasteur/Musée Pasteur.

In 1894, Émile Roux with Louis Martin (1864–1946), a former student of Joseph Grancher (1843–1907), his colleague of Institut Pasteur, reported with great details the preparation of the diphtheria toxin and the immune sera [34]. It contrasted with the vagueness maintained by Pasteur on the preparation of his vaccine against anthrax, which was at odds with the scientific attitude of the very rigorous Dr. Roux. With Dr. Auguste Chaillou (1866–1915), from Necker Hospital, they published the treatment of 300 children, ending with a 50% decreased mortality [35]. The official announcement of the success of the experiment was made by Roux on 5 September 1894 in Budapest during the eighth International Congress of Hygiene and Demography (Figure 3). The results had a great impact in the lay press (Figure 4). "Le Figaro" of which the director was Gaston Calmette (1858–1914), the brother of the Pasteurian Albert Calmette (1863–1933), echoed this success and launched for a call for donations that helped to collect more than 612,000 francs and to gather numerous retired horses. Sarah Bernardt, a star actress, also raised funds for the Institut Pasteur, offering a hundred seats up for auction, at the theater de la Renaissance, for the 1894 première of Gismonda.

Figure 3. Members of the Budapest congress in 1894, lectures on the work on diphtheria. From left to right: Georges Gabritschevsky (1860–1907), Alphonse Laveran (1845–1922), Émile Roux (1853–1933), Léon Perdrix (1859–1917), Edmond Nocard (1850–1903), George Nuttal (1862–1937) and Elie Metchnikoff (1845–1916). ©Institut Pasteur/Musée Pasteur.

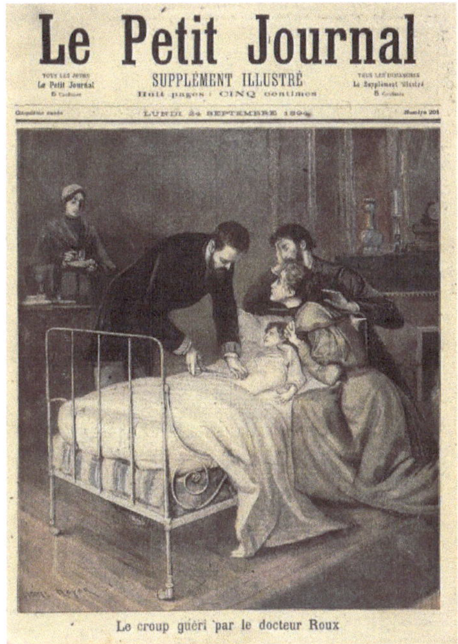

Figure 4. The front page of the "Petit Journal" (24 September 1894) showing Émile Roux saving a child from diphtheria thanks to serotherapy. ©Institut Pasteur/Musée Pasteur.

However, because Roux had compared the outcome of patients treated with immune serum in one hospital with that of untreated patients in another hospital, some physicians were not fully convinced. Among those, were Johannes A. G. Fibiger (1867–1928), a junior physician, working in the ward of Søren Thorvald Sørensen (1849–1928) at Blegdamshospitalet in Copenhagen. In 1896–1897, Fibiger conducted the very first randomized controlled trial. Treatment allocation depended on the day of admittance, and patients received either standard treatment or standard treatment plus serotherapy. Only patients for whom the diphtheria bacterium was then identified were kept in the study. Eight out of 239 patients (3.3%) in the serum treated group, and 30 out of 245 (12.2%) in the control group died [36,37]. Of note, later, Fibiger was awarded with a Nobel Prize (1926), for his discovery of "*Spiroptera carcinoma*" (*Gongylonema neoplasticum*, its current name), a nematode to which he attributed a role in the development of gastric cancer in rats. Systematic reanalyses of Fibiger's data led to conclude that his specific finding was found erroneous [38], despite the links between certain pathogens and cancer were then fully recognized.

Gaston Ramon (1886–1963) (Figure 5) was a veterinarian, hired by Roux to prepare the horse antisera at the annex of Institut Pasteur. In 1923, he showed that the treatment of the diphtheria toxin with formalin and heat resulted in an immunizing molecule that had lost its toxicity, but retained its antigenicity as shown in guinea pigs and horses [39]. Ramon coined the word anatoxin. The same year Alexander Thomas Glenny (1882–1965) from the Wellcome Research Laboratories, Beckenham (Kent) called toxoid his preparation that was only obtained after formalin treatment and needed to be mixed with anti-toxin to be used as an immunizing agent [40]. In France, the vaccination of infants against diphtheria with the Ramon anatoxin has been compulsory since the law of 24 June 1938. Among the other major contributions of Ramon, let us mention the discovery of the adjuvants: "*It is necessary to involve an inflammatory reaction at the site of antigen injection to enhance the immune response.*" [41], the famous «dirty little secret of the immunologists» as claimed by Charles Janeway (1943–2003), the father of the rebirth of innate immunity. Ramon had observed that if a small abscess was occurring at the site of injection of his anatoxin, the antibody titer was greatly enhanced. Then, on purpose he created an inflammation by mixing different substances with his vaccine, including tapioca, his favorite one. Gaston Ramon has been nominated 155 folds for the Nobel prize and is #1 of the scientists with the greatest number of nominations without being awarded. Of note, with 115 nominations Émile Roux is #2 on this podium!

Figure 5. Gaston Ramon (1886–1963) who discovered the diphtheria anatoxin and the adjuvants ©Wikipedia.

4. Tetanus Toxin

The bacillus of tetanus (*Clostridium tetani*) was first described in 1884 in Berlin by Arthur Nicolaier (1862–1942). In 1890, three groups reported the presence of a released toxin: In Denmark, Knud Faber (1862–1956), chief physician at Frederiks Hospital and later at Rigshospitalet and professor of clinical medicine at the University of Copenhagen [42]; In Italy, at the University of Bologna, Giuseppina Cattani (1859–1914) and Guido Tizzoni (1853–1932), medical doctors and bacteriologists [43] and Alessandro Bruschettini (1868–1932) [44]; and in France, Louis Vaillard (1850–1935), and Hyacinthe Jean Vincent (1862–1950) both professors at the military medical school (Val-de-Grâce) showed that filtered cultures of the tetanus bacillus could induce the disease and kill mice, guinea pigs and rabbits [45]. Émile

Roux and Louis Vaillard, following the precursor works of von Behring and Kitasato [27]; initiated a collaboration to further define the preparation of the toxin and its treatment with iodine to obtain an appropriate less toxic immunogen to generate in mice's, rabbit's, guinea-pig's, sheep's, cow's and horse's protective immune sera [46]. They showed that in rabbits the protective immunity lasted more than two years, although they advocated to perform regular boosts. Most interestingly, they reported that the cow milk was a source of anti-toxin in agreement with the ingenious experiments of Ehrlich [29]. In agreement with Metchnikoff's observations, they suggested that the immune sera act on leukocytes to favor the phagocytosis of the tetanus germs, a phenomenon known as opsonization. They showed that the immune sera could be protective even when administered after the deadly bacteria. They also report their first attempts in humans with horse immune sera. Five patients died but two survived. Of course, an inappropriate timing may explain these results. The death of an eleven-year-old child after the extraction of two teeth reminds us that by the end of the 19th century, tetanus could happen in unexpected settings. In 1896, Dr. Eugene Tracey, a British physician reported how he saved a little girl with tetanus by injecting her the immune serum of Drs. Cattani & Tizzoni [47]. In 1897, Élie Metchnikoff compared the sensitivity of different animal species to the tetanus toxin and their respective capacity to produce anti-toxin. He reported that scorpions, beetles, carps, axolotl, tortoise were insensitive while hens, guinea-pigs, frog and caiman when maintained at 32–37 °C, were producing anti-toxin [48]. Auguste-Charles Marie (1864–1935), working in Elie Metchnikoff's laboratory localized the toxin after its injection within the blood compartment, the nervous system and in other organs of frogs, guinea-pigs, rabbits, mice and dogs [49].

Edmond Nocard played again a major role to favor the use of horses for tetanus serotherapy [50]. Institut Pasteur was engaged in a central role during world war I to provide antisera to the armies. In 1914, the Institut Pasteur had 300 horses and was preparing 80,000 vials of antisera a month. In 1918 there were 1462 horses, allowing the preparation of 600,000 vials per month.

After the successful approach by Ramon to prepare the diphtheria anatoxin, the same experimental methodology was developed in 1925 to prepare the tetanus anatoxin, by Pierre Descombey (1895–1930), a young pasteurian who passed away when he was only 35. He successfully immunized horses and guinea pigs [51]. The following year after Albert Lafaille (1891–1963), Ramon's colleague had tested the safety of the injection on himself, Ramon and his colleague Christian Joseph Zoeller (1888–1934), professor at the Val de Grâce military hospital, tested the tetanus anatoxin on a hundred humans and obtained strong neutralizing activity of their sera after three injections [52]. In France, a law of 15 August 1936 rendered the anti-tetanus vaccination mandatory within the armies, and the association of both anatoxins was officialized in 1940 for children after Ramon and Zoeller had shown the feasibility of such association [53].

5. Cholera Toxin

In 1886, The Neapolitan physician, Antonio Cantani, was the first to presume that the cholera toxin was linked to the body substance of the bacteria [54]—a concept which had been further developed by others, especially by Gamaleïa in 1892 [15].

At Institut Pasteur, Metchnikoff compared the sensitivity of different animal species to *Vibrio cholera*. He found that guinea pigs were more sensitive than rabbits, which were more sensitive than mice, while pigeons and hens were insensitive [55]. In addition, he demonstrated that the toxicity was due to a toxin and that the antitoxin immunity was protective. For his demonstration he placed live bacteria in a bag preventing the bacteria's dispersal, and then he implanted the bag into the peritoneal cavity of guinea pigs. Most of them died, thus illustrating that a diffusible product was responsible of the poor outcome. When he placed dead bacteria or culture medium in the bag, the animals survived. Then, he selected the few animals that survived the bags containing the live bacteria and re-injected them with a lethal dose of *V. cholerae*. Not only the animals were protected, but they were able to resist further injections of up to 16 lethal doses. Afterwards, Metchnikoff prepared the cholera toxin and successfully immunized guinea pigs, rabbits, goats and horses demonstrating that their sera displayed a protective activity. It is admirable that the father of cellular immunity made key experiments demonstrating the importance of humoral immunity against bacterial toxins.

Despite these precursors works, the discovery of the cholera toxin is often attributed to an Indian investigator from Calcutta, Sambhu Nath De (1915–1985) who reported in 1959 the toxicity of bacteria-free culture filtrate of *Vibrio cholerae* in rabbits [56]. He showed that the cholera toxin could kill the rabbit, inducing a fall of blood pressure, heart edema, an increased permeability of the capillaries of the intestinal mucosa and an alteration of the kidneys.

6. Endotoxins

Richard Pfeiffer (1858–1945) coined the word endotoxin in 1892 [13]. Pfeiffer had worked as Robert Koch's assistant in the Institute of Hygiene in Berlin. In 1894, he reported the process of in vivo bacteriolysis (1894), known as "the Pfeiffer phenomenon" which was further deciphered by Jules Bordet at Institut Pasteur [57,58]. Pfeiffer accompanied Koch in 1897 in India to investigate the plague epidemics. Pfeiffer thought that the bacteria were containing such a toxic substance, before it was recognized that endotoxins are present on the surface of Gram-negative bacteria and are regularly released by growing bacteria [59]. Two major achievements were obtained by Pasteurians in the field of endotoxins.

Alexandre Besredka (1870–1940) was born in Odessa and had Metchnikoff as a teacher. He moved to Paris in 1893 to study medicine and joined Metchnikoff's laboratory. At his master's death in 1916, Besredka, having spent more than 20 years at his side, became the head of the laboratory Metchnikoff had founded at Institut Pasteur. While Pfeiffer was mistaken in his appreciation of the inability of endotoxins to induce neutralizing antibodies, it was Alexandre Besredka who made the decisive discovery that antisera raised against intravenously injected bacteria developed endotoxin-neutralizing properties. A horse injected with typhoid vaccine generated antibodies against typhoid endotoxin that were able to protect guinea-pigs injected with 30 lethal doses of endotoxin. Once again, it is quite fascinating that in the laboratory of the father of innate cellular immunity, such a key experiment in the field of humoral immunity had been achieved [60].

André Boivin (1895–1949) joined the Institut Pasteur annex in Garches in 1936, working with Gaston Ramon, pursuing his research on smooth and rough endotoxins he had initiated together with Ion and Lydia Mesrobeanu, while he was at the Cantacuzene Institute in Bucharest [61,62]. Boivin and Mesrobeanu had deciphered the biochemical nature of the endotoxins, they initially called "antigène glucido-lipidique", before it acquired his popular name of lipopolysaccharide (LPS).

7. The First Pasteurian Books on Toxins

In 1911, Etienne Burnet (1873–1960) published a book entitled «Microbes and Toxins» (translated in English in 1912) with a preface of Metchnikoff. Burnet was a physician who first joined the École Normale Supérieure (ENS) where Pasteur made most of his career. He entered at Institut Pasteur in 1904 as an assistant in the laboratory of Amédée Borrel (1867–1936) and from 1907–1919, he worked in the laboratory of Metchnikoff. His book is not only a book of bacteriology, dealing with the nature of microbes, but he also addressed the host response, immunity, inflammation, phagocytosis, anaphylaxis, and vaccines. In his chapter devoted to the toxins, he not only focused on bacterial toxins, but he also mentioned the vegetal ones, including ricin. Of note, Jan Danysz (1860–1928), a biologist from Poland who worked with Jules Bordet in South Africa, successfully proposing serotherapy to fight bovine plague [58], investigated the interaction of ricin with its anti-toxin antibodies [63]. Burnet compared the soluble toxins with diastases (enzymes) and also described the endotoxins, paying tribute to Besredka who was the first to report anti-endotoxins.

The second book to be mentioned has been published in 1919. Entitled "Toxines et antitoxines", it has never been translated in English. It addresses vegetal toxins, those found in animal venoms and produced by bacteria, emphasizing on the pathophysiological consequences of their injections through different routes in experimental animals, and the acquired immunity associated with the protective role of antibodies. It was written by three authors: Maurice Nicolle (1862–1932), Emile Césari (1876–1956) and Constant Jouan (1877–1949). Maurice Nicolle is the oldest brother of Charles Nicolle (1866–1936) who received the Nobel prize in 1928 for his work performed while he was director of the Institut Pasteur in Tunis, demonstrating that lice transmit typhus. In 1893, Maurice Nicolle was appointed to replace Waldemar Haffkine (1860–1930) who had developed the cholera and the plague vaccines [64], as preparator of the microbiology course at the Institute. Soon, Louis Pasteur sent him to Turkey at the request of Sultan Abdul Hamid II. There, he directed the Imperial Institute of Bacteriology of Constantinople. From 1896 to 1915, Maurice Nicolle reported numerous investigations on the preparation of toxins, their preservation, and the properties of antisera. Back at the Institut Pasteur Institute (1902–1926), he attracted a number of young researchers and students around him, including Emile Césari who became his assistant in 1909. Their work on toxins and antitoxins, especially on the mutual precipitation of antigens and antibodies, allowed them to develop a method for measuring the in vitro activity of diphtheria and tetanus toxins on the one hand, and on the other hand, the antitoxic potency of anti-diphtheria and anti-tetanus sera. At the request of E. Roux and A. Calmette, he became head of the anti-venomous serotherapy department. The third co-author, Constant Jouan started in 1893 as a preparator in the Department of microbiology applied to hygiene and vaccinations, directed by Charles Chamberland. In 1909, he became assistant laboratory head, responsible for the manufacture of vaccines under the responsibility of Emile Roux. He shared this function with Ernest Fernbach (1875–1962) and came in charge of the laboratory of anthrax vaccines. In 1909, he registered a patent for an apparatus allowing the centrifugation of liquids. In 1919 when war had just ended in Europe, Constant Jouan, decided to devote all of his time to the creation of a new laboratory instruments company. His ingenuity and creativity ensured the rapid growth of the Jouan company. The first to adapt an electric motor to a centrifuge (until then manually operated), he opened new perspectives in research and diversified towards chemistry and biology, making a consortium with other older companies (Maison Adnet (Paris, France) for sterilization and hospital devices; and Maison Mathieu (Paris, France) for surgical instruments). The Nantes (France)-based Jouan company was acquired in September 2003 by its major competitor, the American Thermo Electron.

8. The French-German Relationships

The war of 1870–1871 between France and the Prussian Empire profoundly traumatized Louis Pasteur. He left Paris, worrying about a possible destruction of his laboratory at the École Normale Supérieure during the siege of Paris. He moved to Arbois, then to Pontarlier, Geneva, Lyon and finally joined Émile Duclaux in Clermond-Ferrand. There, he worked in the close by city of Chamalières, at the Kuhn breweries to improve the local production of beers. In June 1871, he patented his process to prepare and preserve beers. In his text, one finds the famous sentence: *"I wish that the beers manufactured with my process carry in France the name of Beer of the National Revenge"*. He sent back his "Honoris causa" degree offered by the University of Bonn and refused to be awarded with the Prussian order of the Royal Crown proposed by the emperor, Guillaume of Prussia (Kronenorden). His words *"science has no homeland, but the scientist has one"* were recalled by Paul Brouardel (1837–1906), the Dean of the Faculty of Medicine during his Jubilé. To his former student, Jules Raulin (1836–1896) Pasteur wrote: *"I try to keep away all these memories and the sight of all our miseries to which I see no salvation except in the despair of an all-out struggle. I would like France to resist until her last man, until her last rampart! I would like the war to be prolonged until the heart of winter so that, the elements coming to our aid, all these vandals would perish from cold, misery and disease. Each of my works until my last day will bear for epigraph: Hate to*

Prussia. Revenge. Revenge." To Luigi Chiozza (1828–1889) a chemist and Italian deputy, in response to an invitation to leave France and to join Italy, Pasteur wrote: *"My dear friend, I received your second letter. How touched I am by these steps and how happy I would be without the misfortunes of my country of these testimonies of esteem granted to my works, I who have always lived for glory."* These last words are a most interesting confession of his motivation. Of course, his hate of Prussia, after the humiliating defeat and the loss of Alsace, a place where he had been professor in his early career (teaching chemistry in Strasbourg), was shared by many of his country people. Among those was Jean-Jacques Henner (1829–1905), an Alsatian artist and friend of his family. He was famous for his allegorical painting "L'Alsace, is waiting" (1871). In the following years, Henner painted the portrait of his daughter, Marie-Louise Pasteur (1876), his daughter-in-law, Jeanne Pasteur (1854–1932), wife of Jean-Baptiste Pasteur (1877), and of Louis Pasteur himself (1877).

Despite his Germanophobic attitude, Pasteur had little influence over his close collaborators. In 1888, Émile Roux sent Alexandre Yersin to visit the laboratory of Robert Koch, Pasteur's best enemy, to follow the courses and to bring back ideas of organization for the laboratories and for teaching purposes. In 1895, Joseph Grancher in his preface of the re-edition of the book "Etude sur les virus" by Jean Hameau (1779–1851) recognized the superiority of the German school in terms of bacterial staining, bacterial cultures on semi-solid media and regarding their identifications [65]. In 1904, Metchnikoff welcomed Robert Koch while he visited the Institut Pasteur. Roux, Metchnikoff and von Behring were friends to the point that Roux and Metchnikoff were the godfathers of Behring's sons. On 4 February 1914, few months before Germany declared war to France, the Pasteurians received Paul Ehrlich and his wife, offering a historical picture of both Nobel prize 1908 standing close to each other (Figure 6). Finally, in March 1914, Roux and Metchnikoff paid tribute to the work of Paul Ehrlich in a paper published in French in a German journal [66]. The complementarity of the works of the Pasteurians and of the German school was illustrated all along this period, when key discoveries in bacteriology and immunology were made on both side of the Rhein.

Figure 6. Visit of Paul Ehrlich and his wife at Institut Pasteur (4 February 1914). From left to right: Camille Delezenne, Constantin Levaditi, Salmon, Jean Danysz, Alphonse Laveran, Félix Mesnil, Mme Ehrlich, Eugène Wollman, Alexandre Besredka, Ernest Fourneau, Emile Roux, Auguste-Charles Marie, Alexandre Salimbeni, Paul Ehrlich, Louis Martin, Elie Metchnikoff, Gabriel Bertrand, Amédée Borrel. ©Institut Pasteur/Musée Pasteur.

Funding: This work received no external funding.

Institutional Review Board Statement: Not applicable.

Informed Consent Statement: Not applicable.

Data Availability Statement: Not applicable.

Acknowledgments: I wish to express my sincere gratitude to Sandra Legout, member of the historical archives at Institut Pasteur, who with her amazing knowledge of the past of this prestigious place, helped me to find all information I needed to write this review.

Conflicts of Interest: The author declares no conflict of interest.

References

1. Cavaillon, J.-M. Historical links between toxinology and immunology. *Pathog. Dis.* **2018**, *76*, fty019. [CrossRef] [PubMed]
2. Cavaillon, J.-M. Exotoxins and endotoxins: Inducers of inflammatory cytokines. *Toxicon* **2018**, *149*, 45–53. [CrossRef] [PubMed]
3. Bergmann, E.V.; Schmiedeberg, O. Ueber das schwefelsaure sepsin (das gift faulender substanzen). *Med. Wiss.* **1866**, *32*, 497–498.
4. Brieger, L. Zur kenntniss des tetanin und des mytilotoxin. *Arch. Pathol. Anat.* **1888**, *112*, 549–551. [CrossRef]
5. Selmi, F. *Sulle Ptomaine od Alcaloidi Cadaverici e Loro Importanza in Tossicologia, Osservazioni del Prof. Francesco Selmi Aggiuntavi una Perizia per la Ricerca Della Morfina*; Presso Nicola Zanichelli: Bologna, Italy, 1878.
6. Cavaillon, J.-M.; Legout, S. Centenary of the death of Elie Metchnikoff: A visionary and an outstanding team leader. *Microbes Infect.* **2016**, *18*, 577–594. [CrossRef] [PubMed]
7. Cavaillon, J.-M.; Legout, S. Duclaux, Chamberland, Roux, Grancher, and Metchnikoff: The five musketeers of Louis Pasteur. *Microbes Infect.* **2019**, *21*, 192–201. [CrossRef]
8. Metchnikoff, E. Études sur la flore intestinale. *Ann. Inst. Pasteur* **1908**, *22*, 929–955.
9. Metchnikoff, E. Poisons intestinaux et scléroses. *Ann. Inst. Pasteur* **1910**, *24*, 755–770.
10. Cavaillon, J.-M. Sir Marc Armand Ruffer and Giulio Bizzozero: The first reports on efferocytosis. *Leukoc. Biol.* **2013**, *93*, 39–43. [CrossRef]
11. Charrin, A.; Ruffer, A. Mécanisme de la fièvre dans la maladie pyocïanique. *C. R. Seances Soc. Biol.* **1889**, *9*, 63–64.
12. Centanni, E. Untersuchungen über das Infectionsfieber. *Dtsch. Med. Wochenschr.* **1894**, *20*, 148–150. [CrossRef]
13. Pfeiffer, R. Untersuchungen über das Choleragift. *Z. Hyg.* **1892**, *11*, 393–412. [CrossRef]
14. Dinarello, C.A. Cytokines as endogenous pyrogens. *J. Infect. Dis.* **1999**, *179*, S294–S304. [CrossRef]
15. Gamaleïa, N. Recherches expérimentales sur les poisons du cholera. *Arch. Méd. Expér. Anat Pathol.* **1892**, *4*, 173–194.
16. Gamaleïa, N. De l'action des ferments solubles sur le poison diphtérique. *C. R. Soc. Biol.* **1892**, *44*, 153–155.
17. Gamaleïa, N. *Les Poisons Bactériens*; Rueff et Cie, J., Ed.; Paris, France, 1892.
18. Pasteur, L.; Joubert, J.-F.; Chamberland, C. La théorie des germes et ses application à la médecine et à la chirurgie. *C. R. Acad. Sci.* **1878**, *86*, 1037–1043.
19. Pasteur, L. Relations de la variole et de la vaccine. Choléra des poules. *Bull. Acad. Méd.* **1880**, *9*, 527–531.
20. Pasteur, L. Études des conditions de la non-récidive de la maladie, et de quelques autres de ses caractères. *Bull. Acad. Med.* **1880**, *44*, 390–401.
21. Pasteur, L. Sur le premier volume des "Annales de L'institut Pasteur", et en particulier sur un mémoire de MM Roux et Chamberland, intitulé "Immmunité contre la septicémie, conférée par des substances solubles". *C. R. Acad. Sci.* **1888**, *106*, 320–324.
22. Smith, K.A. Louis pasteur, the father of immunology? *Front. Immunol.* **2012**, *3*, 68. [CrossRef]
23. Roux, E.; Chamberland, C. Immunité contre la septicémie conférée par des substances solubles. *Rec. Med. Vét.* **1888**, *7e série, Tome 5*, 133–145.
24. Chauveau, A. Le mécanisme de l'immunité acquise. *Rec. Med. Vét.* **1888**, *7e série, Tome 5*, 145–159.
25. Roux, E.; Yersin, A. Contribution à l'étude de la diphtérie. *Ann. Inst. Pasteur* **1888**, *2*, 629–661; **1889**, *3*, 273–288; **1890**, *4*, 385–426.
26. Roux, E. De l'immunité—Immunité acquise et immunité naturelle. *Ann. Inst. Pasteur* **1891**, *5*, 517–533.
27. Behring, E.; Kitasato, S. Ueber das Zustandekommen der Diphtherie-Immunitat und der Tetanus-Immunität bei Thieren. *Dtsch. Med. Wochenschr.* **1890**, *16*, 1113–1114.
28. Baginsky, A. *Die Serumtherapie Der Diphtherie Nach Den Beobachtungen Im Kaiser und Kaiserin Friedrich Kinderkrankenhause in Berlin*; Hirschwald: Berlin, Germany, 1895.
29. Ehrlich, P. Ueber immunität durch vererbung und säugung. *Z. Hyg. Infekt. Krankh.* **1892**, *2*, 183–203. [CrossRef]
30. Von Ehrlich, P.; Kossel, H.; Wassermann, A. Ueber gewinnung und verwendung des diphtherieheilserums. *Dtsch. Med. Wochenschr.* **1894**, *20*, 353–355. [CrossRef]
31. Héricourt, J.; Richet, C. De la transfusion péritonéale et de l'immunité qu'elle confère. *C. R. Acad. Sci.* **1888**, *107*, 748–750.
32. Lahaie, Y.-M.; Watier, H. Contribution of physiologists to the identification of the humoral component of immunity in the 19th century. *Mabs* **2017**, *9*, 774–780. [CrossRef]
33. Simon, J. The origin of the production of diphtheria antitoxin in France, between philanthropy and commerce. *Dynamis* **2007**, *27*, 63–82.
34. Roux, E.; Martin, L. Contribution à l'étude de la diphtérie (sérum-thérapie). *Ann. Inst. Pasteur* **1894**, *8*, 609–639.

35. Roux, E.; Martin, L.; Chaillou, A. Trois cents cas de diphtérie traités par le sérum anti-diphtériques. *Ann. Inst. Pasteur* **1894**, *8*, 640–661.
36. Fibiger, J. Om serumbehandling af difteri. *Hospitalstidende* **1898**, *6*, 309–325; 337–350.
37. Hróbjartsson, A.; Gøtzsche, P.C.; Gluud, C. The controlled clinical trial turns 100 years: Fibiger's trial of serum treatment of diphtheria. *Br. Med. J.* **1998**, *317*, 1243–1245. [CrossRef] [PubMed]
38. Stolt, C.-M.; Klein, G.; Jansson, A.T.R. An analysis of a wrong Nobel Prize—Johannes Fibiger, 1926: A study in the Nobel archives. *Adv. Cancer Res.* **2004**, *92*, 1–12. [PubMed]
39. Ramon, G. Sur le pouvoir floculant et sur les propriétés immunisantes d'une toxine diphtérique rendue anatoxique (anatoxine). *C. R. Acad. Sci.* **1923**, *177*, 1338–1340.
40. Glenny, A.T.; Hopkins, B.E. Diphtheria toxid as an immunizing agent. *Br. J. Exp. Pathol.* **1923**, *4*, 283–288.
41. Ramon, G. Procédé pour accroître la production des antitoxines. *Ann. Inst. Pasteur* **1926**, *40*, 1–10.
42. Faber, K. Die pathogenie des tetanus. *Berl. Klin. Wochenschr.* **1890**, *27*, 710–717.
43. Tizzoni, G.; Cattani, G. Über das Tetanus gift. *Zentralbl. Bakt.* **1890**, *8*, 69–73.
44. Bruschettini, A. Recherches préliminaires sur la diffusion du poison du tétanos dans l'organisme. *Ann. Microgr.* **1890**, *3*, 83–87.
45. Vaillard, L.; Vincent, H.J. Sur le poison tétanique. *C. R. Soc. Biol.* **1890**, *42*, 634–636.
46. Roux, E.; Vaillard, L. Contribution à l'étude du tétanos. Prévention et traitement par le serum anti-toxiique. *Ann. Inst. Pasteur* **1893**, *7*, 65–140.
47. Tracey, H.E. Case of chronic tetanus treated by Tizzoni's antitoxin; recovery. *Lancet* **1896**, *147*, 287. [CrossRef]
48. Metchnikoff, E. Recherches sur l'influence de l'organisme sur les toxines. *Ann. Inst. Pasteur* **1897**, *11*, 801–809.
49. Marie, A. Recherche sur la toxine tétanique. *Ann. Inst. Pasteur* **1897**, *11*, 591–599.
50. Nocard, E. Sur la sérothérapie du tétanos. *Receuil Méd. Vét.* **1897**, *8e série, tome 4*, 481–490.
51. Descombey, P. Vaccination du cheval par l'anatoxine tétanique. *Ann. Inst. Pasteur* **1925**, *39*, 485–504.
52. Ramon, G.; Zoeller, C. De la valeur antigène de l'anatoxine tétanique chez l'homme. *C. R. Acad. Sci.* **1925**, *182*, 245–247.
53. Ramon, G.; Zoeller, C. Les "vaccins associés" par union d'une anatoxine et d'un vaccin microbien (TAB) ou par mélanges d'anatoxines. *C. R. Soc. Biol.* **1926**, *94*, 106–109.
54. Cantani, A. Giftigkeit der Cholerabacillen. *Dtsch. Med. Wochenschr* **1886**, *45*, 789–793. [CrossRef]
55. Metchnikoff, E.; Roux, E.; Taurelli-Salimbeni, A. Toxine et antitoxine cholérique. *Ann. Inst. Pasteur* **1896**, *10*, 257–282.
56. De, S.N.; Sarkar, J.K.; Tribedi, B.P. An experimental study of the action of cholera toxin. *J. Pathol. Bacteriol.* **1951**, *63*, 707–717. [CrossRef] [PubMed]
57. Bordet, J. Les leucocytes et les propriétés actives du sérum chez les vaccinés. *Ann. Inst. Pasteur.* **1895**, *9*, 462–506.
58. Cavaillon, J.-M.; Sansonetti, P.; Goldman, M. 100th Anniversary of Jules Bordet's Nobel Prize: Tribute to a founding father of immunology. *Front. Immunol.* **2019**, *10*, 2114. [CrossRef]
59. Rietschel, E.T.; Cavaillon, J.-M. Endotoxin and anti-endotoxin—The contribution of the schools of Koch and Pasteur: Life, milestone-experiments and concepts of Richard Pfeiffer (Berlin) and Alexandre Besredka (Paris). *J. Endotoxin Res.* **2002**, *8*, 3–16; 71–82. [CrossRef]
60. Besredka, A. De l'anti-endotoxine typhique et des anti-endotoxines en général. *Ann. Inst. Pasteur* **1906**, *20*, 149–154.
61. Boivin, A.; Mesrobeanu, I.; Mesrobeanu, L. Technique pour la préparation des polysaccharides microbiens spécifiques. *C. R. Soc. Biol.* **1933**, *113*, 490–492.
62. Cavaillon, J.-M. André Boivin: A pioneer in endotoxin research and an amazing visionary during the birth of molecular biology. *Innate Immun.* **2020**, *26*, 165–171. [CrossRef]
63. Danysz, J. Contribution à l'étude des propriétés et de la nature des mélanges des toxines avec leurs antitoxines. *Ann. Inst. Pasteur* **1902**, *16*, 331–345.
64. Cavaillon, J.-M.; Osuchowski, M.F. COVID-19 and earlier pandemics, sepsis, and vaccines: A historical perspective. *J. Intensive Med.* **2021**, *1*, 4–13. [CrossRef]
65. Cavaillon, J.-M.; Legout, S. Louis Pasteur: Between myth and reality. *Biomolecules* **2022**, *12*, 596. [CrossRef] [PubMed]
66. Roux, E.; Metchnikoff, E. L'œuvre de Paul Ehrlich. *Berl. Klin. Wochenschr.* **1914**, *71*, 529–531.

Review

Bacillus anthracis, "la maladie du charbon", Toxins, and Institut Pasteur

Pierre L. Goossens

Yersinia, Institut Pasteur, 28 rue du Dr Roux, 75015 Paris, France; pierre.goossens@pasteur.fr

Abstract: Institut Pasteur and *Bacillus anthracis* have enjoyed a relationship lasting almost 120 years, starting from its foundation and the pioneering work of Louis Pasteur in the nascent fields of microbiology and vaccination, and blooming after 1986 following the molecular biology/genetic revolution. This contribution will give a historical overview of these two research eras, taking advantage of the archives conserved at Institut Pasteur. The first era mainly focused on the production, characterisation, surveillance and improvement of veterinary anthrax vaccines; the concepts and technologies with which to reach a deep understanding of this research field were not yet available. The second period saw a new era of *B. anthracis* research at Institut Pasteur, with the anthrax laboratory developing a multi-disciplinary approach, ranging from structural analysis, biochemistry, genetic expression, and regulation to bacterial-host cell interactions, *in vivo* pathogenicity, and therapy development; this led to the comprehensive unravelling of many facets of this toxi-infection. *B. anthracis* may exemplify some general points on how science is performed in a given society at a given time and how a scientific research domain evolves. A striking illustration can be seen in the additive layers of regulations that were implemented from the beginning of the 21st century and their impact on *B. anthracis* research. *B. anthracis* and anthrax are complex systems that raise many valuable questions regarding basic research. One may hope that *B. anthracis* research will be re-initiated under favourable circumstances later at Institut Pasteur.

Keywords: *Bacillus anthracis*; anthrax; *anthracis* toxins; Institut Pasteur; vaccines; regulations; societal control

Key Contribution: The history of *B. anthracis* and Institut Pasteur have been intertwined for almost 120 years, ever since the pioneering work of Louis Pasteur and the foundation of the Institute that carries his name. For the celebration of 200 years since his birth, this historical review will give an overview of how *B. anthracis* research contributed to the renown of Institut Pasteur.

Citation: Goossens, P.L. *Bacillus anthracis*, "la maladie du charbon", Toxins, and Institut Pasteur. *Toxins* **2024**, *16*, 66. https://doi.org/10.3390/toxins16020066

Received: 8 November 2023
Revised: 25 December 2023
Accepted: 30 December 2023
Published: 26 January 2024

1. Introduction

This special issue deals with the contributions of the scientists at Institut Pasteur in the field of toxins. This essay will essentially focus on work performed on *Bacillus anthracis*, its toxins, and "la maladie du charbon" (anthrax) at Institut Pasteur from its origins to the present time. From time to time, crucial contributions from outside Institut Pasteur will be mentioned, but not extensively, as this is not an exhaustive review of anthrax research. This will tend to be a historical essay, a personal point of view (not an opinion), that is sometimes subjective, as it gives my own perception as a scientist who lived through a crucial change in the way science is performed in our rapidly evolving society.

In the first part, this review will present the history of anthrax research at Institut Pasteur from the end of the 19th century to the 1970s, which mainly covers the development and production of the first anthrax vaccines.

This section mainly relies on the archives kept at the Centre de ressources en information scientifique (CeRIS) at Institut Pasteur. As archival research is never completed, there is

an opportunity for further studies in this domain (for editorial reasons, the footnotes have been inserted into the text; they can be read or skipped according to readers' preference).

The second part will concentrate on the 1986–2015 period, exemplifying the richness of original research approaches to microbiology, toxins, and therapeutics.

B. anthracis, the bacterium responsible for "la maladie du charbon" (in French) (Milzbrand in German and anthrax in English) has, from the very early stages, a common history with Louis Pasteur. *B. anthracis* was first observed by Rayer and Davaine in 1850 [1]. It then became the centre of an intense controversy between Louis Pasteur and Robert Koch.

Pasteur and Koch were engaged, at the time, in a fierce scientific competition in the nascent field of microbiology. Many factors interfered, leading to this clash of personalities: their age, scientific recognition, and the language barrier with unfortunate consequences ranging from a lack of knowledge of prior publications to deep misunderstandings (such as during the September 1882 Geneva congress), not forgetting the political context after the Franco-Prussian 1870 war. These details come from the highly informative book on the interactions between Pasteur and Koch, which is available in French [2] with a German translation [3]; an English translation would be invaluable for the scientific community.

Together, they proved that the bacterium was responsible for anthrax and that it could produce spores that account for the periodic resurgence of the disease in the so-called "cursed fields" ("champs maudits" in French [2,4]). On the basis of his work on anthrax and then later on tuberculosis, Koch later proposed his famous "postulates" that link a putative pathogen to a given infectious disease. At the time, anthrax made a strong impression through the first publicised bacterial vaccination by Louis Pasteur in 1881 at Pouilly le Fort (Figure 1) [5]. At the end of the 19th century, the world of bacteria was being discovered and in full expansion; microbes were shown to be the causal agents of numerous diseases.

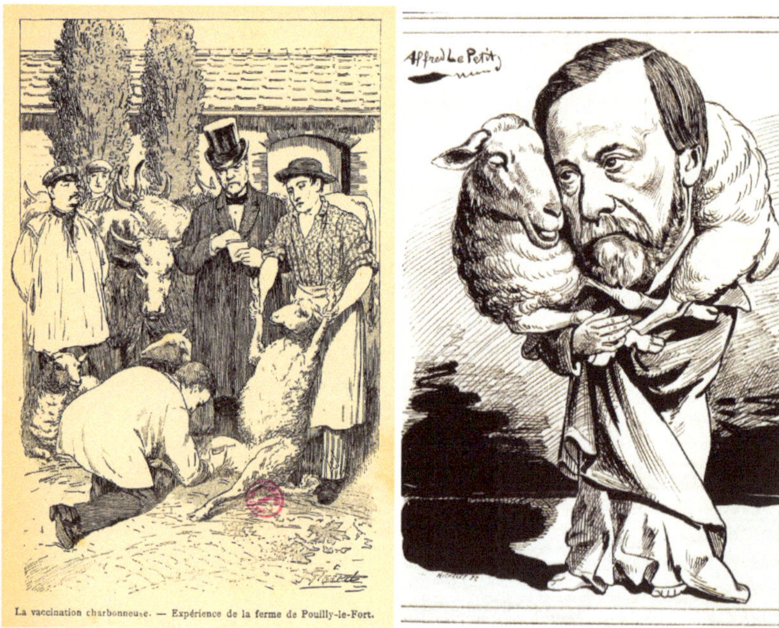

Figure 1. First anthrax vaccination in 1881 at Pouilly le Fort. (left) Emile Roux vaccinating sheep in the presence of Louis Pasteur. Drawing by J Girard, 1887. ©Institut Pasteur/Musée Pasteur. **(right)** Drawing by Alfred Le Petit in Le Charivari, 27 April 1882. https://gallica.bnf.fr/ark:/12148/bpt6k3 073477c#, (accessed on 5 November 2023).

Following the success and the immense interest aroused by the first human vaccination against rabies in July 1885.

Rabies was a frightening disease that struck people's imagination. As a zoonosis transmissible to humans, it was an ideal research field for Louis Pasteur for the application of the notion of pathogen attenuation to vaccination. This research was developed in Pasteur's laboratory, then in École Normale rue d'Ulm in Paris, from 1880 until the first human vaccinations of Joseph Meister and Jean-Baptiste Jupille [6].

Pasteur reported on 1 March 1886 at the Académie des Sciences that 350 people had been vaccinated with only one failure. The rooms at the École Normale were becoming too limited in space to accommodate the increasing number of patients [6,7]. A project for a centre for vaccination against rabies was proposed. This led to the foundation and inauguration of Institut Pasteur at its current location, rue Dutot (now rue du Docteur Roux), on 14 November 1888.

2. The First Golden Age of *B. anthracis* Research: The Vaccines

B. anthracis was known to be responsible for anthrax in domestic and wild herbivorous animals, causing a significant economic problem in the 19th century and the beginning of the 20th century, with massive livestock deaths, along with human infections, especially in wool sorters [4,8–10]. The discovery and the use of attenuated strains by Pasteur led to the introduction of effective vaccination, beginning with chicken cholera. The anthrax vaccine was a live attenuated vaccine first produced at École Normale and then at Institut Pasteur.

Pasteur initiated his research on B. anthracis in 1877 at École Normale. From the archives at Musée Pasteur, it appears that the "vaccin charbonneux" was produced as early as 1882 on premises rue Vauquelin close to the École Normale (AIP PAS. G1 46).

A document from 1883 shows Louis Pasteur as a business manager (Figure 2) annotating each page of the expenditure statement:

"This statement shows that all expenses relating to the vaccine were paid from the vaccine fund. What remains in the fund forms a net profit which is divided into five parts: Mr. Pasteur reserves two parts for himself; he allocates two parts, i.e., an equal sum, to his collaborators. The fifth part constitutes a reserve fund".

"The annual sum allocated to my laboratory by the *Ministère de l'Instruction Publique* is ten thousand francs. At the end of the year, this is barely enough to pay for gas and heating".

"In 1882, 10,000 fr. of the 50,000 remain to be spent. This credit of 50,000 is just enough to cover the laboratory's working expenses. 24 May 1883".

At this time, vaccines relied on the empirical process of the attenuation of pathogens (as will be discussed later, the basis of the pathogenesis of *B. anthracis* was unknown—the toxins were only fully described in 1954 [11], and the plasmids carrying the genes coding for the toxins and capsule were reported in 1983 and 1985 [12–14]). The stability of the attenuation level was far from mastered, leading to the low, but existing, frequency of adverse effects in vaccinated animals (AIP SVV.1). Furthermore, as the vaccine was alive, it needed to be maintained through *in vitro* (rarely *in vivo*) passages and its attenuation controlled.

Historically (AIP SVV.1 & SVV.2), the anthrax vaccines were developed and produced in the "Service des Vaccins Vétérinaires", initially by Charles Chamberland from 1889 to 1904; the vaccine against "Rouget du porc" (swine erysipelas, *Erysipelothrix rhusiopathiae*) was the other major bacterial vaccine produced during this period.

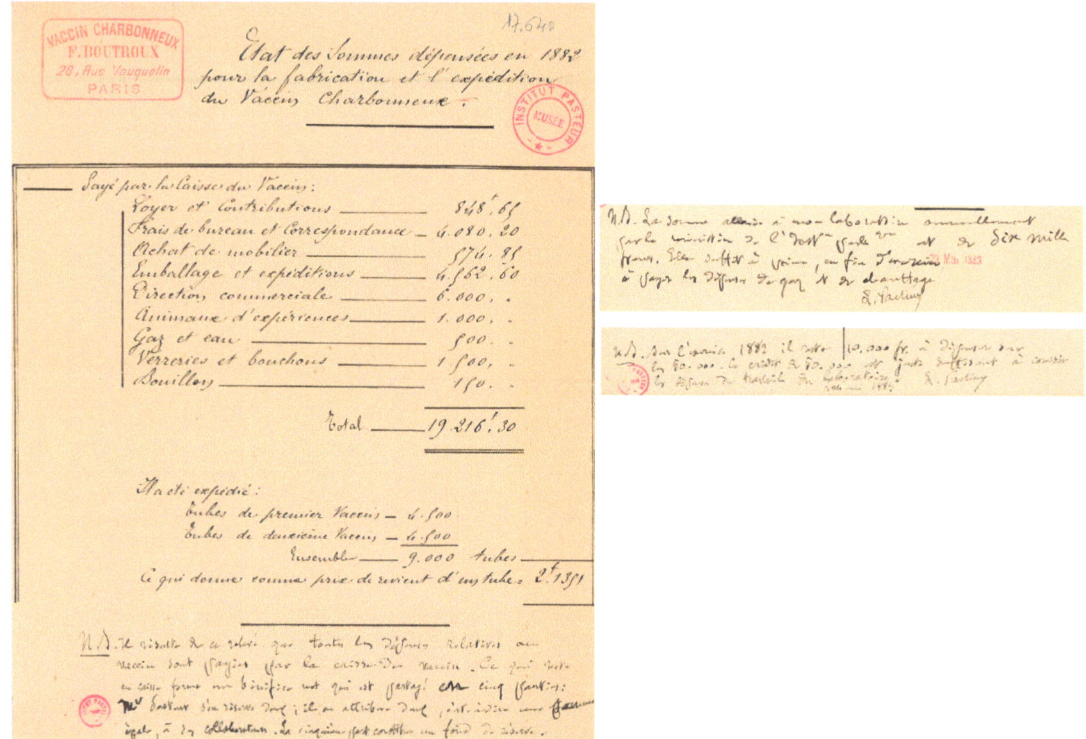

Figure 2. **Statement of manufacturing and shipping costs in 1882 for the "Vaccin Charbonneux".** At the bottom of the page, Louis Pasteur's autograph can be read; two others on two further pages of this document are shown on the right (english translation in the text). French transcription: *"Il résulte de ce relevé que toutes les dépenses relatives au vaccin sont payées par la caisse du vaccin. Ce qui reste en caisse forme un bénéfice net qui est partagé en cinq parties: Mr Pasteur s'en réserve deux; il en attribue deux, c'est-à-dire une somme égale, à ses collaborateurs. La cinquième part constitue un fonds de réserve".* *"La somme allouée à mon laboratoire annuellement par le ministère de l'Instuction publique est de dix mille francs. Elle suffit à peine, en fin d'exercice, à payer les dépenses de gaz et de chauffage".* *"Sur l'exercice 1882 il reste 10.000 fr. à dépenser sur les 50.000. Ce crédit de 50.000 est juste suffisant à couvrir les dépenses de travail du laboratoire. Le 24 mai 1883".* AIP PAS. G1 46. ©Institut Pasteur/Musée Pasteur.

The anthrax vaccine was distributed in many countries; the production was relocated to produce and distribute the vaccine directly on site. An interesting document gives some insights into this aspect. In 1886, a contract was signed between Charles Chamberland and Henri Lefebvre de Sainte Marie—defined as "député, sous-directeur, laboratoire de M. Pasteur" and "ancien inspecteur général de l'agriculture", respectively; it aimed to entrust the production of the anthrax vaccine outside France (and its colonies) for 30 years (Figure 3). Some countries were excluded, as other exclusivity contracts were already running; these were Argentina, Paraguay, Uruguay, and Austria-Hungary. The selling prices were defined. In another instance, a little later (1888–1891), Louis Pasteur sent Adrien Loir, his nephew, to Australia to confirm that Cumberland disease, which affected cattle on this continent, was, in fact, anthrax. Adrien Loir then set up a vaccine production unit in Sydney [15].

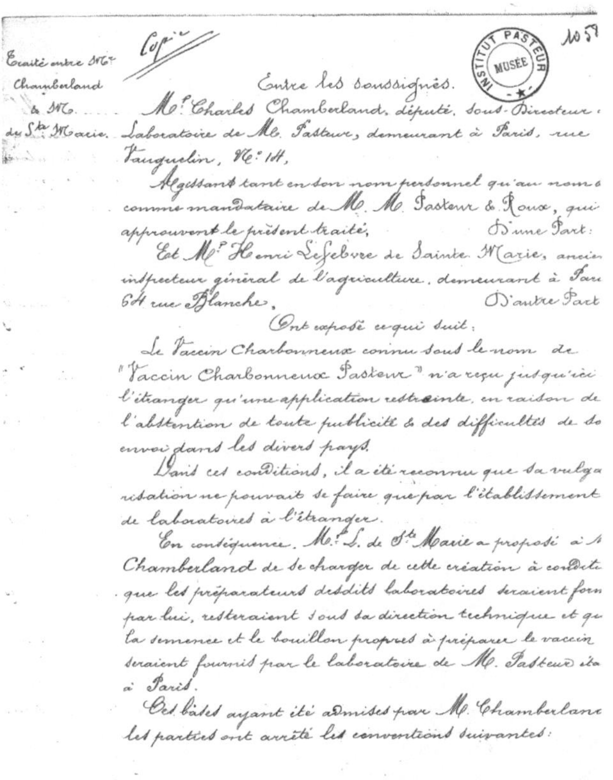

Figure 3. **Contract for the production of the anthrax vaccine in foreign countries (first page),** signed in 1886 between Charles Chamberland and Henri Lefebvre de Sainte Marie. Chamberland represented Pasteur, Roux, and himself: "The anthrax vaccine known as "Vaccin Charbonneux Pasteur" has so far received only limited application abroad, due to the absence of any publicity and the difficulties of shipping it to various countries. Under these conditions, it was recognized that its wider distribution could only be achieved through the establishment of laboratories abroad. Consequently, Mr. L. de Ste Marie proposed to Mr. Chamberland that he take charge of the creation of such laboratories, on condition that the assistants of the said laboratories, trained by him, would remain under his technical direction, and that the seed and broth suitable for preparing the vaccine would be supplied by Mr. Pasteur's laboratory in Paris. These terms having been accepted by Mr. Chamberland, the parties entered into the following agreements:" (english translation: Dominique Goossens). *"Le vaccin charbonneux connu sous le nom de "Vaccin Charbonneux Pasteur" n'a reçu jusqu'ici à l'étranger qu'une application restreinte, en raison de l'abstention de toute publicité & des difficultés de son envoi dans les divers pays. Dans ces conditions, il a été reconnu que sa vulgarisation ne pouvait se faire que par l'établissement de laboratoires à l'étranger. En conséquence, Mr L. de Ste Marie a proposé à Mr Chamberland de se charger de cette création à condition que les préparateurs desdits laboratoires seraient formés par lui, resteraient sous sa direction technique et que la semence et le bouillon propres à préparer le vaccin seraient fournis par le laboratoire de M. Pasteur étant à Paris. Ces bases ayant été admises par M. Chamberland, les parties ont arrêté les conventions suivantes:"* AIP PAS.G1 33 ©Institut Pasteur/Musée Pasteur.

After Chamberland's illness and death (1908), the "Service des Vaccins Vétérinaires" was headed by Émile Roux as "Chef de service" and Constant Jouan as "Chef de laboratoire adjoint"; Jouan actually managed the service, as Émile Roux had his own directorial activities. Jouan's presence at Institut Pasteur can be traced back to 1893; he was "préparateur" in the "Service de Microbiologie appliquée à l'hygiène et des vaccinations", managed by Chamberland. Then Jouan left the Institut Pasteur (before 1925).

The exact date has not yet been found in the Archives at Institut Pasteur. Jouan then created the well-known laboratory equipment enterprise that still carries his name (see the following link for the 1933 catalogue: http://www.bium.univ-paris5.fr/histmed/medica/ cote?extaphpin014, accessed on 5 November 2023). (All the information in this paragraph was researched and kindly communicated by Sandra Legout, CeRIS, Institut Pasteur)

Victor Frasey, the director of the "écuries d'Alleray" (stables at Alleray, premises not far from the Institut Pasteur campus) then assumed responsibility, with Charles Truche and André Staub as "adjoints". Truche later headed the service until 1934. André Staub was an "assistant" from October 1906 onwards; from 1934, he managed the "Service des Vaccins Vétérinaires" and maintained this activity until 1951, the date of his retirement (Figure 4).

Figure 4. André Staub (*circa* 1935) worked on *B. anthracis* from 1906 to 1951. Other pathogens were also studied, such as swine erysipelas, avian influenza, chicken cholera, contagious bovine pleuropneumonia, fowl typhoid, and classical swine fever. His laboratory notebooks (spanning 1901–1951), which are kept in the Archives of Institut Pasteur, give a rare glimpse into the functioning of a research laboratory at the beginning of the 20th century. ©Institut Pasteur/Musée Pasteur.

When perusing the laboratory notebooks and the official reports of these structures, it emerges that the main research on *B. anthracis* focused on the production, characterisation, surveillance, and improvement of the vaccines. This type of approach was similar to that used for the vaccines against chicken cholera, swine erysipelas, or rabies.

Anthrax vaccination usually required two to three inoculations of the attenuated vaccine strains with increasing virulence, with the "first vaccine" being the most attenuated; thus, so-called first, second, and, sometimes, third vaccines were produced, as is apparent from the laboratory notebooks (AIP SVV.1&2). Their degree of residual virulence was regularly tested in mice, guinea pigs, and rabbits, resulting in the modification of their use. For instance in an entry on 15 September 1934 (Figure 5): "trial of the second anthrax vaccine (used as third vaccine from 22 September 1934) . . . trial of the third anthrax vaccine (used as second vaccine from 22 September 1934)" or "second anthrax vaccine . . . too strong, to be used as third vaccine (vials labelled 23 April 1940)" (translation PLG).

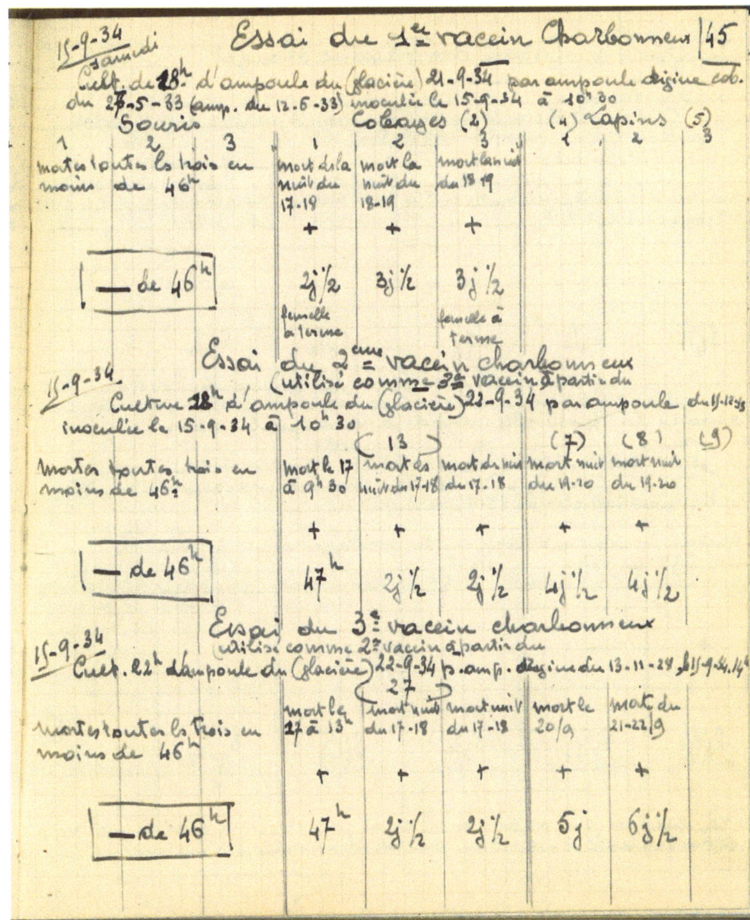

Figure 5. Detail of André Staub's laboratory notebook, illustrating how, in 1934, the *B. anthracis* "Pasteur vaccines" were checked and adapted. ©Institut Pasteur/Archives—Fonds Service des vaccins vétérinaires.

One of the particulars was that each animal species to be vaccinated showed different susceptibilities to anthrax: some are highly susceptible (sheep), others much less so (bovines); hence, the residual virulence of the less attenuated vaccines in susceptible species. Accidents of vaccination regularly occurred in the vaccinated animals, prompting additional research (AIP SVV.1&2). There were repeated attempts to obtain a "vaccin unique", a single vaccine that could protect all animal species. However, this did not actually succeed, and *in fine* there were several "vaccin unique" that were either specific to sheep, bovines, or goats ("vaccin unique mouton", "vaccin unique bovin", and "vaccin unique chèvre"; AIP SVV.1&2). Usually, the bovine vaccine was a two-fold dose of the sheep vaccine.

In other cases, following instances of some vaccine batches having suspected low efficiency or in areas where there was a high level of *B. anthracis* spore contamination, the production of the vaccine strain was tailored by adapting its attenuation levels, with such denominations as "vaccin fort" (i.e., strong) and "vaccin special" found in the laboratory notebooks.

An intriguing point is what type of attenuation the Pasteur vaccines harboured, as the substratum of virulence was then unknown. Did they lose one of the plasmids, thus becoming atoxinogenic or unencapsulated? On 31 March 1922, André Staub explored this aspect (Figure 6), and the following is what he observed:

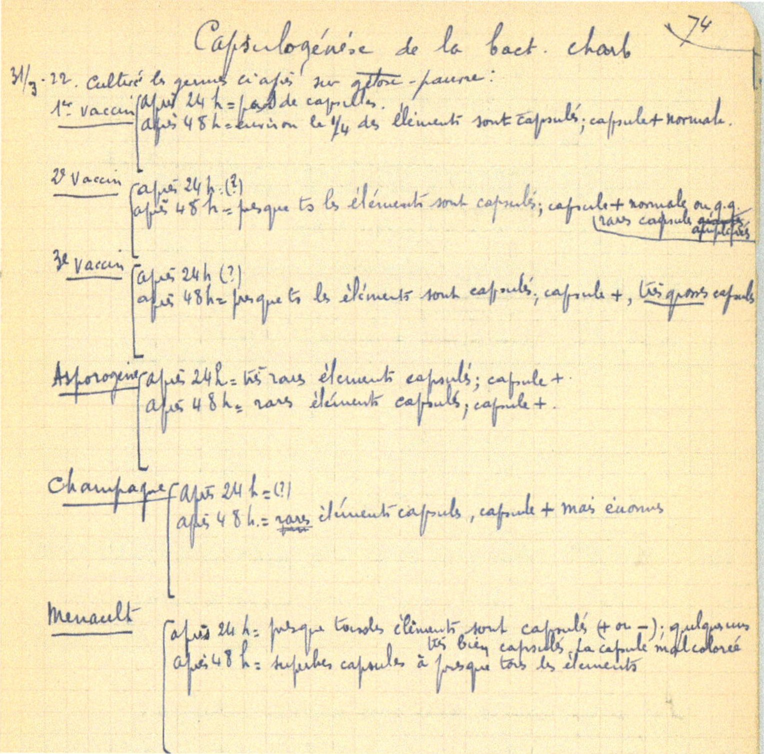

Figure 6. In March 1922, André Staub tested the capsulation of the Pasteur vaccine strains. The "asporogene" strain was related to assays developing another type of attenuated vaccinal strain; "Champagne" and "Menault" refer to two other strains that were addressed to the laboratory. ©Institut Pasteur/Archives—Fonds Service des vaccins vétérinaires.

"First vaccine: around $\frac{1}{4}$ of encapsulated bacteria, normal capsule
Second vaccine: almost all bacteria are encapsulated, normal capsule, rare 'amplified' capsule
Third vaccine: almost all bacteria are encapsulated, very thick capsule" (my own translation).
Clearly, the "Pasteur" strains used were encapsulated but at different levels, either in terms of the percentage of the entire bacterial population or the quantity of capsular material per bacterium. They were most probably toxinogenic; testing toxinogenesis was, however, not available at this time, so this question will most probably remain unanswered. One hope would be to sequence the vaccine strains. Some (many?) strains have been labelled "Pasteur strains" in various laboratories throughout the world, originating from strain exchanges between laboratories. As they have been stored for many years, how much they reflect the original strains with the minimum of genetic changes during storage and cultivation remains to be evaluated. Some of these "Pasteur" strains were later tested for capsulation, showing capsulation heterogeneity and suggesting that some form of encapsulation reversal could occur [13,14]. Our current knowledge of genetics and gene regulation might provide a more specific basis for the phenotype of these strains (these approaches have tentatively been applied to some of the "Pasteur" strains [16,17]).

One last opportunity would be to exploit the ancient Pasteur first and second vaccine strains stored in sealed vials in the Institut Pasteur museum that have been preserved until the present day (Figure 7).

Figure 7. Ancient vials of Pasteur vaccines form part of the museum collection; exposed in the "salle des souvenirs scientifiques du Musée Pasteur". Photo credit: Pierre L. Goossens.

> *The conservation of biological samples in a museum is an interesting topic, raising ethical concerns in terms of human sources and the regulatory concerns relevant to specific regulations (see Section 5). Different approaches are usually considered, ranging from destruction—and the loss of biological patrimony— to storage and access according to regulations. Valorisation could be a key mission for a museum [18,19]; in this respect, the collection in the Musée Pasteur might provide valuable data. Constant vigilance should, nevertheless, be exerted, following the evolution of the regulations and the consequences for biological patrimony through the degree of stringency of their application/implementation in each institution, Institut Pasteur included) to avoid irreversible decisions that might be regretted later.*

Due to the advances in sequencing and genetic analysis, their characterisation should bring interesting insights into this old question; one advantage is the fact that the vaccines exist in the form of spores that are highly resistant. Spores are hard to break, and it may prove challenging to extract enough DNA for meaningful sequencing in these limited, precious samples; they may also have accumulated some mutations due to cosmic radiation [20–22]. Let us be imaginative and optimistic.

In the 1930s, unencapsulated strains were reported by Nicolas Stamatin in Romania and Max Sterne in South Africa [23–25]. When used for vaccination, these strains gave lower mortality/morbidity and were safer for use on cattle [26]. At the time, nobody understood the basis of this attenuation, as the plasmids had not yet been described. The Sterne strain arrived at Institut Pasteur on 29 December 1947 through André Staub (Figure 8, AIP SVV.1) and was tested for its protective efficiency [27]. No report considering its potential use as an alternative vaccine was found in the laboratory notebooks.

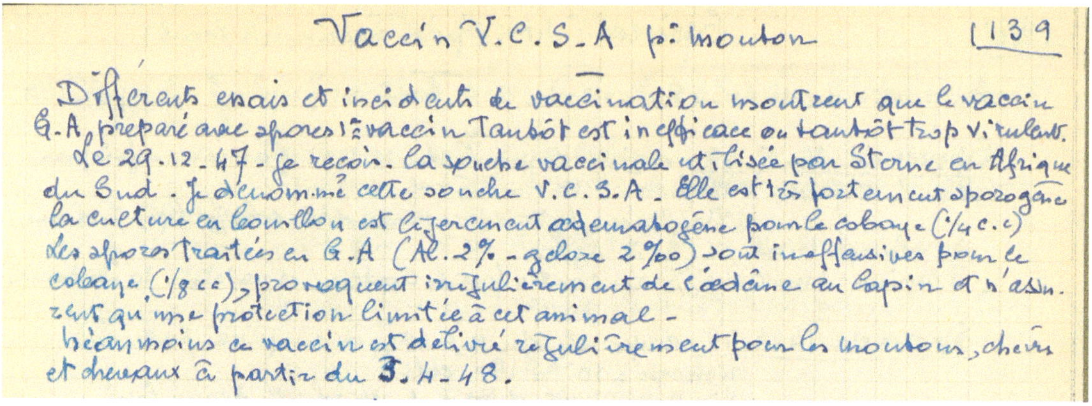
Vaccin V.C.S.A p: mouton 139

Différents essais et incidents de vaccination montrent que le vaccin G.A, préparé avec spores 1er vaccin tantôt est inefficace ou tantôt trop virulent.
Le 29.12.47 - je reçois la souche vaccinale utilisée par Sterne en Afrique du Sud. Je dénomme cette souche V.C.S.A. Elle est très fortement sporogène la culture en bouillon est légèrement oedematogène pour le cobaye (1/4 c.c) Les spores traitées en G.A (Al.2% - gélose 2‰) sont inoffensives pour le cobaye (1/8 cc), provoquent irrégulièrement de l'oedème au lapin et n'assurent qu'une protection limitée à cet animal -
Néanmoins ce vaccin est délivré régulièrement pour les moutons, chèvres et chevaux à partir du 3.4-48.

Figure 8. The unencapsulated strain isolated by Max Sterne in South Africa reached the Institut Pasteur on 29 December 1947, according to André Staub's laboratory notebook entry. "Vaccine V.C.S.A. for sheep. Various trials and accidents of vaccination show that the G.A. vaccine prepared from spores of the first vaccine is sometimes inefficacious or sometimes too virulent. On 29 December 1947 I receive the vaccine strain used by Sterne in South Africa. I designate this strain V.C.S.A. It is highly sporogenic, the culture in broth medium is slightly edematogen for the guinea pig (1/4 cm^3). The spores treated in G.A. (Al.2%–gelose 2‰) are innocuous for the guinea pig (1/8 cm^3), irregularly provoke edema in rabbit and provide only a limited protection for this animal. However this vaccine is regularly delivered for sheep, goat and horses from 3 April 1948" (my own translation). G = gélose; A = Alun [28]. The meaning of the acronym V.C.S.A. is unknown, but may be guessed as Vaccin Charbonneux South Africa, awaiting further findings. ©Institut Pasteur/Archives—Fonds Service des vaccins vétérinaires.

Apart from the vaccines, immune sera were produced from rabbits and horses for diagnosis ("sérum précipitant anticharbonneux pour réaction d'Ascoli") and for the experimental testing of protection transfer, with no great success achieved (AIP SVV.1).

When André Staub retired in 1951, the Service des Vaccins Vétérinaires was fused with the Service de Microbiologie Animale, which exerted its activity into research on viruses (an emerging domain at this time) under Henri Jacotot, André Vallée, and Bernard Virat as successive heads, assuring the permanence of *B. anthracis* activity until the beginning of the 1970s. In 1954, for instance, Bernard Virat assessed the longevity of *B. anthracis* spores from samples ranging from 1884 to 1900, gathered in the museum ([29] and AIP SMA.1 for the original data); only 4 out of 100 samples could be revived and three of them had kept their initial virulence but were unable to protect rabbits against a virulent challenge.

Looking back at the activity of anthrax vaccine production at the Institut Pasteur provides some hints as to the number of doses delivered. In 1914, for the 25th anniversary of the Institut Pasteur, Émile Roux mentioned the following:

"The oldest of our practical departments is that of the anthrax vaccine, it goes back to the famous experiment at Pouilly-le-Fort, in 1881, and was organised by Chamberland. Soon the vaccine for swine erysipelas was also developed and for the past thirty-two years, the department has delivered 41,649,592 doses of anthrax vaccine and 10,716,906 doses of swine erysipelas vaccine. Messrs. Jouan and Staub who ensure the preparation of these vaccines deserve the recognition of farmers" (english translation Dominique Goossens).

> *"Le plus ancien de nos services pratiques est celui des vaccins charbonneux, il date de la célèbre expérience de Pouilly-le-Fort, en 1881, et fut organisé par Chamberland. Bientôt le vaccin du rouget des porcs vint s'ajouter à celui du charbon et depuis trente-deux ans que le service fonctionne, il a délivré 41 649 592 doses de vaccin charbonneux et 10 716 906 doses de vaccin du rouget. MM. Jouan et Staub, qui assurent la préparation de ces*

vaccins, ont droit à la reconnaissance des agriculteurs " (discours de M le Docteur Roux, Le XXVe Anniversaire de l'Institut Pasteur. In: Revue internationale de l'enseignement, tome 67, Janvier–Juin 1914. pp. 60–82).

https://www.persee.fr/doc/revin_1775%E2%80%936014_1914_num_67_1_6822 (accessed on 5 November 2023).

In 1936, anthrax Pasteur vaccine production was in the range of 120,000 annual doses for sheep and 90,000 doses for bovine; in 1950, this rose to around 300,000 and 100,000 annual doses, respectively (Figure 9, AIP SMA.1).

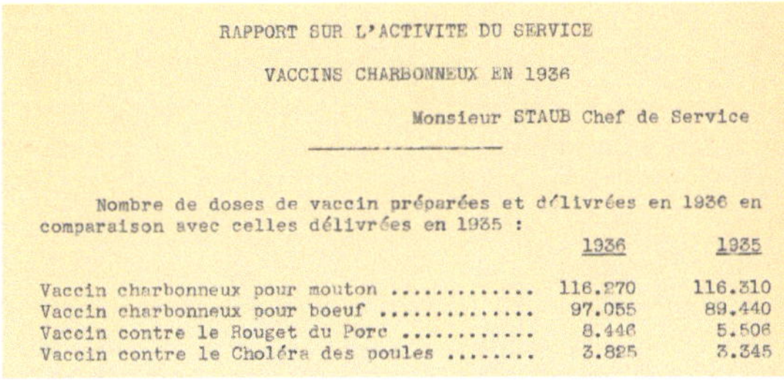

RAPPORT SUR L'ACTIVITE DU SERVICE

VACCINS CHARBONNEUX EN 1936

Monsieur STAUB Chef de Service

Nombre de doses de vaccin préparées et délivrées en 1936 en comparaison avec celles délivrées en 1935 :

	1936	1935
Vaccin charbonneux pour mouton	116.270	116.310
Vaccin charbonneux pour boeuf	97.055	89.440
Vaccin contre le Rouget du Porc	8.446	5.506
Vaccin contre le Choléra des poules	3.825	3.345

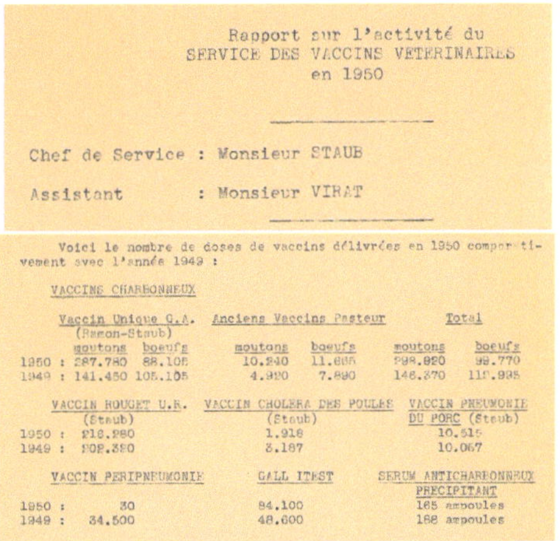

Rapport sur l'activité du
SERVICE DES VACCINS VETERINAIRES
en 1950

Chef de Service : Monsieur STAUB

Assistant : Monsieur VIRAT

Voici le nombre de doses de vaccins délivrées en 1950 comparativement avec l'année 1949 :

VACCINS CHARBONNEUX

	Vaccin Unique Q.A. (Ramon-Staub)		Anciens Vaccins Pasteur		Total	
	moutons	boeufs	moutons	boeufs	moutons	boeufs
1950 :	287.780	88.105	10.240	11.665	298.820	99.770
1949 :	141.450	105.105	4.920	7.890	146.370	112.995

	VACCIN ROUGET U.R. (Staub)	VACCIN CHOLERA DES POULES (Staub)	VACCIN PNEUMONIE DU PORC (Staub)
1950 :	218.280	1.918	10.515
1949 :	202.380	3.187	10.067

	VACCIN PERIPNEUMONIE	GALL ITEST	SERUM ANTICHARBONNEUX PRECIPITANT
1950 :	30	84.100	165 ampoules
1949 :	34.500	48.600	168 ampoules

Figure 9. **Details from annual reports on vaccine production in the Service des vaccins vétérinaires** in 1935–36, and 1949–50, showing the number of doses of the various anthrax vaccines produced at Institut Pasteur, along with the other manufactured vaccines against animal pathogens. ©Institut Pasteur/Archives—Fonds Service des vaccins vétérinaires.

In the 1960s, Institut Pasteur was experiencing a delicate financial period; during his directorship (1971–1976), Jacques Monod obtained an increase in financial support from the French government. Let us recall that, from its creation, Institut Pasteur was a private enterprise, in order to retain its independence. The downside was that it had an obligation to find financial sources, for example, through the industrial commercialisation of its products (vaccines, diagnostics, antisera, antitoxins, etc.). The acceptance of this public financial contribution was tied to the separation of the research activities and the production/diagnostic activities. This led to the creation of Institut Pasteur Production (1972–73), which was later split in two: Diagnostic Pasteur and Pasteur Vaccins. The latter was then fused within Institut Merieux where the Sterne vaccine was produced. The Pasteur anthrax vaccines, which were already in decline, then disappeared, and this was the end of an era [30].

It seems from the Archives that the vaccine was produced until the early 1970s when Bernard Virat deposited the *B. anthracis* strains in the collection of the Institut Pasteur. The activity around *B. anthracis* was quite low, mainly covering the functions of a current national reference center. Work on *B. anthracis* was then interrupted till 1986 (perusing the litterature shows that some experiments were performed in other laboratories, using *B. anthracis* as a tool/target for antimicrobial therapy assays (for instance, [31]).

3. Anthrax Toxins: The Puzzle of a Complex Research Domain

In the previous section, toxins produced by *B. anthracis* are not mentioned. This is not because they were not looked for. Other toxins were described as early as 1889–90 for diphtheria or tetanus, for example ([32–35], see this special issue [36,37]); toxins were even mentioned by Louis Pasteur for chicken cholera caused by *Pasteurella multocida* [38], though no confirmation could be obtained thereafter.

At Institut Pasteur, André Staub searched for the toxins produced by *B. anthracis* as soon as 1909, as can be read from his laboratory notebooks, and this carried on regularly going forward (for example in 1911 and 1920; AIP SVV.1&2). He did not, however, obtain clear and tangible results on toxic activities. Anne-Marie Staub, his daughter, and Pierre Grabar followed up this query later in the 1940s at Institut Pasteur, taking advantage of the progress in antibody purification techniques to explore which *B. anthracis* antigens could be involved in toxicity [39,40]; no clear demonstration could be achieved, the time was not ripe for such a breakthrough.

So why were so many years necessary to reach the basis of our current knowledge on *B. anthracis* toxins (initially acquired by Harry Smith from 1954 onwards, with a series of accompanying papers in the following years characterising the system)? Harry Smith has given an excellent, highly readable, and vivid account describing the conditions of this discovery [11].

Let us first begin with an *a posteriori* brief overview of what is currently known of *B. anthracis* toxins before turning to the potential reasons for this delayed description.

Two main toxic activities are produced by *B. anthracis*, which were considered (for many years) as (1) a lethal toxin leading to cellular death and (2) an edema toxin easily observed through the characteristic edema in anthrax or in animal models. However, these two activities are mediated through a third component, named protective antigen (PA), that ensures the entry of the catalytic components edema factor (EF) and lethal factor (LF) into the cell. PA multimerises—classically an heptamer—at the cell surface after interaction with a cell receptor. EF and LF then interact with two adjacent molecules of PA, with the heptamer thus accomodating three molecules of EF/LF (octamerisation has also been reported, thus accomodating four molecules of EF and/or LF [41]).

Historically, two toxins were described: edema and lethal toxins (ET and LT, respectively EF+PA and LF+PA). The current view is to name this complex a tripartite toxin, which can exert two toxic activities. After internalisation and intracellular trafficking (the complex events are out of the scope of this review; for extended notions, see specialised reviews, for example [42–45]), EF and LF are translocated into the cytosol, where they exert their enzymatic toxic activity.

In 1982, Steve Leppla showed that EF is a calmodulin-dependent adenylate cyclase [46]; LF enzymatic activity (zinc metalloproteinase) and cellular targets (the majority of mitogen-activated protein kinase kinases, MAPKKs) were unknown until 1994 and 1998, respectively [47–49].

Now, what could be the reasons for this lengthy delay before discovery when compared to other toxins such as diphtheria or tetanus toxin?

1. First, as just mentioned, the *B. anthracis* toxin(s) is a multi-component toxin, an AB toxin. Such multi-molecular architecture necessitates the purification of at least two components to be able to produce an active toxin that can be tested *in vitro* or *in vivo*. The detection of direct toxicity that mimics the pathology of the infection *per se* through the inoculation of filtered bacterial extracts or culture medium was, indeed, central to diphtheria or tetanus toxin discovery [32–35]. Furthermore, to follow the presence of a toxin in a given sample is much easier when its enzymatic activity is known (but not necessarily, as tetanus toxin enzymatic activity was discovered many years after its toxicity (1992 vs. 1889, see [50]). For *B. anthracis*, the edema was a pathognomical sign both in humans and animals; it could be easily followed *in vivo* during production and purification. However, *in vitro* experiments were hampered until calmodulin dependence was recognised [46].

2. Second, the production of the edema and lethal toxins by *B. anthracis* necessitates specific induction conditions [51] (toxin and capsule expression are co-ordinately regulated [52]). Many years were needed to finally unravel these *in vitro* conditions (bicarbonate + CO_2), and thus mimic the *in vivo* environment to obtain sufficient levels

of toxins. Harry Smith freed himself from these constraints and unknowns, as he purified the toxins directly from serum and peritoneal liquids from infected animals.

3. Third, another crucial aspect is the availability of biochemistry techniques used for the isolation and purification of biological molecules. The majority were developed and became available during the second half of the 20th century (electrophoresis, column purification, and immunotechniques, among others [53]) and could, thus, not be applied to in-depth analyses of the composition of the biological milieu (either procaryotic or eucaryotic). The main technique available at the turn of the 20th century was filtration (filtre de Chamberland).

Let us also keep in mind that the vaccines developed, initially at École Normale and later at Institut Pasteur, then internationally, were quite effective. When combined with the emergence of the live attenuated unencapsulated vaccines on the one hand [23,24], and the advent of antibiotics after the second world war on the other hand, veterinary anthrax was efficiently controlled. The economic pressure was less urgent, and this research domain was no longer a priority. However, as a consequence of the existence of programs for biological weaponry development, research was pursued in military/army laboratories, such as in Porton Down, UK, where Harry Smith encountered favourable conditions to develop his basic science project on *B. anthracis* toxins.

Taken together, this provides some clues as to why the *B. anthracis* toxins required a considerable amount of time before being unequivocally detected, purified, and characterised.

4. The Second Golden Age of *B. anthracis* Research

As mentioned above, the separation of the research activity from the diagnostics and vaccine activities at the beginning of the 1970s led to the interruption of research on *B. anthracis* at Institut Pasteur. Work on *B. anthracis* was revived from 1986 onwards at Institut Pasteur by Michèle Mock, a scientist at the CNRS, a national French research organisation (Centre National de la Recherche Scientifique). During her doctoral and post-doctoral formation, she specialised in colicins, which can be considered plasmid-borne bacterial toxins that are directed against other bacteria [54]. Her postdoctoral training in John Collier's "toxin" laboratory (then in Los Angeles) made her fluent in the new techniques in molecular biology such as cloning and sequencing among many others [55]. The period was, indeed, blooming with new technologies in molecular biology. This was also applied to *B. anthracis*, hence the discovery of the genetic substratum of the toxins; the pXO1 "toxin plasmid" was first described in 1983 [12] and the "capsule plasmid" in 1985 [13,14], paving the way for future avenues of research.

Michèle Mock became aware of Steve Leppla's work on *B. anthracis* edema factor and its calmodulin-dependent adenylate cyclase toxin activity [46] and decided to initiate a new route in her research career as she was keen to explore bacterial toxins; *B. anthracis* was now known to produce toxins. The scientific environment at the Institut Pasteur was favourable at that time. For instance, another bacterium that also produces a calmodulin-dependent adenylate cyclase, *Bordetella pertussis*, was actively studied in Agnès Ullmann's laboratory on the floor below, and the scientific exchanges between the two laboratories were key to the successful emergence of the *B. anthracis* project. Furthermore the immense opportunity of the genetic tools developed by Patrick Trieu-Cuot that ultimately enabled the heterogramic transfer of genetic material between *Escherichia coli* and *B. anthracis* [56], allowed the generation of *B. anthracis* mutants. Everything was, thus, perfect for initiating a new era in *B. anthracis* research at Institut Pasteur, all the more as the senior scientists then in charge of the decisions gave their green light.

Interestingly, Michèle Mock's *primum movens* was to understand the contribution of the *B. anthracis* virulence factors to the infection and, thus, explore the *in vivo* effect of inactivating each toxin—not only focusing on the genetics. Complementary expertise was, thus, introduced in the laboratory, leading to a multi-disciplinary approach, ranging from structural analysis, biochemistry, genetic expression and regulation, to bacterial-host cell interactions, *in vivo* pathogenicity in various animal models, and therapy development.

The laboratory was a scientific hub for many colleagues, leading to scientific discussions, training, and the exchange of materials (almost impossible nowadays with our current regulations in France, see Section 5).

Summarising almost 25 years of *B. anthracis* original research from the anthrax Pasteur laboratory is quite a challenge. Some of the key points selected from this abundant research output are presented below:

4.1. On Toxins

If one summarises these years of research on the genetics of *B. anthracis*, the laboratory was a pioneer in terms of the construction of bacterial and plasmid systems for generating mutants, cloning the toxin genes (*cya*, *lef*, and *pagA* for the EF, LF, and PA moieties, respectively), and their subsequent inactivation through the insertion of antibiotic cassettes or point mutations [57–60].

In addition to exploring the contribution of each toxin component in virulence and pathogenicity, the inactivation of each gene enabled better purification of the remaining toxin for cellular or *in vivo* experiments, the aim being to purify them directly from *B. anthracis* and not from recombinant *E. coli*, as those toxins were contaminated with LPS, with its confounding multiple effects.

Similarly, another practical aim was to increase the production of toxins to increase purification yields; hence, the initial interest in regulation of toxin gene production. The regulation facet, which was, of course, also investigated for its fundamental scientific interest, was successfully developed by Agnès Fouet and her group, leading to the exploration of some of the central regulatory networks (*atxA*, *pagR*, and *codY*, among others [52,61–65]).

4.2. On Bacterial Cell Surface

The purified toxin components were initially contaminated with high molecular weight proteins from the vegetative cells; the *sap*/*eag* system of the S-layer, thus, became a research focus, both at the genetic (gene organisation, regulation) and structural level. Inactivating them was a means to allow better purification of the toxin components but, at the same time, opened the way to unravelling the regulation of the production of this surface layer of the vegetative cells [66–69].

Pursuing the studies on *B. anthracis* cell surface led to the exploration of the structure and regulation of the other main major virulence factor, the pseudoproteic poly-gamma-D-glutamate (PDGA) capsule [67,70–72]. Sortases and cell surface-anchored proteins also became a focus of interest [73,74].

In parallel to the exploration of the vegetative cell surface, the spore surface was extensively studied, as it was of interest for vaccine development; furthermore, spore surface composition and structure is central to sporulation and germination, hence its implication in successful colonisation during infection, this being an obvious target for vaccines [75,76].

4.3. On Pathophysiology

If one wishes to have a global view of the *B. anthracis* infectious process, the crucial point is to remember that anthrax is a toxi-infection with two facets: (1) one is the infection *per se*, i.e., the encapsulated bacteria disseminate systemically and multiply, leading to major terminal septicemia, and (2) the other is the deleterious effects of the toxins secreted by the multiplying bacteria on multiple organs and cellular systems (for this aspect, many excellent reviews are available, with just a few cited here [42–45]).

The studies at the anthrax Pasteur laboratory were initiated on the unencapsulated toxinogenic attenuated Sterne background [77–79], as it was easier and safer to manipulate. One of the drawbacks of this is that they reproduce only the toxin arm of the infection and necessitate specific animal models susceptible to toxin-only effects [80].

The exploration of the infection arm, i.e., infection with encapsulated non-toxinogenic strains, was developed in the laboratory from 2000 and developed by Pierre L. Goossens and his group [81,82]. Applying the bioluminescence technology to follow real-time *B. anthracis* infection *in vivo*, Ian J. Glomski deciphered the dynamics of bacterial dissemination from the portal of entry in cutaneous, inhalational, and gastric infections, either with encapsulated non-toxinogenic strains (exploring how the bacteria disseminate in the absence of toxins) or with toxinogenic unencapsulated strains (exploring how toxins interfere with the host defense mechanisms). The absence or presence of the poly-gamma-D-glutamate capsule strikingly modifies *B. anthracis* dissemination, pinpointing how crucial the animal model used could be to recapitulating anthrax [80–84].

A dual pattern of *in vivo* bacterial behaviour was unravelled during inhalational infection with the wild-type strain—both encapsulated and toxinogenic. Infection in each infected host progresses along two patterns of dissemination, either one mimicking what occurs when only ET (i.e., EF + PA) is expressed or another when only LT (i.e., LF + PA) is expressed [85]. This raises the intriguing possibility that the bacteria at the site of entry may initially preferentially express either ET or LT, with each toxin inducing different patterns of subsequent colonisation and dissemination. Is this a stochastic event at the bacterial level, or is this related to variations in the milieu that surrounds the bacteria at the portal of entry, thus influencing the ratio of EF/LF expression? A similar pattern of a temporal balance of EF/LF local secretion levels could be deduced from histological observations in the spleen (where depending on the size of the infectious foci—hence their "age"—an initial LF histological effect was followed by predominant edema provoked by EF) [85]. This describes a complex pattern of *B. anthracis in vivo* behaviour that will depend on local EF/LF production ratios and on the parameters in the local tissular micro-environment, such as the O_2/CO_2 balance (nasopharynx vs. lung, spleen, or liver) and temperature (cutaneous vs. deep organs).

> *The influence of the local tissular micro-environment on pathogen-host interactions occurs in other diseases, such as cutaneous vs. visceral Leishmaniasis, in terms of temperature or Mycobacterium tuberculosis colonisation in different areas of the lungs, depending on the O_2 tension: top vs. posterior areas for biped vs. quadruped behaviour (Gilles Marchal and Geneviève Milon, personal communication).*

Interestingly, although it is known that *B. anthracis* is a tripartite toxin, scientists in the anthrax field still reason as if there were two distinct toxins. As Mahtab Moayeri and Steve Leppla reflect: "The combinatorial toxins to this day remain named after the early observations made about their *in vivo* effects (lethality and edema)" [43]. Since the PA-heptamer binds three molecules of EF and/or LF, the ratio of EF/LF produced in the bacterial micro-environment will most probably influence the ratio of EF/LF molecules bound to PA, hence the quantity of each toxic moiety a cell is exposed to. Another question then emerges: if three molecules of EF and/or LF are bound to a PA-multimer, does each EF/LF moiety have the same probability of being translocated into the cytosol? In other terms, what is the probability of translocation for each remaining molecule? Is it equivalent for each, or is there a decrease of efficiency after each translocation event? The less favourable issue (for the bacterium) is that only one molecule can be translocated, implying that the PA-multimer would be trapping two potentially toxic molecules—an interesting and stimulating concept.

If one further pushes consideration on the cellular model, the complexity of the toxin effects increases; a single cell will bind various quantities of PA-multimers depending on cell surface receptor density, with each enabling the translocation of various quantities of EF and LF. As the intracellular pathways affected by each toxin are interdependent, synergistic or antagonistic effects will, in the end, lead to complex consequences for the cellular pathway functions, not forgetting that little is known about any possible disparity in *in vivo* EF vs. LF intra- and extra-cellular half-lives (extracellular proteases, proteasomes, etc.), hence the consequences on the intracellular EF/LF ratio and subsequent toxicity. All these points are still unclear and warrant further research.

4.4. On Therapeutics

When looking back at what anthrax represented in the 19th century, it was mainly a veterinary concern for the society of the time, particularly from an economic point of view; epidemics in livestock had dire consequences [8,9]. The antibiotic era had not yet begun, and the sole manner of control was knowledge of contaminated areas and carcass management. Antibiotherapy and vaccination drastically changed the philosophy of veterinary anthrax control; infected animals are usually disposed of or treated with antibiotics when needed, and vaccination protects the remaining livestock from any further extension of the epidemic. Research on veterinary anthrax is no longer a key research domain—veterinary viral infections are more deadly and pose more of a threat to global health.

Human anthrax usually develops from direct contact with infected animals or products [80]: cutaneous, through the manipulation of infected animals; inhalational—wool sorters' disease (now, though rarely, through the resuspension of spores from contaminated skin used for making drums [86]); and digestive, through the ingestion of insufficiently cooked meat from infected animals. Human cases are extremely rare in developed countries due to adequate management. Anecdotally, an epidemic occurred in 2009 in drug users due to the injection of contaminated heroin [87]. There is no aerial human-to-human transmission, and basic protective measures (avoiding contact with potentially contaminated biological material) are usually sufficient. Whatever the origin, antibiotic therapy is the treatment of choice [88].

However, *B. anthracis* may have been used as a bioweapon for a long time from what is usually mentioned in the literature [4]. During the 20th century, programs in some countries have been carried out for such nefarious uses [89–91]. The anthrax letter events in the USA in 2001 [89,92] exemplified its potential effectiveness for malevolent purposes. The defence authorities in many countries, thus, increased their interest and support for anthrax research. Due to its expertise and central position both in France and internationally, the anthrax laboratory at Institut Pasteur had, indeed, been contacted in the 1990s for counselling to increase threat responsiveness; research programs were, thus, developed—therapeutic approaches in particular. They took advantage of the knowledge acquired throughout the years of *B. anthracis* research, expanding and suggesting novel avenues of basic research in return, with their potential future applications for disease control. This dialogue between applied and basic research was a major characteristic of the Institut Pasteur from its birth (and even earlier) when looking back on how Louis Pasteur developed his research axes, ranging from the tartrate studies to rabies vaccination [38,93].

These demands of the French defence authorities led to the establishment of a long-term collaboration with the IRBA (Institut de Recherche Biomédicale des Armées, the medical research structure of the army). The transfer of knowledge was central between the two research structures, and complementarity in the scientific approaches was key to this successful collaboration (for example, Jean-Nicolas Tournier was a pioneer of exploiting the high level technology of bi-photon imaging to follow in real time the dynamics of *B. anthracis* inhalational infection in the lung both *in vitro* and *in vivo* [94]), collaboration illustrated by a number of co-publications [95–101]. Similarly, collaborations were set up with our colleagues in the CEA (Commissariat à l'énergie atomique et aux énergies alternatives) for the development of rapid and sensitive detection technologies, either for toxins or spores [102–105].

4.4.1. Preventive Therapies

The prevention of anthrax in humans relies on vaccination. The current human vaccines are based on PA being present in the bacterial culture supernatants of strains equivalent to the Sterne strain; these vaccines are produced in the US and UK. Their main aim is to target inhalational anthrax, should such an epidemic occur during warfare or bioterrorism action. These vaccines have answered the official requirements for vaccine development in humans.

However, as human anthrax is a rare disease, it means that one relies on experimental laboratory data and few epidemiological evaluations. Some concerns have been raised about the actual efficiency of the anthrax human vaccine [106]. Here emerges the problem of the choice of the animal model for testing therapeutics in general and, more specifically, for anthrax [80]. The animal model used will explore (to various degrees) each facet of the anthrax toxi-infection (see above). Thus, the PA-based vaccines relying on toxin neutralisation will mainly be tested in animals that are susceptible to the toxins. In contrast, animal models highly susceptible to the infection facet of anthrax will never be used for testing anti-toxin therapies, as the afforded protective effects would be masked by the overwhelming infectious process. In particular, the PA-based vaccines do not protect mice against an infection with a fully virulent wild-type strain, whatever the route of infection (and especially in inhalational anthrax, which is the most difficult infection to control).

The availability of a human vaccine in France could be problematic if urgently needed to protect given populations; hence, vaccination design for better protection in humans was also a main concern of the French MoD for the Pasteur anthrax laboratory, leading to the tentative development of a phase I human anthrax vaccine based on its experimental data.

As will become apparent in Section 5, the development of an anthrax human vaccine is subject to many hurdles; manipulating B. anthracis is not an easy task under current regulations. Drawing on our experience in such development, it seems judicious and reasonable to favour working on a strain belonging to the B. cereus group outside anthracis to produce spores, specific spore antigens, or the capsular poly-gamma-D-glutamate through the expression of its biosynthesis operon; unpublished experiments in the anthrax Pasteur laboratory found that B. cereus spores could replace B. anthracis spores, albeit with a small decrease in efficiency. Recombinant PA and LF as key vaccine components are already produced in E. coli. Due to the 500 bp rule of the French regulations governing any samples that may contain genetic material from B. anthracis (see below), such vaccine development concerning a human vaccine seems, for the time being, probably more easily (while safe) managed at the European level outside France.

In order to summarise the almost 15-year vaccine project, the main aim was to target both arms of the infection through the addition of (i) formaldehyde-inactivated spores (FIS) and (ii) the poly-gamma-D-glutamate capsule naturally coupled to the peptidoglycan (PGC) to the PA-based vaccines. The end results saw full protection against subcutaneous infection and a high level of protection after an inhalational challenge with a fully virulent *B. anthracis* strain in the notoriously hard-to-protect mouse model ([57,70,95,107] and Fabien Dumetz, unpublished data, Figure 10A). Most interestingly, the anti-toxin humoral immune response could play a protective role even without toxin neutralisation; adding a toxin component to the spore immunisation led to a significant increase in protection against non-toxinogenic encapsulated strains (86.5% vs. 47.5%, Figure 10B; [57] and unpublished results). Both the humoral and CD4-T cell arms of the immune response could be involved by creating a micro-environment unfavourable to initial bacterial growth.

It is usually considered that humans are more sensitive to the toxemia (than to the infection) arm of anthrax and should be protected with the current PA-based vaccines [80]; so what was the rationale for this focus on the highly susceptible mouse model of anthrax? One key point is that experiments are usually performed on adult animals in the absence of infection, stress, or any other pathologies (except when stresses or age are the central aim of the study); the data are then translated in terms of humans in the same physiological state. The challenge is to ensure that the same level of protection is reached under a stress condition, be it physical, psychological, or in the presence of concurrent viral, bacterial, or parasitic infections. Anthrax vaccination interest relies, at least initially, on protecting defence personnel in the case of malevolent use as a biological weapon; the field conditions would most probably vastly differ from a normal physiological state. Will a human exposed to such stress conditions be "turned into a mouse" in terms of immune response and defence mechanisms, thus becoming highly susceptible to *B. anthracis* infection? A vaccine formula targeting both facets of the anthrax toxi-infection, thus, seems safer and warrants further scientific efforts. This is now apparent from the wider international interest in the use of capsular material as a co-vaccination antigen and in developing vaccines targeting spores, toxins, and bacillus [108,109].

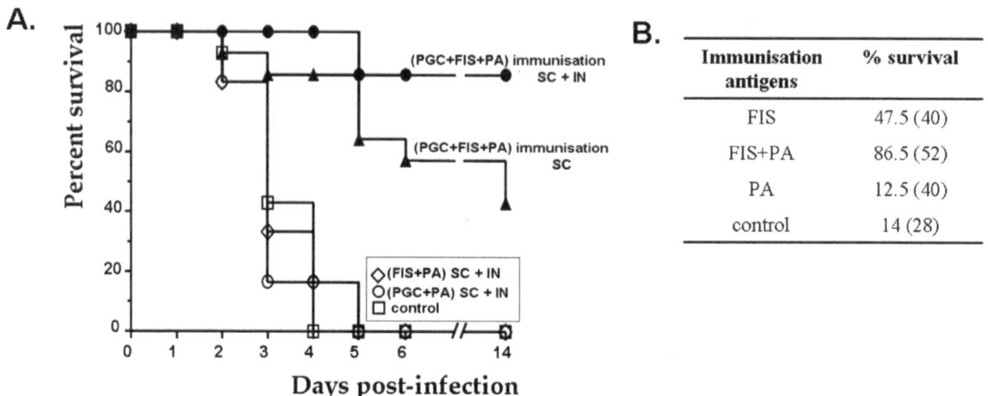

Immunisation antigens	% survival
FIS	47.5 (40)
FIS+PA	86.5 (52)
PA	12.5 (40)
control	14 (28)

Figure 10. (**A**) **Significant protection against** *B. anthracis* **inhalational infection in the hard-to-protect mouse model.** Capsular poly-gamma-D-glutamate naturally anchored to the peptidoglycan (PGC) combined with formaldehyde-inactivated spores (FIS) and protective antigen (PA) were used as immunogens; double immunisation—systemic (SC) and mucosal (IN)—increased the afforded level of protection compared to single SC immunisation. Notably, the FIS + PA vaccine (as initially developed in [57] that fully protects against a cutaneous challenge) did not afford protection against inhalational infection, even after a combined SC + IN immunisation regimen. Immunisation protocols were as per [57] for FIS + PA, [70] for PGC, and [95] for the intranasal immunisation; the inhalational challenge dose of the fully virulent wild-type *B. anthracis* 9602 strain was $5.84 \pm 0.10 \log_{10}$ spores for outbred OF1 mice. The results were synthesised from four independent experiments; unpublished data from the postdoctoral work of Fabien Dumetz. (**B**) **The protection provided by anti-PA immunisation is mediated through neutralisation-independent mechanisms.** The immune response directed against the PA toxin moiety (FIS + PA) increases the protection afforded by spore-only (FIS) immunisation against a challenge with the non-toxinogenic encapsulated *B. anthracis* 9602P strain. The protocols of immunisation and challenge of OF1 mice were as per [57]. The results were synthesised from five independent experiments, with the number of animals in brackets.

4.4.2. Curative Therapies

In addition to prevention through vaccination, curative therapies targeting the toxins or the bacterium were also studied in the Pasteur anthrax laboratory. Since their discovery, the *B. anthracis* toxins were considered the main virulence factors to be neutralised in anthrax. Monoclonal anti-PA, -LF and -EF antibodies were developed as soon as the technology was accessible [110]. These tools enabled the mapping of the domains of the toxin components, the quantitative detection and assay of the toxins, and the follow-up of their presence in experimental samples [77,111].

The therapeutic potential of inhibiting EF enzymatic activity with adefovir was shown through a collaboration with Wei-Jen Tang [85].

A collaboration with Gilles Guichard saw the proposal of a novel alternative to antimicrobial peptides (AMP), i.e., oligoureas, non-natural peptidomimetics that mimic the structure of AMP. They were as efficiently bactericidal as AMP, with the advantage of not being cleaved *in vivo*, thus increasing their half-life [112,113].

Another highly efficient molecule, both *in vitro* and *in vivo*, is secretory phospholipase A_2-IIa (sPLA$_2$-IIa), an anti-microbial agent well-studied in *Staphylococcus aureus*; it was anthracidal at the nano-molar level (collaboration Lhousseine Touqui [114–116]). Intriguingly, a correlation may exist between sPLA$_2$-IIa levels and the degree of resistance to the infection arm of *B. anthracis* infection in different species. The human enzymatic molecule might, thus, be considered an interesting and highly valuable therapeutic in humans in some cases if needed. Anecdotally, we observed that allicin, the component mediating the pungent odour of garlic, was anthracidal (unpublished, following a query from Pr David Mirelman, who kindly provided the purified molecule [117,118]).

5. "MOT" (Special Agents) Regulation

B. anthracis is on the list of agents that can be used as a biological weapon. After the 2001 anthrax events in the USA [89,92], a flurry of regulations was implemented internationally. They aimed to regulate the biological activities in the laboratories, controlling the safety/security aspects of their use, including the dual-use notion, while keeping track of them and of the scientists manipulating them. Now, after more than 10 years of implementation, it seems worthwhile to evaluate how this radical change at the societal level has impacted research activities on *B. anthracis* (and all other concerned agents) over the longer term.

For its part, France, one of three countries with highly stringent regulations (with Canada and Singapore, as mentioned in a working document), decided on drastically restrictive regulations linked to judicial sanctions (MicroOrganismes et Toxines regulations, MOT).

Links showing some of the evolutions of the French regulation (accessed on 5 November 2023):

2004: https://www.legifrance.gouv.fr/codes/id/LEGIARTI000006690180/2004-08-11/

2012: https://www.legifrance.gouv.fr/codes/id/LEGIARTI000025104612/2012-05-01/

2014: https://www.legifrance.gouv.fr/codes/id/LEGIARTI000028352295/2014-02-01/

2023: https://www.legifrance.gouv.fr/loda/id/LEGIARTI000047600565/2023-05-29/

and https://www.legifrance.gouv.fr/codes/section_lc/LEGITEXT000006072665/LEGISCTA0 00019217188/2023-08-04/

The sanctions in 2023 are up to 5–7 years' imprisonment and a EUR 375,000–750,000 fine

For example, one of the most restrictive rules was the "500 bp" rule, initially linked to the control of ricin gene manipulation but applied to all MOT items. Thus, any sample that may harbour genetic material of more than 500 bp in length originating from *B. anthracis* is considered MOT and should be declared, tracked, and stored in specific access-restricted locations.

Historically, when these regulations were first implemented and their implication not yet mastered, any longer than 500 bp of DNA purified from B. anthracis was considered "MOT", even if the same exact sequence could be found in another non-"MOT" microorganism; similarly, any molecule purified from B. anthracis could be considered MOT, even if it was a molecule found in any living cells such as ATP. Fortunately, scientific soundness rapidly prevailed, and the regulations were more precisely specified.

The MOT regulations led to the destruction of part of the biological patrimony in France—a study to approach this aspect would be of interest; any destructions were, of course, perfectly managed alongside all the safety concerns and procedures that already existed. Many laboratories and scientists did not wish to enter an indefinite, complex, and time-consuming circle of tracking and surveillance. Furthermore, from impromptu discussions, it emerged that scientists felt they were specifically targeted by this suspicion, and they questioned whether potential malevolent users would effectively be deterred. Let us not forget that *B. anthracis* is naturally present in France and in many countries, both at the endemic and epidemic level [88], and can easily enter our countries (refer to the inhalational and gastrointestinal anthrax cases originating from drum skins [86,119,120] among other reports).

The application of the regulations is sometimes so time- and mind-consuming, e.g., conforming to all procedures or waiting for the required official authorisations, that it can impede crucial scientific activity. For diagnosis, forensic, or therapeutic aims, the delay in receiving a strain or testing the efficacy of given therapeutics, such as vaccination, should be the shortest possible; strict adherence to regulation may have consequences, thus requiring a framework for specific emergency circumstances. This raises the points of preparedness in times of health urgency, societal or political pressure, and how to balance the need for effective control vs. efficiency in responsiveness to an emergency. Some positive evolutions have recently been observed, but more is needed to find alignment with the spirit of the intended aims of these necessary regulations, which is a titanic task (one may wonder whether the letter was followed more than the spirit of their intended aims, due to the potential judicial consequences).

Three additional layers of regulations were implemented from the beginning of the 21st century: animal ethics regulating animal use, safety regulations for BSL3 (biosafety level) laboratory and animal facilities, and regulations concerning genetically modified organisms. Research on *B. anthracis* is directly concerned with all counts. In some ways, all these regulations may interfere with the *bona fide* proposals of scientific projects and concepts, although illustrating the genuine and legitimate concerns of our current societies (from Galileo to Frankenstein and the sorcerer's apprentice, scientists have been part of the collective imagination of society). It is delicate for us to think well or ill of the current situation, as we are intrinsically involved; it will depend on our future colleagues within their societies to evaluate the soundness of the way ours has decided to rein in our scientific activity (one may worry whether our expertise, preparedness, and reactivity could currently be made fragile if confronted with the use of *B. anthracis* as a biological weapon or, more generally, to any agent on the MOT list; let us remain optimistic).

6. Some Points of Discussion

Before concluding this review, some scientific considerations will be proposed to the thoughtful imagination of our colleagues in the anthrax field.

- First, let us stress one technical (although seeming trivial at first sight), in fact, a crucial aspect for a valid interpretation of the anthracidal effects of a given anti-bacterial agent, i.e., sporicidal vs. bactericidal. Usually, a certain incubation time for the tested molecule is required, during which germination may occur and the antibacterial molecule exerts its effect. Data interpretation should, thus, be drawn carefully, distinguishing between direct sporicidal activity and killing of germinated spores.
- Through our studies, we observed that the cellular effects of the toxins were mediated *in fine* through epigenetic modifications of the intoxinated cell [121]. This opens the path of alternative therapeutics aimed at restoring the functionality of the intoxinated cells in reversing these epigenetic modifications. Such "resuscitation" therapeutics are not confined to anthrax and have been considered in other pathologies where epigenetic memory is thought to be part of the persistence of homeostasis perturbation [122].
- Lung *B. anthracis* infection is a secondary infection from within capillaries, with the blood-circulating encapsulated bacteria being trapped due to their large diameter [123]. This phenomenon was already suggested at the beginning of the 20th century (cited in [124]). Conceptually, this may represent a more general way for a pathogen to reach the lungs; as a secondary event, this is an interesting concept to consider (for example, for encapsulated *Streptococcus pneumoniae* or *Yersinia pestis* at a given dissemination stage).
- *B. anthracis* toxins and spores could be detected in the blood, liver, and/or spleen as early as 1 h after intra-nasal delivery [99]. Such rapid crossing of the respiratory epithelium might be a more general way for a pathogen to interact with its host. Protein delivery through the respiratory epithelium has, indeed, been well-studied for therapeutic means [125,126], though the exact mechanisms are still debated (through pores, between adjacent cells, transcytosis, etc.).
- A crucial challenge is the induction of a respiratory immune response that will control inhalational anthrax. A recent report using BCG [127,128] raises the interesting concept

that an immune response generated from within the lung capillaries could trigger a more effective protective immunity than from the aerial space. *B. anthracis* could be a valid candidate for such exploration.

7. Conclusions

In summary, until now, *Bacillus anthracis* and Institut Pasteur enjoyed almost 120 years, starting from the pioneering work of Louis Pasteur and his "lieutenants" in the nascent fields of microbiology and vaccination [2,6,38] and blooming between 1986 and 2010 following the molecular biology/genetic revolution (exemplified by the Nobel prize attributed to André Lwoff, Jacques Monod, and François Jacob at Institut Pasteur). The 1986–2010 anthrax laboratory was one of the first that was able to genetically manipulate *B. anthracis* and explore its many components—especially toxins—and their role in pathogenicity, leading to the comprehensive unravelling of many facets of anthrax toxi-infection. The tools and mutants developed in the laboratory were exchanged internationally and exploited in various laboratories before they were strictly controlled through the current regulations.

This second golden age of *B. anthracis* research is exemplified by the high-level scientific recognition and the many international collaborations and meeting participations that occurred; a special thought can be given to the International Conference on Anthrax beginning in 1989 in Winchester (UK), then which happened in 2003 in parallel with the third International Workshop on the molecular biology of *Bacillus cereus*, *Bacillus thuringiensis*, and *Bacillus anthracis* (expanded thereafter every two years as the BACT series, regrouping scientists from the *cereus*, *thuringiensis*, and *anthracis* fields, as these bacteria belong to the same group, i.e., *Bacillus cereus* [129] (Figure 11).

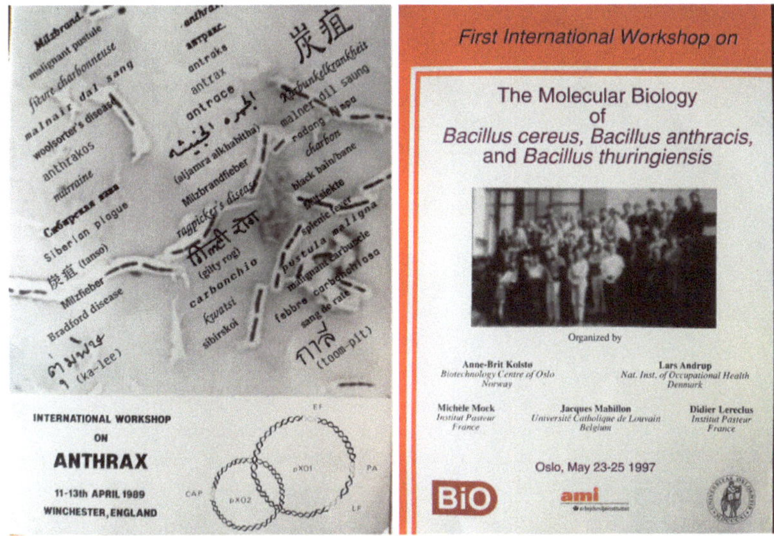

Figure 11. Cover page of the two first international congresses on anthrax in 1989 (Winchester, UK) and on the molecular biology of *the B. cereus* group in 1997 (Oslo, Norway). ©Michèle Mock, private archives.

Indeed, in recent years, reports have emerged on *B. cereus* harbouring similar virulence factors as those found in *B. anthracis* (toxins and capsule), causing anthrax-like pathologies in humans and great apes [130–132]. In the course of the anthrax Euronet 2004–2006 European project [133], a highly stimulating collaboration on these *B. cereus* biovar *anthracis* from apes was developed between our colleagues from the Robert Koch Institute and our laboratory at Institut Pasteur, putting together our complementary and synergistic expertise [134]. A scientific collaboration was finally taking place between our two institutions more than 100 years after the fierce competition between Louis Pasteur and Robert Koch in the 1880s [2].

The *B. anthracis* case could serve to illustrate some general points on how science is performed in a given society at a given time and how a scientific research domain evolves.

Many parameters obviously shape the way a given scientific activity can develop, such as the available technologies, the amount of investment in times of lack of financial resources, their mode of attribution/allocation, the degree of scientific openness and the way society aims to control it through specific regulations. The evolution of the global organisation and function of scientific structures in the 21st century led to a shortening of laboratory life duration and the advent of short-to-middle-term projects (typically two to three years).

All this aspects obviously influence the quality and orientation of research at the level of society. Moreover it questions their impact on the emergence of new concepts, technologies, or applications, as their basic and applied significance can be unpredictable, unexpected, and often unrealised until circumstances are suitable for their development.

The 1990s and early 2000s were ideal for anthrax research at Institut Pasteur in Paris, capitalising on novel concepts, available technologies, scientific critical mass, positive scientific support, grants, and access to biological facilities for this pathogen, both at the laboratory and animal levels. The decision to terminate *B. anthracis* research at the Institut Pasteur was taken in 2015 at the institutional level. This raises the issue of a loss of knowledge and expertise in the anthrax field at the basic and applied levels in France.

B. anthracis and anthrax as a toxi-infection are complex systems that raise many valuable questions in basic research. One may hope that *B. anthracis* research will be initiated *de novo* later at Institut Pasteur with new technologies, mindsets, and ways of exploring the domain. In between, expertise will be lost and will need to be rediscovered through the data published by our predecessors; this is a challenging feat. Indeed, not all technical and experimental details are available in the published papers, regardless of the efforts and implications at the author, editor, and reviewer levels. A part that is missing is the transfer of tacit knowledge, wherein there always remain some unmentioned details [135].

Having experienced the *B. anthracis* field for many years, I noticed the many, often unwritten, hints given around in various anthrax laboratories, such as growth conditions or spore activation steps. The loss of laboratory memory is rapid, and information will need to be rediscovered (a waste of time, energy, and money, but at the same time, a great joy for the discoverer of a previously known and forgotten fact). Another missing part relates to the nature of language (even scientific) that plays a prominent role in this relative failure. On the one hand, an author cannot describe everything exhaustively.

The French author Georges Perec explored this intriguing aspect in some of his works, such as "Tentative d'épuisement d'un lieu parisien" (An Attempt at Exhausting a Place in Paris), where the impossibility of describing absolutely everything was more than obvious. Michael Baxandall analysed similar attempts at describing pictures (Pattern of intentions, Introduction, Yale University Press) and reached the same conclusion.

On the other hand, our language cannot describe everything; some blanks always remain [136].

Over these past years, regulation constraints have been implemented both at the national and institutional levels. Our young colleagues entering the anthrax field (or any MOT) experience a scientific world full of constraints (this could be called "the Matrix effect", as they will live in a constrained world shaped by our society without sensing how it could be outside). On the positive side, such constraints might counterintuitively stimulate creativity.

Georges Perec was keen to exploit constraints when stimulating creation in his works. Among the many playing with constraints, two are impressive: "la disparition" (trans-

*lated as "a void"), around 300 pages without the vowel E, or "Les Revenentes" (translated
as "The Exeter Text: Jewels, Secrets, Sex") a novel of around 60 pages with only the
vowel E, excluding all the others. In another artistic domain, Nadia Boulanger gave the
same advice to jazzman Quincy Jones "Mieux il saura écrire avec des contraintes, plus il
deviendra libre" ("the more he would know how to write with constraints, the freer he
will be";*

*https://www.radiofrance.fr/francemusique/podcasts/musicopolis/un-americain-a-paris-
quincy-jones-entre-barclay-et-boulanger-1192593 (accessed on 5 November 2023)*

We can be confident in the ability of future scientists to take advantage of these
constraints and be creative while keeping an open mind. As Robert Goossens wrote
in 1957:

"The autonomy of life reveals itself unpredictably, the important point is to seize
upon the apparently contradictory element of prevalent observation and to integrate it
into knowledge instead of rejecting it as an aberration of nature [. . .] since in experimental
biology, the essential is not the single-minded pursuit of our initial aim, but the capacity of
also seeing what we were not seeking".

*("L'autonomie de la vie se révèle imprévisiblement, le tout est de saisir le fait apparemment
contradictoire à l'observation commune et de l'intégrer dans la connaissance au lieu de
le rejeter comme une aberration de la nature [. . .] car en expérimentation biologique,
l'essentiel n'est pas de poursuivre toujours le but que l'on cherchait, mais de voir aussi ce
qu'on n'y cherchait pas" p. 140)* [137]

Let us dream of a third golden age of *B. anthracis* research at Institut Pasteur.

Funding: This research received no external funding.

Institutional Review Board Statement: Not applicable.

Informed Consent Statement: Not applicable.

Data Availability Statement: All documents of this historical review are available in the Archives of
Institut Pasteur.

Acknowledgments: This contribution covers a domain that possesses a long history. My respect and
admiration go to our historical predecessors. Their names are present at the top of their publications,
often freely available in numeric form. From the present, I will be unable to mention all my colleagues
in Michèle Mock's lab; suffice to recognise them as they are present as co-authors on all the published
papers from the lab; let me mention more specifically Thomas Candela, Michel 'TCM' Haustant and
the memory of Fabien Brossier. I would like to recognise the pleasure of working with the scientists
in formation or post-formation to whom I hope I was able to transmit a constructive view of science
and help to develop their, each unique, potential, they gave me so much in return, Alex, Manue, Ian,
Benoît, Emilie, Maria, Clémence, Fabien, Mathieu and Guillain.
In the *Bacillus anthracis* field (and *cereus!*), I had the pleasure of interacting with many colleagues
during all these years; among them, a specific thought is paid to Art Friedlander, Ann-Brit Kolsto,
Anne Moir, Silke Klee, and Mahtab Moyaeri. A dear memory is the encounter and collective
discussion with Harry Smith at the Santa Fe BACT-2005 meeting; live history is always priceless.
I wish to thank my colleagues from the Anthrax Euronet group, among them, especially, Wolfgang
Beyer, Bassam Hallis, Michael Hudson, Silke Klee, and Cesare Montecucco. This European grant
brought together vastly different expertise, initiating discussions and collaborations. One of these
collaborations was with Silke Klee and Susann Dupke at the RKI, with Roland Grunow and Fabian
Leendertz. It was such a great pleasure to collaborate on a thrilling subject, share our expertise, and
start a scientific collaboration between Institut Pasteur and Robert Koch Institute, which, due to
historical reasons, did not happen at the time of Louis Pasteur and Robert Koch.
One of the great pleasures in science relies on the collaborations that sometimes emerge unexpectedly;
I especially remember Wei-Jen Tang in Chicago, Gilles Guichard in Bordeaux; Lhousseine Touqui at
Institut Pasteur, and Grégory Jouvion and Michel Huerre when histopathology expertise was still
available at Institut Pasteur; François Bécher, Eric Ezan, Daniel Gillet, and Nathalie Morel at CEA. A
specific collaboration flourished with IRBA; I cannot cite all colleagues during these almost 20 years

in the immunology, bacteriology, and virology fields. They should know I deeply value our scientific interactions and hope I was able to forward what I received from my colleagues during my career. A special thought for Jean-Nicolas Tournier, with whom a collaboration developed from almost the time of my arrival in Michèle Mock's lab; our scientific discussions and sharing of experiments, technologies, and friendship were precious. I fully appreciated the exchanges with Patrice Binder, his vast experience and his support for the anthrax field. A warm remembrance of the late Jacques Viret, from our first encounter in 1982 at CRSSA in Clamart to the last at IRBA in Grenoble, his soft spot for Jung and René Thom and our discussions on many other stimulating topics.

This contribution deals with history; I would have been unable to delve into old and less-old documents in the Institut Pasteur archives and photothèque without the help of especially Sandra Legout, Chantal Pflieger, Christian Sany, Catherine Cécilio and Michaël Davy. I would like to stress the invaluable contribution of Sandra Legout, who has developed a deep knowledge of Institut Pasteur and its personnel in addition to her other archival research interests.

I wish to acknowledge my immense gratitude to Javier Pizarro-Cerda, and Olivier Dussurget. Javier accepted me in his lab at a delicate period. I sincerely hope that I have been able to contribute the knowledge and scientific vision I was able to acquire during my career to my lab colleagues. I am immensely grateful to Geneviève Milon, who supported me in difficult times at Pasteur, and for our impromptu discussions; as Marivaux wrote *"Dans ce monde, il faut être un peu trop bon pour l'être assez"*. I also remember my life as a newborn "pasteurien" in her and Gilles Marchal's group in Robert Fauve's lab. Their vision of science and transmission was enlightening. A wonderful memory is, thanks to Gilles and Geneviève, my CIBA foundation bursary and the following short period in Ralph Steinman's lab in Zanvil Cohn's department; this was an incredible scientific period full of discoveries including the fascinating mobile dendritic cell. Last but not least, I am deeply indebted to Michèle Mock for these wonderful years of research; she accepted my sometimes-unorthodox queries, creating a trusting environment where everything could be considered and discussed without prejudice. She never ceased to discuss and critically read my manuscripts even after the anthrax lab closure; the current one included—many thanks also to Maxime Schwartz for his constructive criticism and remarks. Dominique Goossens has always been by my side during all these years, "I have forgot why I did call thee back—Let me stand here till thou remember it—I shall forget, to have thee still stand there, Rememb'ring how I love thy company—And I'll still stay, to have thee still forget".

Conflicts of Interest: The author declares no conflict of interest.

Abbreviations

AIP	Archives de l'Institut Pasteur
AMP	Anti-Microbial Peptides
EF	Edema Factor
ET	Edema Toxin
FIS	Formaldehyde-Inactivated Spores
LF	Letal Factor
LT	Letal Toxin
PA	Protective Antigen
PDGA	Poly-gamma D-Glutamic Acid
PGC	Capsule naturally coupled to PeptidoGlycan
SMA	Service de Microbiologie Animale
SVV	Service des Vaccins Vétérinaires

References

1. Davaine, C. Recherches relatives à l'action de la chaleur sur le virus charbonneux. *C. R. Acad. Sci.* **1873**, *77*, 726–729.
2. Perrot, A.C.; Schwartz, M. *Pasteur et Koch: Un Duel de Géants Dans le Monde des Microbes*; Ed. Odile Jacob: Paris, France, 2014.
3. Perrot, A.C.; Schwartz, M. *Robert Koch und Louis Pasteur: Duell Zweier Giganten*; Ed. Theiss: Darmstadt, Germany, 2015.
4. Schwartz, M. Jekyll and Mr. Hyde: A short history of anthrax. *Mol. Asp. Med.* **2009**, *30*, 347–355. [CrossRef]
5. Pasteur, L.; Chamberland, C.; Roux, E. Compte rendu sommaire des expériences faites à Pouilly-le-Fort, près Melun, sur la vaccination charbonneuse. In *Comptes Rendus Hebdomadaires des Séances de l'Académie des Sciences*; 1881; Volume XCII, p. 1378.
6. Perrot, A.; Schwartz, M. *Pasteur et ses Lieutenants*; Ed. Odile Jacob: Paris, France, 2013.
7. Delaunay, A. *L'Institut Pasteur des Origines à Aujourd'hui*; Ed. France-Empire: Paris, France, 1962.
8. Beyer, W.; Turnbull, P.C. Anthrax in animals. *Mol. Aspects Med.* **2009**, *30*, 481–489. [CrossRef]
9. Hugh-Jones, M.; Blackburn, J. The ecology of *Bacillus anthracis*. *Mol. Asp. Med.* **2009**, *30*, 356–367. [CrossRef]

10. Mock, M.; Montecucco, C. Editorial. *Mol. Asp. Med.* **2009**, *30*, 345–346. [CrossRef]
11. Smith, H. Discovery of the anthrax toxin: The beginning of studies of virulence determinants regulated *in vivo*. *Int. J. Med. Microbiol.* **2002**, *291*, 411–417. [CrossRef]
12. Mikesell, P.; Ivins, B.E.; Ristroph, J.D.; Dreier, T.M. Evidence for plasmid-mediated toxin production in *Bacillus anthracis*. *Infect. Immun.* **1983**, *39*, 371–376. [CrossRef]
13. Green, B.D.; Battisti, L.; Koehler, T.M.; Thorne, C.B.; Ivins, B.E. Demonstration of a capsule plasmid in *Bacillus anthracis*. *Infect. Immun.* **1985**, *49*, 291–297. [CrossRef] [PubMed]
14. Uchida, I.; Sekizaki, T.; Hashimoto, K.; Terakado, N. Association of the encapsulation of *Bacillus anthracis* with a 60 megadalton plasmid. *J. Gen. Microbiol.* **1985**, *131*, 363–367. [CrossRef] [PubMed]
15. Perrot, A.; Schwartz, M. *Le Neveu de Pasteur: Ou la vie Aventureuse d'Adrien Loir, Savant et Globe-Trotter (1862–1941)*; Ed. Odile Jacob: Paris, France, 2020.
16. Antwerpen, M.; Beyer, W.; Bassy, O.; Ortega-Garcia, M.V.; Cabria-Ramos, J.C.; Grass, G.; Wolfel, R. Phylogenetic Placement of Isolates within the Trans-Eurasian Clade A.Br.008/009 of *Bacillus anthracis*. *Microorganisms* **2019**, *7*, 689. [CrossRef] [PubMed]
17. Girault, G.; Blouin, Y.; Vergnaud, G.; Derzelle, S. High-throughput sequencing of *Bacillus anthracis* in France: Investigating genome diversity and population structure using whole-genome SNP discovery. *BMC Genom.* **2014**, *15*, 288. [CrossRef]
18. Nakahama, N. Museum specimens: An overlooked and valuable material for conservation genetics. *Ecol. Res.* **2021**, *36*, 13–23. [CrossRef]
19. Yeates, D.; Zwick, A.; Mikheyev, A. Museums are biobanks: Unlocking the genetic potential of the three billion specimens in the world's biological collections. *Curr. Opin. Insect Sci.* **2016**, *18*, 83–88. [CrossRef] [PubMed]
20. Lewis, K.; Epstein, S.; Godoy, V.G.; Hong, S.H. Intact DNA in ancient permafrost. *Trends Microbiol.* **2008**, *16*, 92–94. [CrossRef]
21. Price, P.B. Microbial life in glacial ice and implications for a cold origin of life. *FEMS Microbiol. Ecol.* **2007**, *59*, 217–231. [CrossRef]
22. Price, P.B. Microbial genesis, life and death in glacial ice. *Can. J. Microbiol.* **2009**, *55*, 1–11. [CrossRef] [PubMed]
23. Stamatin, N.; Stamatin, L. Le pouvoir immunisant de souches acapsulogènes du *B. anthracis*. *C. R. Soc. Biol.* **1936**, *122*, 491–493.
24. Sterne, M. The use of anthrax vaccines prepared from avirulent (unencapsulated) variants of *Bacillus anthracis*. *Onderstepoort J. Vet. Sci. Anim. Ind.* **1939**, *13*, 307–312.
25. Shlyakhov, E.; Vata, A.; Platkov, E. Immunological diagnosis *in vivo* and acapsular anthrax vaccine: A pioneer achievement of romanian medical science. *J. Prev. Med.* **2001**, *9*, 52–56.
26. Stamatin, N. L'immunisation anticharbonneuse au moyen d'une souche de *Bacillus anthracis* acapsulogène chez le mouton. *C.R. Soc. Biol. Paris* **1937**, *125*(pt6), 90–92. Available online: https://gallica.bnf.fr/ark:/12148/bpt6k6541878j/f96.item (accessed on 5 November 2023).
27. Staub, A.M.; Virat, B.; Levaditi, J. Pouvoir létal de certaines souches acapsulogènes de *B. anthracis*. *Ann. Inst. Pasteur* **1955**, *88*, 244–246.
28. Ramon, G.; Staub, A. Essais sur l'immunisation contre le charbon. Sur une nouvelle formule de vaccination charbonneuse. *Rev. d'immunologie* **1936**, *2*, 401–414. [CrossRef]
29. Jacotot, H.; Virat, B. Longevity of *B. anthracis* spores (Pasteur's first vaccine). *Ann. Inst. Pasteur* **1954**, *87*, 215–217.
30. Virat, B. La vaccination anticharbonneuse. *Econ. Méd. Anim.* **1967**, *8*, 17–21.
31. Fauve, R.M.; Delaunay, A. Résistance cellulaire à l'infection bactérienne. V. Modifications cytoplasmiques observées *in vitro*, dans des macrophages de souris, après injection de bactéries capables ou non de multiplication intracellulaire [Cellular resistance to bacterial infection. V. Cytoplasm modifications *in vitro* in mouse macrophages after injection of bacteria capable or incapable of intracellular multiplication]. *Ann. Inst. Pasteur* **1966**, *111*, 78–84.
32. von Behring, E.; Kitasato, S. Ueber das zustandekommen der diphtherie-immunität und der tetanus-immunität bei thieren. *Dtsch. Med. Wochenschr.* **1890**, *49*, 1–6.
33. Kitasato, S. Ueber den tetanusbacillus. *Zeitschr. Hyg.* **1889**, *7*, 225–234. [CrossRef]
34. Nocard, E. Sur la sérothérapie du tétanos: Essais de traitement préventif. *Bull. Soc. Cent. Méd. Vét.* **1895**, 407–418.
35. Roux, E.; Yersin, A. Contribution à l'étude de la diphtérie. *Ann. Inst. Pasteur* **1888**, *12*, 629–661.
36. Cavaillon, J.-M. From Bacterial Poisons to Toxins: The Early Works of Pasteurians. *Toxins* **2022**, *14*, 759. [CrossRef] [PubMed]
37. Popoff, M.R.; Legout, S. Anaerobes and Toxins, a Tradition of the Institut Pasteur. *Toxins* **2023**, *15*, 43. [CrossRef]
38. Maquart, B. *Louis Pasteur, Le Visionnaire, Sous la Direction d'Annick Perrot et Maxime Schwartz*; Ed. La Martinière: Paris, France, 2017.
39. Grabar, P.; Staub, A.M. Recherches immunochimiques sur la bactéridie charbonneuse: II.-Les fractions protéidiques du liquide d'oedème charbonneux et des extraits de *B. anthracis*. *Ann. Inst. Pasteur* **1944**, *70*, 12–143.
40. Staub, A.M.; Grabar, P. Recherches immunochimiques sur la bactéridie charbonneuse: I.-Le liquide d'oedème de cobaye et les polyosides. *Ann. Inst. Pasteur* **1944**, *7*, 16–32.
41. Kintzer, A.F.; Thoren, K.L.; Sterling, H.J.; Dong, K.C.; Feld, G.K.; Tang, I.I.; Zhang, T.T.; Williams, E.R.; Berger, J.M.; Krantz, B.A. The protective antigen component of anthrax toxin forms functional octameric complexes. *J. Mol. Biol.* **2009**, *392*, 614–629. [CrossRef]
42. Brossier, F.; Mock, M. Toxins of *Bacillus anthracis*. *Toxicon* **2001**, *39*, 1747–1755. [CrossRef] [PubMed]
43. Moayeri, M.; Leppla, S.H. Cellular and systemic effects of anthrax lethal toxin and edema toxin. *Mol. Asp. Med.* **2009**, *30*, 439–455. [CrossRef]

44. Moayeri, M.; Leppla, S.H.; Vrentas, C.; Pomerantsev, A.P.; Liu, S. Anthrax Pathogenesis. *Annu. Rev. Microbiol.* **2015**, *69*, 185–208. [CrossRef]

45. Tournier, J.N.; Rossi Paccani, S.; Quesnel-Hellmann, A.; Baldari, C.T. Anthrax toxins: A weapon to systematically dismantle the host immune defenses. *Mol. Asp. Med.* **2009**, *30*, 456–466. [CrossRef] [PubMed]

46. Leppla, S.H. Anthrax toxin edema factor: A bacterial adenylate cyclase that increases cyclic AMP concentrations of eukaryotic cells. *Proc. Natl. Acad. Sci. USA* **1982**, *79*, 3162–3166. [CrossRef]

47. Duesbery, N.S.; Webb, C.P.; Leppla, S.H.; Gordon, V.M.; Klimpel, K.R.; Copeland, T.D.; Ahn, N.G.; Oskarsson, M.K.; Fukasawa, K.; Paull, K.D.; et al. Proteolytic inactivation of MAP-kinase-kinase by anthrax lethal factor. *Science* **1998**, *280*, 734–737. [CrossRef]

48. Klimpel, K.R.; Arora, N.; Leppla, S.H. Anthrax toxin lethal factor contains a zinc metalloprotease consensus sequence which is required for lethal toxin activity. *Mol. Microbiol.* **1994**, *13*, 1093–1100. [CrossRef] [PubMed]

49. Vitale, G.; Pellizzari, R.; Recchi, C.; Napolitani, G.; Mock, M.; Montecucco, C. Anthrax lethal factor cleaves the N-terminus of MAPKKS and induces tyrosine/threonine phosphorylation of MAPKS in cultured macrophages. *J. Appl. Microbiol.* **1999**, *87*, 288. [CrossRef] [PubMed]

50. Rossetto, O.; Scorzeto, M.; Megighian, A.; Montecucco, C. Tetanus neurotoxin. *Toxicon* **2013**, *66*, 59–63. [CrossRef] [PubMed]

51. Meynell, E.; Meynell, G.G. The Roles of Serum and Carbon Dioxide in Capsule Formation by *Bacillus anthracis*. *J. Gen. Microbiol.* **1964**, *34*, 153–164. [CrossRef] [PubMed]

52. Sirard, J.C.; Mock, M.; Fouet, A. The three *Bacillus anthracis* toxin genes are coordinately regulated by bicarbonate and temperature. *J. Bacteriol.* **1994**, *176*, 5188–5192. [CrossRef]

53. Morange, M. *A History of Biology*; Princeton University Press: Princeton, NJ, USA, 2021.

54. Mock, M.; Schwartz, M. Mutations which affect the structure and activity of colicin E3. *J. Bacteriol.* **1980**, *142*, 384–390. [CrossRef] [PubMed]

55. Mock, M.; Miyada, C.G.; Collier, R.J. Genetic analysis of the functional relationship between colicin E3 and its immunity protein. *J. Bacteriol.* **1984**, *159*, 658–662. [CrossRef]

56. Trieu-Cuot, P.; Gerbaud, G.; Lambert, T.; Courvalin, P. *In vivo* transfer of genetic information between gram-positive and gram-negative bacteria. *EMBO J.* **1985**, *4*, 3583–3587. [CrossRef]

57. Brossier, F.; Levy, M.; Mock, M. Anthrax spores make an essential contribution to vaccine efficacy. *Infect. Immun.* **2002**, *70*, 661–664. [CrossRef]

58. Cataldi, A.; Labruyere, E.; Mock, M. Construction and characterization of a protective antigen-deficient *Bacillus anthracis* strain. *Mol. Microbiol.* **1990**, *4*, 1111–1117. [CrossRef]

59. Labruyere, E.; Mock, M.; Surewicz, W.K.; Mantsch, H.H.; Rose, T.; Munier, H.; Sarfati, R.S.; Barzu, O. Structural and ligand-binding properties of a truncated form of *Bacillus anthracis* adenylate cyclase and of a catalytically inactive variant in which glutamine substitutes for lysine-346. *Biochemistry* **1991**, *30*, 2619–2624. [CrossRef]

60. Pezard, C.; Duflot, E.; Mock, M. Construction of *Bacillus anthracis* mutant strains producing a single toxin component. *J. Gen. Microbiol.* **1993**, *139*, 2459–2463. [CrossRef] [PubMed]

61. Chateau, A.; van Schaik, W.; Six, A.; Aucher, W.; Fouet, A. CodY regulation is required for full virulence and heme iron acquisition in *Bacillus anthracis*. *FASEB J.* **2011**, *25*, 4445–4456. [CrossRef] [PubMed]

62. Fouet, A. AtxA, a *Bacillus anthracis* global virulence regulator. *Res. Microbiol.* **2010**, *161*, 735–742. [CrossRef]

63. Fouet, A.; Mock, M. Regulatory networks for virulence and persistence of *Bacillus anthracis*. *Curr. Opin. Microbiol.* **2006**, *9*, 160–166. [CrossRef]

64. Mignot, T.; Couture-Tosi, E.; Mesnage, S.; Mock, M.; Fouet, A. *In vivo Bacillus anthracis* gene expression requires PagR as an intermediate effector of the AtxA signalling cascade. *Int. J. Med. Microbiol.* **2004**, *293*, 619–624. [CrossRef] [PubMed]

65. van Schaik, W.; Chateau, A.; Dillies, M.A.; Coppee, J.Y.; Sonenshein, A.L.; Fouet, A. The global regulator CodY regulates toxin gene expression in *Bacillus anthracis* and is required for full virulence. *Infect. Immun.* **2009**, *77*, 4437–4445. [CrossRef]

66. Fouet, A.; Mesnage, S. *Bacillus anthracis* cell envelope components. *Curr. Top. Microbiol. Immunol.* **2002**, *271*, 87–113. [CrossRef]

67. Fouet, A.; Mesnage, S.; Tosi-Couture, E.; Gounon, P.; Mock, M. *Bacillus anthracis* surface: Capsule and S-layer. *J. Appl. Microbiol.* **1999**, *87*, 251–255. [CrossRef]

68. Mesnage, S.; Tosi-Couture, E.; Gounon, P.; Mock, M.; Fouet, A. The capsule and S-layer: Two independent and yet compatible macromolecular structures in *Bacillus anthracis*. *J. Bacteriol.* **1998**, *180*, 52–58. [CrossRef]

69. Mignot, T.; Mesnage, S.; Couture-Tosi, E.; Mock, M.; Fouet, A. Developmental switch of S-layer protein synthesis in *Bacillus anthracis*. *Mol. Microbiol.* **2002**, *43*, 1615–1627. [CrossRef]

70. Candela, T.; Dumetz, F.; Tosi-Couture, E.; Mock, M.; Goossens, P.L.; Fouet, A. Cell-wall preparation containing poly-gamma-D-glutamate covalently linked to peptidoglycan, a straightforward extractable molecule, protects mice against experimental anthrax infection. *Vaccine* **2012**, *31*, 171–175. [CrossRef]

71. Candela, T.; Fouet, A. *Bacillus anthracis* CapD, belonging to the gamma-glutamyltranspeptidase family, is required for the covalent anchoring of capsule to peptidoglycan. *Mol. Microbiol.* **2005**, *57*, 717–726. [CrossRef]

72. Candela, T.; Mock, M.; Fouet, A. CapE, a 47-amino-acid peptide, is necessary for *Bacillus anthracis* polyglutamate capsule synthesis. *J. Bacteriol.* **2005**, *187*, 7765–7772. [CrossRef] [PubMed]

73. Aucher, W.; Davison, S.; Fouet, A. Characterization of the sortase repertoire in *Bacillus anthracis*. *PLoS ONE* **2011**, *6*, e27411. [CrossRef] [PubMed]

74. Davison, S.; Couture-Tosi, E.; Candela, T.; Mock, M.; Fouet, A. Identification of the *Bacillus anthracis* (gamma) phage receptor. *J. Bacteriol.* **2005**, *187*, 6742–6749. [CrossRef] [PubMed]
75. Sylvestre, P.; Couture-Tosi, E.; Mock, M. A collagen-like surface glycoprotein is a structural component of the *Bacillus anthracis* exosporium. *Mol. Microbiol.* **2002**, *45*, 169–178. [CrossRef] [PubMed]
76. Sylvestre, P.; Couture-Tosi, E.; Mock, M. Contribution of ExsFA and ExsFB proteins to the localization of BclA on the spore surface and to the stability of the *Bacillus anthracis* exosporium. *J. Bacteriol.* **2005**, *187*, 5122–5128. [CrossRef] [PubMed]
77. Brossier, F.; Weber-Levy, M.; Mock, M.; Sirard, J.C. Role of toxin functional domains in anthrax pathogenesis. *Infect. Immun.* **2000**, *68*, 1781–1786. [CrossRef] [PubMed]
78. Pezard, C.; Berche, P.; Mock, M. Contribution of individual toxin components to virulence of *Bacillus anthracis*. *Infect. Immun.* **1991**, *59*, 3472–3477. [CrossRef]
79. Sirard, J.C.; Guidi-Rontani, C.; Fouet, A.; Mock, M. Characterization of a plasmid region involved in *Bacillus anthracis* toxin production and pathogenesis. *Int. J. Med. Microbiol.* **2000**, *290*, 313–316. [CrossRef]
80. Goossens, P.L. Animal models of human anthrax: The Quest for the Holy Grail. *Mol. Asp. Med.* **2009**, *30*, 467–480. [CrossRef] [PubMed]
81. Glomski, I.J.; Piris-Gimenez, A.; Huerre, M.; Mock, M.; Goossens, P.L. Primary involvement of pharynx and peyer's patch in inhalational and intestinal anthrax. *PLoS Pathog.* **2007**, *3*, e76. [CrossRef] [PubMed]
82. Piris-Gimenez, A.; Corre, J.P.; Jouvion, G.; Candela, T.; Khun, H.; Goossens, P.L. Encapsulated *Bacillus anthracis* interacts closely with liver endothelium. *J. Infect. Dis.* **2009**, *200*, 1381–1389. [CrossRef] [PubMed]
83. Glomski, I.J.; Corre, J.P.; Mock, M.; Goossens, P.L. Noncapsulated toxinogenic *Bacillus anthracis* presents a specific growth and dissemination pattern in naive and protective antigen-immune mice. *Infect. Immun.* **2007**, *75*, 4754–4761. [CrossRef] [PubMed]
84. Glomski, I.J.; Dumetz, F.; Jouvion, G.; Huerre, M.R.; Mock, M.; Goossens, P.L. Inhaled non-capsulated *Bacillus anthracis* in A/J mice: Nasopharynx and alveolar space as dual portals of entry, delayed dissemination, and specific organ targeting. *Microbes Infect.* **2008**, *10*, 1398–1404. [CrossRef] [PubMed]
85. Dumetz, F.; Jouvion, G.; Khun, H.; Glomski, I.J.; Corre, J.P.; Rougeaux, C.; Tang, W.J.; Mock, M.; Huerre, M.; Goossens, P.L. Noninvasive imaging technologies reveal edema toxin as a key virulence factor in anthrax. *Am. J. Pathol.* **2011**, *178*, 2523–2535. [CrossRef] [PubMed]
86. Bennett, E.; Hall, I.M.; Pottage, T.; Silman, N.J.; Bennett, A.M. Drumming-associated anthrax incidents: Exposures to low levels of indoor environmental contamination. *Epidemiol. Infect.* **2018**, *146*, 1519–1525. [CrossRef]
87. Price, E.P.; Seymour, M.L.; Sarovich, D.S.; Latham, J.; Wolken, S.R.; Mason, J.; Vincent, G.; Drees, K.P.; Beckstrom-Sternberg, S.M.; Phillippy, A.M.; et al. Molecular epidemiologic investigation of an anthrax outbreak among heroin users, Europe. *Emerg. Infect. Dis.* **2012**, *18*, 1307–1313. [CrossRef]
88. WHO. *Anthrax in Humans and Animals*; WHO: Geneva, Switzerland, 2008; ISBN 9789241547536.
89. Guillemin, J. Anthrax: The investigation of a deadly outbreak. *N. Engl. J. Med* **2000**, *343*, 1198. [CrossRef]
90. Guillemin, J. *Detecting Anthrax: What We Learned from the 1979 Sverdlovsk Outbreak*; Springer: Dordrecht, The Netherlands, 2001; Volume 35, pp. 75–85.
91. Guillemin, J. *Biological Weapons: From the Invention of State-Sponsored Programs to Contemporary Bioterrorism*; Columbia University Press: New York, NY, USA, 2004.
92. Turner, J.N. *Amérithrax*; Ed. de l'aube/L'aube Noire: La Tour d'Aigues, France, 2014.
93. Cossart, P.; Schwartz, M. Pasteur, a visionary: *Comptes Rendus. Biologies* **2022**, *345*, 1–5. Available online: https://comptes-rendus.academie-sciences.fr/biologies/item/10.5802/crbiol.97.pdf (accessed on 5 November 2023). [CrossRef] [PubMed]
94. Fiole, D.; Deman, P.; Trescos, Y.; Mayol, J.F.; Mathieu, J.; Vial, J.C.; Douady, J.; Tournier, J.N. Two-photon intravital imaging of lungs during anthrax infection reveals long-lasting macrophage-dendritic cell contacts. *Infect. Immun.* **2014**, *82*, 864–872. [CrossRef] [PubMed]
95. Gauthier, Y.P.; Tournier, J.N.; Paucod, J.C.; Corre, J.P.; Mock, M.; Goossens, P.L.; Vidal, D.R. Efficacy of a vaccine based on protective antigen and killed spores against experimental inhalational anthrax. *Infect. Immun.* **2009**, *77*, 1197–1207. [CrossRef] [PubMed]
96. Klezovich-Benard, M.; Corre, J.P.; Jusforgues-Saklani, H.; Fiole, D.; Burjek, N.; Tournier, J.N.; Goossens, P.L. Mechanisms of NK cell-macrophage *Bacillus anthracis* crosstalk: A balance between stimulation by spores and differential disruption by toxins. *PLoS Pathog.* **2012**, *8*, e1002481. [CrossRef]
97. Le Gars, M.; Haustant, M.; Klezovich-Benard, M.; Paget, C.; Trottein, F.; Goossens, P.L.; Tournier, J.N. Mechanisms of Invariant NKT Cell Activity in Restraining *Bacillus anthracis* Systemic Dissemination. *J. Immunol.* **2016**, *197*, 3225–3232. [CrossRef]
98. Rougeaux, C.; Becher, F.; Ezan, E.; Tournier, J.N.; Goossens, P.L. *In vivo* dynamics of active edema and lethal factors during anthrax. *Sci. Rep.* **2016**, *6*, 23346. [CrossRef]
99. Rougeaux, C.; Becher, F.; Goossens, P.L.; Tournier, J.N. Very Early Blood Diffusion of the Active Lethal and Edema Factors of *Bacillus anthracis* After Intranasal Infection. *J. Infect. Dis.* **2020**, *221*, 660–667. [CrossRef]
100. Tournier, J.N.; Quesnel-Hellmann, A.; Mathieu, J.; Montecucco, C.; Tang, W.J.; Mock, M.; Vidal, D.R.; Goossens, P.L. Anthrax edema toxin cooperates with lethal toxin to impair cytokine secretion during infection of dendritic cells. *J. Immunol.* **2005**, *174*, 4934–4941. [CrossRef]

101. Trescos, Y.; Tessier, E.; Rougeaux, C.; Goossens, P.L.; Tournier, J.N. Micropatterned macrophage analysis reveals global cytoskeleton constraints induced by *Bacillus anthracis* edema toxin. *Infect. Immun.* **2015**, *83*, 3114–3125. [CrossRef]

102. Chenau, J.; Fenaille, F.; Caro, V.; Haustant, M.; Diancourt, L.; Klee, S.R.; Junot, C.; Ezan, E.; Goossens, P.L.; Becher, F. Identification and validation of specific markers of *Bacillus anthracis* spores by proteomics and genomics approaches. *Mol. Cell. Proteom.* **2014**, *13*, 716–732. [CrossRef]

103. Chenau, J.; Fenaille, F.; Ezan, E.; Morel, N.; Lamourette, P.; Goossens, P.L.; Becher, F. Sensitive detection of *Bacillus anthracis* spores by immunocapture and liquid chromatography-tandem mass spectrometry. *Anal. Chem.* **2011**, *83*, 8675–8682. [CrossRef]

104. Duriez, E.; Goossens, P.L.; Becher, F.; Ezan, E. Femtomolar detection of the anthrax edema factor in human and animal plasma. *Anal. Chem.* **2009**, *81*, 5935–5941. [CrossRef] [PubMed]

105. Morel, N.; Volland, H.; Dano, J.; Lamourette, P.; Sylvestre, P.; Mock, M.; Creminon, C. Fast and sensitive detection of *Bacillus anthracis* spores by immunoassay. *Appl. Environ. Microbiol.* **2012**, *78*, 6491–6498. [CrossRef]

106. Sykes, A.; Brooks, T.; Dusmet, M.; Nicholson, A.G.; Hansell, D.M.; Wilson, R. Inhalational anthrax in a vaccinated soldier. *Eur. Respir. J.* **2013**, *42*, 285–287. [CrossRef] [PubMed]

107. Mock, M.; Fouet, A. Anthrax. *Annu. Rev. Microbiol.* **2001**, *55*, 647–671. [CrossRef]

108. Chabot, D.J.; Ribot, W.J.; Joyce, J.; Cook, J.; Hepler, R.; Nahas, D.; Chua, J.; Friedlander, A.M. Protection of rhesus macaques against inhalational anthrax with a *Bacillus anthracis* capsule conjugate vaccine. *Vaccine* **2016**, *34*, 4012–4016. [CrossRef]

109. Wang, J.Y.; Roehrl, M.H. Anthrax vaccine design: Strategies to achieve comprehensive protection against spore, bacillus, and toxin. *Med. Immunol.* **2005**, *4*, 4. [CrossRef]

110. Brossier, F.; Levy, M.; Landier, A.; Lafaye, P.; Mock, M. Functional analysis of *Bacillus anthracis* protective antigen by using neutralizing monoclonal antibodies. *Infect. Immun.* **2004**, *72*, 6313–6317. [CrossRef]

111. Brossier, F.; Sirard, J.C.; Guidi-Rontani, C.; Duflot, E.; Mock, M. Functional analysis of the carboxy-terminal domain of *Bacillus anthracis* protective antigen. *Infect. Immun.* **1999**, *67*, 964–967. [CrossRef]

112. Antunes, S.; Corre, J.P.; Mikaty, G.; Douat, C.; Goossens, P.L.; Guichard, G. Effect of replacing main-chain ureas with thiourea and guanidinium surrogates on the bactericidal activity of membrane active oligourea foldamers. *Bioorg. Med. Chem.* **2017**, *25*, 4245–4252. [CrossRef]

113. Teyssieres, E.; Corre, J.P.; Antunes, S.; Rougeot, C.; Dugave, C.; Jouvion, G.; Claudon, P.; Mikaty, G.; Douat, C.; Goossens, P.L.; et al. Proteolytically Stable Foldamer Mimics of Host-Defense Peptides with Protective Activities in a Murine Model of Bacterial Infection. *J. Med. Chem.* **2016**, *59*, 8221–8232. [CrossRef]

114. Gimenez, A.P.; Wu, Y.Z.; Paya, M.; Delclaux, C.; Touqui, L.; Goossens, P.L. High bactericidal efficiency of type IIa phospholipase A2 against *Bacillus anthracis* and inhibition of its secretion by the lethal toxin. *J. Immunol.* **2004**, *173*, 521–530. [CrossRef]

115. Piris-Gimenez, A.; Paya, M.; Lambeau, G.; Chignard, M.; Mock, M.; Touqui, L.; Goossens, P.L. *In vivo* protective role of human group IIa phospholipase A2 against experimental anthrax. *J. Immunol.* **2005**, *175*, 6786–6791. [CrossRef] [PubMed]

116. Raymond, B.; Ravaux, L.; Memet, S.; Wu, Y.; Sturny-Leclere, A.; Leduc, D.; Denoyelle, C.; Goossens, P.L.; Paya, M.; Raymondjean, M.; et al. Anthrax lethal toxin down-regulates type-IIA secreted phospholipase A_2 expression through MAPK/NF-κB inactivation. *Biochem. Pharmacol.* **2010**, *79*, 1149–1155. [CrossRef]

117. Ankri, S.; Mirelman, D. Antimicrobial properties of allicin from garlic. *Microbes Infect.* **1999**, *1*, 125–129. [CrossRef]

118. Harris, J.C.; Cottrell, S.L.; Plummer, S.; Lloyd, D. Antimicrobial properties of *Allium sativum* (garlic). *Appl. Microbiol. Biotechnol.* **2001**, *57*, 282–286. [CrossRef] [PubMed]

119. Gastrointestinal anthrax after an animal-hide drumming event—New Hampshire and Massachusetts, 2009. *MMWR Morb. Mortal Wkly. Rep.* **2010**, *59*, 872–877.

120. Marston, C.K.; Allen, C.A.; Beaudry, J.; Price, E.P.; Wolken, S.R.; Pearson, T.; Keim, P.; Hoffmaster, A.R. Molecular epidemiology of anthrax cases associated with recreational use of animal hides and yarn in the United States. *PLoS ONE* **2011**, *6*, e28274. [CrossRef] [PubMed]

121. Raymond, B.; Batsche, E.; Boutillon, F.; Wu, Y.Z.; Leduc, D.; Balloy, V.; Raoust, E.; Muchardt, C.; Goossens, P.L.; Touqui, L. Anthrax lethal toxin impairs IL-8 expression in epithelial cells through inhibition of histone H3 modification. *PLoS Pathog.* **2009**, *5*, e1000359. [CrossRef] [PubMed]

122. Yang, J.H.; Hayano, M.; Griffin, P.T.; Amorim, J.A.; Bonkowski, M.S.; Apostolides, J.K.; Salfati, E.L.; Blanchette, M.; Munding, E.M.; Bhakta, M.; et al. Loss of epigenetic information as a cause of mammalian aging. *Cell* **2023**, *186*, 305–326.e327. [CrossRef]

123. Jouvion, G.; Corre, J.P.; Khun, H.; Moya-Nilges, M.; Roux, P.; Latroche, C.; Tournier, J.N.; Huerre, M.; Chretien, F.; Goossens, P.L. Physical Sequestration of *Bacillus anthracis* in the Pulmonary Capillaries in Terminal Infection. *J. Infect. Dis.* **2016**, *214*, 281–287. [CrossRef]

124. Bloom, W.L.; Mc, G.W. Studies on infection with *Bacillus anthracis*; physiological changes in experimental animals during the course of infection with *B. anthracis*. *J. Infect. Dis.* **1947**, *80*, 137–144. [CrossRef]

125. Brain, J.D. Inhalation, deposition, and fate of insulin and other therapeutic proteins. *Diabetes Technol. Ther.* **2007**, *9* (Suppl. S1), S4–S15. [CrossRef]

126. Hastings, R.H.; Folkesson, H.G.; Matthay, M.A. Mechanisms of alveolar protein clearance in the intact lung. *Am. J. Physiol. Lung Cell. Mol. Physiol.* **2004**, *286*, L679–L689. [CrossRef]

127. Darrah, P.A.; Zeppa, J.J.; Maiello, P.; Hackney, J.A.; Wadsworth, M.H., 2nd; Hughes, T.K.; Pokkali, S.; Swanson, P.A., 2nd; Grant, N.L.; Rodgers, M.A.; et al. Prevention of tuberculosis in macaques after intravenous BCG immunization. *Nature* **2020**, *577*, 95–102. [CrossRef]
128. Larson, E.C.; Ellis-Connell, A.L.; Rodgers, M.A.; Gubernat, A.K.; Gleim, J.L.; Moriarty, R.V.; Balgeman, A.J.; Ameel, C.L.; Jauro, S.; Tomko, J.A.; et al. Intravenous Bacille Calmette-Guérin vaccination protects simian immunodeficiency virus-infected macaques from tuberculosis. *Nat. Microbiol.* **2023**, *8*, 2080–2092. [CrossRef] [PubMed]
129. Helgason, E.; Okstad, O.A.; Caugant, D.A.; Johansen, H.A.; Fouet, A.; Mock, M.; Hegna, I.; Kolsto, A.B. *Bacillus anthracis, Bacillus cereus*, and *Bacillus thuringiensis*—One species on the basis of genetic evidence. *Appl. Environ. Microbiol.* **2000**, *66*, 2627–2630. [CrossRef] [PubMed]
130. Hoffmaster, A.R.; Hill, K.K.; Gee, J.E.; Marston, C.K.; De, B.K.; Popovic, T.; Sue, D.; Wilkins, P.P.; Avashia, S.B.; Drumgoole, R.; et al. Characterization of *Bacillus cereus* isolates associated with fatal pneumonias: Strains are closely related to *Bacillus anthracis* and harbor *B. anthracis* virulence genes. *J. Clin. Microbiol.* **2006**, *44*, 3352–3360. [CrossRef]
131. Hoffmaster, A.R.; Ravel, J.; Rasko, D.A.; Chapman, G.D.; Chute, M.D.; Marston, C.K.; De, B.K.; Sacchi, C.T.; Fitzgerald, C.; Mayer, L.W.; et al. Identification of anthrax toxin genes in a *Bacillus cereus* associated with an illness resembling inhalation anthrax. *Proc. Natl. Acad. Sci. USA* **2004**, *101*, 8449–8454. [CrossRef] [PubMed]
132. Klee, S.R.; Ozel, M.; Appel, B.; Boesch, C.; Ellerbrok, H.; Jacob, D.; Holland, G.; Leendertz, F.H.; Pauli, G.; Grunow, R.; et al. Characterization of *Bacillus anthracis*-like bacteria isolated from wild great apes from Cote d'Ivoire and Cameroon. *J. Bacteriol.* **2006**, *188*, 5333–5344. [CrossRef]
133. Hudson, M.J.; Beyer, W.; Böhm, R.; Fasanella, A.; Garofolo, G.; Golinski, R.; Goossens, P.L.; Hahn, U.; Hallis, B.; King, A.; et al. *Bacillus anthracis*: Balancing innocent research with dual-use potential. *Int. J. Med. Microbiol.* **2008**, *298*, 345–364. [CrossRef] [PubMed]
134. Brezillon, C.; Haustant, M.; Dupke, S.; Corre, J.P.; Lander, A.; Franz, T.; Monot, M.; Couture-Tosi, E.; Jouvion, G.; Leendertz, F.H.; et al. Capsules, toxins and AtxA as virulence factors of emerging *Bacillus cereus* biovar *anthracis*. *PLoS Negl. Trop. Dis.* **2015**, *9*, e0003455. [CrossRef] [PubMed]
135. Feder, T. Q&A: Harry Collins on acquiring and using scientific knowledge. *Phys. Today* **2021**. [CrossRef]
136. Lefebvre, J.; Testenoire, P.-Y. Blancs de l'écrit, blancs de l'écriture. *Linguistique L'écrit* **2019**, *1*, 1–16. [CrossRef]
137. Goossens, R. Les Fermentations Antibiotiques: Essai sur le Déterminisme Expérimental dans les Biosynthèses. Ph.D. Thesis, University of Paris, Paris, France, 1957. Available online: http://cadic.pasteur.fr/exl-php/vue-consult/ipp_recherche/ISO00015 109 (accessed on 5 November 2023).

Review

Pasteurian Contributions to the Study of *Bordetella pertussis* Toxins

Camille Locht

Univ. Lille, CNRS, Inserm, CHU Lille, Institut Pasteur de Lille, U1019-UMR 9017-CIIL-Centre for Infection and Immunity of Lille, F-59000 Lille, France; camille.locht@pasteur-lille.fr

Abstract: As a tribute to Louis Pasteur on the occasion of the 200th anniversary of his birth, this article summarizes the main contributions of scientists from Pasteur Institutes to the current knowledge of toxins produced by *Bordetella pertussis*. The article therefore focuses on publications authored by researchers from Pasteur Institutes and is not intended as a systematic review of *B. pertussis* toxins. Besides identifying *B. pertussis* as the causative agent of whooping cough, Pasteurians have made several major contributions with respect to the structure–function relationship of the *Bordetella* lipo-oligosaccharide, adenylyl cyclase toxin and pertussis toxin. In addition to contributing to the understanding of these toxins' mechanisms at the molecular and cellular levels and their role in pathogenesis, scientists at Pasteur Institutes have also exploited potential applications of the gathered knowledge of these toxins. These applications range from the development of novel tools to study protein–protein interactions over the design of novel antigen delivery tools, such as prophylactic or therapeutic vaccine candidates against cancer and viral infection, to the development of a live attenuated nasal pertussis vaccine. This scientific journey from basic science to applications in the field of human health matches perfectly with the overall scientific objectives outlined by Louis Pasteur himself.

Keywords: *Bordetella*; lipo-oligosaccharide; adenylyl cyclase toxin; pertussis toxin; vaccines

Key Contribution: Scientists at Pasteur Institutes have made major contributions to the understanding of the mechanisms of *Bordetella pertussis* toxins, their role in pathogenesis and applications based on *B. pertussis* toxins.

Citation: Locht, C. Pasteurian Contributions to the Study of *Bordetella pertussis* Toxins. *Toxins* **2023**, *15*, 176. https://doi.org/10.3390/toxins15030176

Received: 31 January 2023
Revised: 17 February 2023
Accepted: 23 February 2023
Published: 25 February 2023

1. Introduction

The *Bordetella* genus contains over a dozen different species, four of which are bona fide pathogens for warm-blooded vertebrates [1]. *Bordetella pertussis* causes whooping cough in humans. The human-adapted lineage of the ovine pathogen *Bordetella parapertussis* cases mild whooping-cough-like symptoms in humans. *Bordetella bronchispetica* is able to infect many mammals and causes rhinitis in pigs and kennel cough in dogs. It may also infect humans and is considered an opportunistic pathogen for humans, as it can cause symptomatic infections in individuals with immune deficiency. Finally, *Bordetella avium* is a bird pathogen. Other *Bordetella* species, such as *Bordetella holmesii*, have also occasionally been associated with human disease, although mostly in immunocompromised individuals. In this article, I briefly review the contribution of scientists from Pasteur Institutes to the understanding of *Bordetella* toxins.

2. From the Identification of the Whooping Cough Agent to the Concept of *Bordetella* Toxins

The first *Bordetella* species that was discovered is *B. pertussis*, initially called *Haemophilus pertussis*, because of a haemolytic halo that can be observed when the organism is grown on blood agar plates, suggesting the presence of a haemolysin toxin. *B. pertussis* was first observed by Jules Bordet in 1900 [2] when he was at the Pasteur Institute in Paris and examined the expectoration of his 5-month-old daughter, who was suffering from typical whooping cough, under a microscope. However, the organism was cultured for the first time only in 1906 and definitively identified as the whopping cough agent by Jules Bordet, at the Institut Pasteur de Bruxelles, together with his brother-in-law, Octave Gengou. It became clear very quickly that successful culturing of *B. pertussis* only occurred very early after the onset of the disease, while the symptoms last for weeks to months after the organism has disappeared. This observation led Bordet and Gengou to postulate the action of toxic compounds and to qualify pertussis as a toxin-mediated disease. The presence of toxic substances was further confirmed by the observation that the intraperitoneal injection of killed *B. pertussis* extracts led to a pleural effusion and eventually the death of guinea pigs [3].

3. Lipo-Oligosaccharide

The first evidence of a *Bordetella* toxin, initially named endotoxin, was presented by Bordet and Gengou in a publication in 1909 [4] that referred to a toxin present in whole-cell lysates with strong toxicity in rabbits and guinea pigs. When injected into mice, rabbits or guinea pigs, it induced dermonecrotic lesions and death of the animals. Toxicity was lost upon heating for 30 min at 56 °C. Because of its heat lability, it is likely that this toxin corresponds to what has later been referred to as heat-labile or dermonecrotic toxin, a 1464 amino-acid-long protein toxin [5].

However, in addition to the heat-labile dermonecrotic toxin, *Bordetella* species also produce a non-proteinaceous endotoxin, which corresponds to the *Bordetella* lipo-oligosaccharide (LOS). This molecule was extensively characterized by Martine Caroff and colleagues [6], and its action on immune cells was investigated by Jean-Marc Cavaillon and colleagues at the Institut Pasteur in Paris. His team studied the structure–function relationship of *Bordetella* LOS and found a specific binding of this molecule to lectin-like receptors on the surface of macrophages [7] and subsequent polyclonal B-cell activation. The polysaccharide moiety of the *Bordetella* LOS further stimulated IL-1 secretion by macrophages and monocytes [8]. The precise LOS structural determinant responsible for the induction of IL-1 secretion was identified as the inner core region consisting of 2-keto-3-deoxy-D-manno-octulosonic acid and heptose [9]. In contrast, the lipid A moiety of the *Bordetella* LOS was found to be 1000 to 10,000 times less effective in inducing IL-1 secretion than the complete LOS molecule [10].

4. Adenylyl Cyclase/Haemolysin Toxin

4.1. Biogenesis

As previously observed by Bordet and Gengou, growth of *B. pertussis* on blood agar plates results in the appearance of a haemolytic halo surrounding the bacterial colonies, suggesting the production and secretion of a haemolytic molecule. This haemolytic activity is carried by the C-terminal domain of a 177.312 kDa protein, which contains an N-terminal catalytic domain expressing potent adenylyl cyclase activity [11]. A highly potent secreted adenylyl cyclase was first identified by Hewlett et al. [12], but the conclusive link of this enzyme with the haemolysin activity of *Bordetella* came from the seminal discovery of its corresponding structural gene by elegant work performed at the Institut Pasteur in Paris by Glaser et al. [13]. It was known that the *Bordetella* adenylyl cyclase needed calmodulin to express its enzyme activity. By generating an *Eschericia coli* strain that overproduced recombinant calmodulin, the teams at the Institut Pasteur identified *B. pertussis* DNA fragments able to complement adenylyl cyclase deficiency in this *E. coli* strain. The isolation and sequence of the full-length gene encoding the adenylyl cyclase, named *cyaA*, revealed an open reading frame of 1706 codons and predicted a protein consisting of four domains, including the N-terminal 400-residue-long calmodulin-sensitive cyclase domain. The C-terminal domain was hypothesized to possess haemolytic properties, which was confirmed by subsequent work by the same team [14]. They uncovered the similarities between this C-terminal domain and the *E. coli* alpha-haemolysin and with the *Pasteurella haemolytica* leukotoxin. Furthermore, they established the parallelism between the *E. coli* alpha-haemolysin secretion mechanism and the secretion mechanism of the *Bordetella* adenylyl cyclase/haemolysin toxin (ACT) by sequencing the *cyaA* downstream region and identifying two genes, named *cyaB* and *cyaD*. These two genes code for a 712- and a 400-residue-long protein, which are similar to *E. coli* HlyB and HlyD, respectively, both of which are involved in the secretion of the *E. coli* alpha-haemolysin. However, the *Bordetella* ACT secretion also required an additional protein, the gene of which was identified downstream of *cyaD* and named *cyaE*. The production of *Bordetella* ACT and its secretion as a single 200 kDa polypeptide chain was confirmed by Bellalou et al. in subsequent studies [15].

In addition to the identification of adenylyl cyclase and haemolysin as a single protein and its secretion apparatus, the team also made important contributions to the understanding of the regulation of its expression. The expression of the *Bordetella* virulence genes is under the control of the two-component BvgA/S signal-transducing system. BvgA is a transcriptional activator, and BvgS is a sensor protein that transmits information on environmental changes to BvgA via a phosphorylation cascade [16]. Laoide and Ullmann [17] found that the *cyaA* gene is under the control of BvgA, while transcription of the downstream *cyaBDE* cistrons, which are separated by 77 nucleotides from the *cyaA* translational stop codon, is independent of BvgA. Using primer-extension analyses, they identified the transcriptional start site of *cyaA* and observed transcription only under BvgA$^+$ conditions. Using the same technology, they also identified the transcriptional start site of the *cyaBDE* operon and found that this operon was transcribed under both BvgA$^+$ and BvgA$^-$ conditions. However, the strength of the latter promoter was 4 to 5 times lower than that of the *cyaA* gene. Steffen et al. [18] then demonstrated that, in contrast to some other BvgA-regulated genes, the *cyaA* gene requires phosphorylation of BvgA in order to be activated by this transcription factor and that phosphorylation of BvgA is sufficient to activate transcription of *cyaA*. This was demonstrated by the development of an in vitro transcription system using the RNA polymerase of *Bordetella*. In a series of runoff transcription experiments, they showed that the *Bordetella* RNA polymerase was able to efficiently drive transcription of the *cyaA* gene in the presence of phosphorylated BvgA. When *E. coli* RNA polymerase was used instead of the *Bordetella* enzyme, transcription levels were markedly reduced, suggesting that the *E. coli* RNA polymerase is less efficient in the formation of the transcription initiation complex than its *Bordetella* counterpart.

4.2. Calmodulin Binding and Catalysis

The structure–function relationship of ACT was also extensively studied by scientists of the Institut Pasteur in Paris. Considering the importance of calmodulin binding for adenylyl cyclase activity, Ladant examined the interaction of the enzyme with calmodulin and found that the two molecules interact with each other in a 1:1 stoichiometry and that the interaction was stronger in the presence of calcium than in its absence [19]. Trypsin treatment of calmodulin-bound adenylyl cyclase resulted in two distinct fragments of 18 and 25 kDa, respectively, both of which interacted with calmodulin. After trypsin cleavage, the two fragments remained catalytically active when bound to calmodulin, whereas in the absence of calmodulin, the two fragments lost their catalytic activity. The 18-kDa fragment was then identified as the main calmodulin-binding domain, and the 25-kDa fragment located at the N-terminal moiety was identified as the catalytic domain [20]. Using a fluorescent ATP analogue, Sarfati et al. [21] demonstrated that in the presence of calcium, calmodulin increased substrate binding of the adenylyl cyclase, probably through the induction of conformational changes. Work by Bouhss et al. [22] narrowed down the essential peptide for calmodulin binding to a 72 amino acid peptide that spans the C terminus of the 25-kDa peptide and the N terminus of the 18-kDa peptide and showed that the hydrophobic helix around Trp-242 is critical for calmodulin binding. However, the N-terminal half of the catalytic 25-kDa domain also contributes to calmodulin binding, albeit to a much lesser degree. Regions critical for calmodulin binding were subsequently identified around Leu-247 and Cys-335 by Ladant et al. [23] using an insertional mutagenesis approach. Further mutagenesis experiments using an alanine substitution approach revealed that replacement of Arg-338, Asn-347 and Asp-360 by alanine dramatically reduced the affinity of adenylyl cyclase to calmodulin, and molecular dynamics simulations suggested that the mutations may have caused large fluctuations of the calcium-binding loops of calmodulin, which may have weakened the interaction of the enzyme with calmodulin and destabilized the catalytic loop of the adenylyl cyclase [24]. More recently, O'Brien et al. [25] reported that calmodulin helps the adenylyl cyclase enzyme to adopt the folding necessary to express its enzymatic activity, as evidenced by a series of small-angle X-ray scattering, hydrogen/deuterium exchange mass spectrometry and synchrotron radiation circular dichroism measurements. In the absence of calmodulin, adenylyl cyclase is structurally disordered, with a 75-residue-long peptide in the 18-kDa C-terminal domain, spanning residues 201 to 275. Binding of calmodulin to this region transitions the domain from a disordered to an ordered conformation and induces allosteric effects to stabilize the distant catalytic site. This intrinsically disordered region may destabilize the enzyme in the absence of calmodulin so that it is only active when needed, i.e., within the target cell where calmodulin is present. The intrinsically disordered structure may also be helpful for the toxin to cross the bacterial secretion system and the target cell membrane.

The catalytic mechanism was investigated by Glaser et al. [26], first by using site-directed mutagenesis to identify critical amino acid residues for enzyme function. Replacement of lysines at position 58 or 65 by glycines strongly reduced the catalytic activity. Later, four additional residues, i.e., Asp-188, Asp-190, His-298 and Glu-301, were found to be important for enzyme activity [27]. The substitution of these amino acids had a relatively minor impact on calmodulin binding. Instead, binding to a photoactivable ATP analogue was strongly affected, especially by substitutions of Asp-190. A mechanism was proposed whereby Asp-188 and Asp-190 of the enzyme interact with Mg^{++}-ATP, thereby stabilizing the transition state. In this model, Lys-65 or Lys-58 interacts with the α-phosphate group of the Mg^{++}-ATP. Another amino acid can then act as a basic catalyst for the cyclization of the ATP.

Another important residue for catalysis is His-63 [28]. While replacement of His-298 by leucine or arginine did strongly reduce enzyme activity, it did not alter the kinetic characteristics of the adenylyl cyclase. In contrast, substitution of His-63 altered its kinetic properties, indicating that this residue is directly involved in catalysis itself, possibly as part of a charge relay system in a general acid/base enzyme mechanism.

4.3. Haemolytic and Receptor-Binding Domain

As shown by Glaser et al. [14] and Bellalou et al. [15], the haemolytic activity of ACT is confined to the 1306 C-terminal residues of the molecule, showing sequence similarities to the *E. coli* α-haemolysin, although its haemolytic activity is weaker than that of the *E. coli* α-haemolysin. By constructing mutant derivatives of ACT devoid of adenylyl cyclase activity or lacking the catalytic N-terminal domain, Sakamoto et al. [29] showed that haemolysis does not require enzyme activity and that the C-terminal moiety of the molecule is sufficient to express full haemolytic activity.

The latter domain is also essential for the transport of the catalytic domain into the target cell and is itself composed of two distinct subdomains. The N-proximal subdomain contains hydrophobic segments, which are essential for haemolytic and toxin activity [15]. By using an artificial planar–lipid bilayer system, Benz et al. [30] demonstrated that ACT forms small ion-permeable pores in lipid bilayers with a very small diameter dependent on the subdomain containing the hydrophobic segments. However, the pore-forming activity of the native protein was substantially higher than that of a recombinant analogue, suggesting the role of a post-translational modification of the toxin.

The post-translational modification of ACT is catalysed by the product of an additional gene, namely *cyaC*, which was discovered by Barry et al. [31] at the Medical College of Virginia. This allowed Sebo et al. [32] to produce fully active ACT by recombinant *E. coli* co-expressing *cyaC*, together with *cyaA* [32]. Through in vitro chemical acylation of lysines on non-modified, inactive, recombinant ACT, Heveker et al. [33] activated ACT by transferring lauric, myristic or palmitic acid chains, as evaluated by both its haemolytic and toxic activities. However, the activity of the chemically modified ACT was still lower than that of native ACT. Nevertheless, this study suggests that activation of ACT by CyaC occurs through acylation of one or more lysine residue(s) of the toxin. The modified lysine residues were identified as Lys-983 and, potentially, Lys-860 by the Hewlett group at the University of Virginia [34,35] and found to be palmitoylated by CyaC.

Initially, it was hypothesized that the domain containing the hydrophobic segments would form channels in the target cell membranes through which the catalytic domain can gain access to the cytoplasm. However, the size of the channels is too small to allow for translocation of a protein domain, even when totally unfolded [30]. Karst et al. [36] identified an additional segment spanning residues 375–485 located at the C-terminal end of the catalytic domain that can also bind to membranes and is able to destabilize lipid bilayers. When this segment was deleted from ACT, toxin activity was abrogated. This segment contains two subsegments spanning residues 414 to 440 and 454 to 484, respectively, that are predicted to form α-helical structures with membrane-interacting potential [37]. A synthetic peptide corresponding to one of these two subsegments was shown to bind to and permeabilize membranes containing anionic lipids. It adopted an α-helical conformation upon interaction with the lipid bilayer and was able to permeabilize lipid vesicles. For the peptide to acquire its secondary structure, the anionic nature of the lipids is important, most likely via the interaction of the lipid with the two positively charged arginine residues of the peptide. More recently, Voegele et al. found that this peptide can translocate across lipid bilayers and then bind to calcium-loaded calmodulin, thereby pulling the entire catalytic domain through the plasma membrane [38]. This interaction with calmodulin appears to be essential for translocation, as substitutions of amino acid residues that interact with calmodulin or the use of calmodulin inhibitors abrogates translocation. These observations indicate that calmodulin plays a double role in ACT action by aiding in membrane translocation of the adenylyl cyclase domain and by stabilizing the active conformation of the enzyme.

Calcium plays an important role in ACT action not only by binding to calmodulin. Both ACT toxicity and haemolysis are calcium-dependent [39], and the toxin can directly bind calcium via several high-affinity and low-affinity binding sites [40]. Binding of calcium to high-affinity sites was proposed to be necessary for the haemolytic activity of the toxin, while binding to the low-affinity sites induces conformational rearrangements of the protein. The latter sites were mapped to the C-terminal part of the protein characterized by Asp-Gly-rich repeats common to the members of the RTX (for repeat in toxin) family. Work by Bauche et al. [41] established that the calcium-binding sites extend beyond the Asp-Gly-rich repeats and include adjacent protein segments. These adjacent segments are essential for the calcium-binding sites to fold into a stable parallel ß helix upon binding to calcium. In the absence of calcium, the Asp-Gly-rich repeat region is intrinsically disordered [42]. Since the calcium concentration is low within the bacterial cytosol, the newly synthesized ACT is likely in an unfolded conformation, which was shown to be important for ACT to be secreted via a type 1 secretion system through the *B. pertussis* cell wall. Once secreted, ACT can bind calcium and thereby adopt its functional conformation. The disorder-to-order transition triggered by calcium binding also reduces the mean net charge of ACT, which influences its ability to multimerize [43]. This calcium effect strongly depends on the C-terminal flanking region of the Asp-Gly repeats. Like many other RTX proteins, ACT tends to form multimers in vitro, such as after a denaturing/renaturing cycle of purified ACT. However, the monomeric form displays substantially stronger haemolytic and toxin activity than the multimeric forms [44], suggesting that the physiologically active form of ACT is monomeric, which is a form favoured by calcium and post-translational acylation.

Nevertheless, the fact that membrane permeabilization by ACT requires oligomerization suggests that the protein may oligomerize once it has partitioned into the membrane.

The Asp-Gly-rich C-terminal region of ACT is also its receptor-binding domain. The ACT receptor was identified by Guermonprez et al. [45] as the $\alpha_M\beta_2$ (CD11b/CD18) integrin expressed on the surface of leukocytes, as evidenced by the ability of anti-CD11b and anti-CD18 antibodies (but not by antibodies to other integrins) to block ACT binding to neutrophils and ensuing increases in intracellular cAMP levels. Although ACT can bind to and invade many different cell types, Chinese hamster ovary cells became highly sensitive to ACT when transfected with CD11b/CD18 but not with other integrins, and binding to CD11b/CD18 strictly depended on the presence of calcium. CD11b/CD18 is mostly expressed on the surface of macrophages, neutrophils and dendritic cells, suggesting a primary action of ACT on these innate immune cells.

4.4. Biological Activities of ACT

The team headed by Nicole Guiso has extensively studied the biological functions of ACT. Thirty years ago, this group showed that macrophages are indeed a major cell type targeted by ACT [46], which is consistent with the presence of the ACT receptor in these cells. Macrophages infected with *B. pertussis* undergo apoptosis, which depends on ACT, as infection with ACT-deficient *B. pertussis* does not induce apoptosis. Both the catalytic and the haemolytic domains are required for apoptosis induction. In vivo, in a murine respiratory challenge model, infection by *B. pertussis* was shown to induce apoptosis of macrophages, along with neutrophils, present in bronchoalveolar lavage fluids and lung tissues [47]. In vivo macrophage apoptosis was not observed when mice were infected with ACT-deficient *B. pertussis* strains, while neutrophil apoptosis was still observed in these mice. ACT-deficient *B. pertussis* strains were also affected in their growth rate during the early phase of infection in a murine respiratory challenge model [48], and both the haemolytic and the catalytic domains were shown to be important for early growth during infection. Furthermore, ACT was found to co-operate with pertussis toxin (PTX) in the infectious process and in the induction of histopathological lesions, as well as in the recruitment of inflammatory cells, including neutrophils, in the lungs [47,49].

In addition to apoptosis of macrophages, ACT also induces more subtle responses of monocytes. The increase in intracellular cAMP levels within human monocytes upon *B. pertussis* infection is inversely correlated with TNF-α production, superoxide anion release and Hsp70 expression [50]. These effects depend on the production of ACT and were not observed when the cells were incubated with an ACT-deficient strain but could be reproduced by incubation in the presence of purified ACT, suggesting that the toxin affects monocyte responses to infection to evade innate antimicrobial defence strategies.

ACT may also play a role in the interaction of *B. pertussis* with epithelial cells. Curiously, ACT-deficient mutants were able to invade human HTE tracheal epithelial cells more efficiently than fully virulent isogenic strains [51], suggesting that ACT prevents invasion of these cells. *B. pertussis* also induces the secretion of IL-6 by HTE cells in an ACT-dependent manner [52], indicating its role in the inflammatory process triggered by *B. pertussis* infection.

The protective effect of ACT as a potential vaccine antigen against pertussis was also tested by the Guiso team using a mouse intranasal infection model. Administration of anti-ACT antibodies to 3-week-old female BALB/c mice significantly protected them from pulmonary and intracerebral lesions, as well as death induced by challenge with high doses of the 18,323 strain of *B. pertussis*, as did active immunization with the purified catalytic domain of ACT [53]. Follow-up studies showed that ACT and its catalytic domain protected mice equally well against bacterial lung colonization in a sublethal model of *B. pertussis* infection, albeit not as well as a whole-cell vaccine comparator [54] due to neutralizing antibodies binding to residues 373–400 of the catalytic domain that block toxin translocation. However, later studies by the same group showed that the C-terminal domain of ACT is essential for protection against *B. pertussis* colonization. Using a series of truncated

ACT derivatives, Betsou et al. [55] found that only immunization with derivatives that contained the Asp-Gly-rich repeat region conferred protection against lung colonization by *B. pertussis*. This region is also the predominant part of the protein that elicits antibodies upon infection by *B. pertussis* in mice, as well as in humans. In addition to this region, the last 217 amino acids of ACT are also essential for antibody recognition and protective activity, most likely because of its role in folding, e.g., to display the protective epitopes. ACT derivatives that lacked the Asp-Gly-rich repeat region were nevertheless able to induce strong ACT-neutralizing antibodies, suggesting that protection induced by ACT may be independent of elicited neutralizing antibodies. CyaC-catalyzed acylation was also found to be important for ACT-mediated protection against *B. pertussis* lung colonization [56]. This was shown by comparing recombinant ACT produced by *E. coli* in the presence or absence of coexpressed *cyaC*. In a sublethal nasal challenge model, recombinant ACT produced in the absence of coexpressed *cyaC* provided no protection at all against lung colonization. In contrast, when the mice were immunized with CyaC-activated ACT, a significant reduction in bacterial colonization of the lungs was observed compared to non-vaccinated mice. Partial activation through low levels of *cyaC* coexpression resulted in intermediate levels of protection. However, ACT produced in *cyaC*-co-expressing *E. coli* was less protective than native ACT, possibly due to differences in the chemical nature of the acyl group between the recombinant and the native forms of ACT or to the presence of small amounts of contaminating antigens in the native ACT preparation.

4.5. Applications

The various unique properties of ACT have prompted the proposal of potential applications in a variety of different fields. Its invasive properties led Sebo et al. [57] to propose ACT as a vehicle to deliver foreign epitopes directly into the cytosol for the induction of CD8$^+$ cytotoxic T cells. When target cells were invaded by a recombinant ACT analogue that carries a CD8$^+$ T-cell epitope from the lymphocytic choriomeningitis virus (LCMV), they were found to be readily lysed by cytotoxic CD8$^+$ T cells specific to this epitope, indicating that ACT may be a good carrier for T-cell epitopes to be presented by the major histocompatibility complex class I molecules. Moreover, active immunization with enzymatically inactive hybrid ACT molecules carrying viral epitopes of LCMV or human immunodeficiency virus (HIV) resulted in antigen-specific class I restricted CD8$^+$ CTL responses that could kill peptide-loaded target cells [58]. Using a mouse model of lethal intracerebral LCMV infection, Saron et al. [59] found that genetically inactivated hybrid ACT molecules carrying the LCMV epitope protected the animals from death through a CD8$^+$ T-cell-dependent mechanism. The use of hybrid ACT analogues for the presentation of epitopes to CD8$^+$ T cells was then extended to other antigens, including tumour antigens [60], which led to the finding that net negative charges are detrimental to the translocation process and the ensuing induction of CD8$^+$ T-cell responses. Furthermore, proof of concept was provided that this approach can stimulate antitumour immunity [61]. The ACT vector was further improved by inserting multiple copies of MHC class I and class II restricted epitopes, which, in addition to CD8$^+$ CTL responses, triggered potent Th1-type immune responses, as evidenced by high production of IL-2 and IFN-γ [62]. The ACT vector was also found to be capable of accommodating several different CD8$^+$ T-cell epitopes, as exemplified by an LCMV polyepitope, an HIV and an ovalbumin epitope inserted simultaneously at three different permissive sites of ACT [63]. After administration to mice, all three were processed and induced CTL responses, and the immunized mice were protected against lethal challenge with LCMV. Importantly, prior immunization with ACT did not appear to prevent the induction of CD8$^+$ T-cell responses to the grafted epitopes by the recombinant ACT analogues.

Given that ACT preferentially binds to CD11b/CD18-expressing cells, including dendritic cells, targeting foreign antigens to these antigen-presenting cells may be an efficient way to not only induce CD8$^+$ CTL responses but also T-helper cell and B-cell responses. This was shown by using a recombinant ACT derivative carrying the full-length

HIV-1 Tat protein [64]. Immunization with this hybrid toxin in the absence of adjuvant induced high levels of long-lasting Tat-specific neutralizing antibodies and Th1-type T-cell responses. The same molecule was also able to induce strong neutralizing antibody and Th1-type T-cell responses in African green monkeys [65], indicating that ACT can also deliver foreign antigens to dendritic cells of primates.

Similarly, full-length and various segments of the E7 oncoprotein from the human papilloma virus 16 were inserted into genetically detoxified ACT and were shown to elicit strong CTL and Th1 responses in mice [66]. The vaccines were also able to trigger tumour regression in mice injected with E7-expressing tumour cells. Tumour regression reached up to 100%, accompanied by 100% survival, which exceeded that of the comparator consisting of the E7 peptide administered together with CpG ODN 1826. This vaccine candidate, combined with a HPV18 E7-ACT protein, is now in clinical development and was shown to be safe and immunogenic in women infected with HPV16 or HPV18 [67].

The observation that the catalytic domain of ACT can be split into two fragments of 18 and 25 kDa, respectively, that can complement each other in the presence of calmodulin [19,20] has led to the development of a two-hybrid system to study protein–protein interactions [68]. Separately, the two fragments do not express adenylyl cyclase activity in the absence of calmodulin. However, when two proteins that interact with each other are fused to the two complementary 18-kDa and 25-kDa ACT fragments, enzyme activity is restored, and high levels of cAMP are produced, even in the absence of calmodulin. cAMP can then bind to the transcriptional activator CAP, which triggers the transcription of catabolic operons, including the lactose and maltose operons. This can easily be monitored phenotypically in an *E. coli* strain that lacks the endogenous adenylyl cyclase by the use of indicator plates. This bacterial two-hybrid system is able to detect interactions between small peptides and between entire proteins. It can also be used in genetic screening to identify proteins that interact with a specific target. In contrast to the yeast two-hybrid system, it can be used to study protein–protein interactions in the cytosol, as well as in the inner membrane.

5. Pertussis Toxin

In addition to ACT, *B. pertussis* also produces PTX. Unlike ACT, which is also produced by *B. parapertussis* and *B. bronchiseptica*, PTX is exclusively produced by *B. pertussis*, although the *ptx* genes are present in the genomes of two other *Bordetella* species [69]. This toxin is composed of five different subunits, named S1 to S5 according to their decreasing molecular weights, arranged in a hexameric structure with a 1S1:1S2:1S3:2S4:1S5 stoichiometry (for review, see [70]). It is a member of the A-B toxin family, in which the A subunit (here S1) expresses enzyme activity and the B oligomers (here S2 to S5) are responsible for target cell receptor binding. Unlike ACT, which can enter the cell directly through the plasma membrane, PTX is taken up by receptor-mediated endocytosis and reaches the endoplasmic reticulum through retrograde transport, where the S1 is translocated into the cytosol. In the cytosol, S1 catalyses the transfer of the ADP–ribosyl moiety of NAD onto the alpha subunit of signal-transducing Gi/o proteins, which is the basis of PTX toxicity and the mechanism of hallmark features of pertussis, such as leukocytosis.

5.1. Structure–Function Relationship of PTX

Before joining the Institut Pasteur de Lille, I was fortunate enough to isolate and sequence the structural gene of PTX [71]. This work established that the five subunits are produced as independent polypeptides encoded by a single polycistronic operon in the order of *ptxABDEC*, coding for S1, S2, S4, S5 and S3, respectively. Each polypeptide is synthesized with a typical cleavable signal peptide, suggesting that they are transported via the Sec apparatus through the inner membrane of *B. pertussis* into the periplasm, where the assembly into the holotoxin occurs. The S1 subunit contains regions of homology to the enzymatically active A subunits of cholera toxin and *E. coli* heat-labile toxin, two other

ADP-ribosylating toxins, consistent with S1 being the catalytic subunit. S2 and S3 share a relatively high degree of sequence similarities. Each one is associated with an S4 subunit.

Subsequent work of my group has shown that the S2/S4 and S3/S4 dimers display a certain degree of specificity in receptor binding [72]. While the S2/S4 dimer binds to haptoglobin, the S3/S4 dimer binds to the toxin receptor on the surface of Chinese hamster ovary cells. Deletions of Asn-105 in S2 and of Lys-105 in S3 reduced haptoglobin and Chinese hamster ovary binding, respectively. When both residues were deleted, PTX-mediated mitogenesis, an ADP-ribosylation-independent activity of the toxin, was abolished. When the S2 or the S3 gene was deleted from the *B. pertussis* chromosome, PTX analogues were produced that contain either two S3 or two S2 subunits [73], indicating that the two subunits can substitute each other, although this does not occur in wild-type *B. pertussis*. Interestingly, the toxin analogues lacking the S2 subunit were less efficiently secreted by *B. pertussis* than the wild-type version, while the deletion of the S3 subunit significantly increased PTX secretion, suggesting that S2 plays a role in the secretion of the toxin. Consistent with differences in receptor recognition by S2 and S3, the toxin analogue containing two S3 subunits was approximately 10 times more efficient in in vivo ADP-ribosylation of Gi proteins in Chinese hamster ovary cells, while the analogue containing two S2 subunits was roughly 100-fold less efficient than the natural toxin. When THP-1 cells were used in the in vivo ADP ribosylation assay, both mutant toxins were less efficient than the wild-type molecule.

The five subunits all contain intrachain disulphide bonds: one in S1, three in S2 and S3 and two in S4 and S5. With Rudy Antoine, we have shown that the disulphide bond in S1 linking Cys-41 to Cys-200 is essential for assembly of S1 with the B oligomer [74]. Removal of each one of the two cysteines of S1 resulted in the secretion of the B oligomer without S1.

5.2. Mechanism of PTX S1 Subunit Enzyme Activity

Once within the cytoplasm of the target cell, the disulphide bond of S1 must be reduced for ADP ribosylation to occur. Deletion of Cys-41 or of the adjacent Ser-30 abolished enzyme activity, while substitutions of Cys-41 by serine or glycine still allowed for detectable activity, albeit at reduced levels [75]. This allowed us to compare enzyme kinetics between the wild-type S1 and its mutant derivative, which indicated that the catalytic rate was not affected by the amino acid substitution but that the K_m of the mutant protein for NAD was increased relative to that of the wild-type enzyme, suggesting that Cys-41 is located close to the NAD-binding site. Of note, Cys-41 is located between two segments of S1 with strong sequence similarities to the A subunits of cholera toxin and *E. coli* heat-labile toxin.

In addition to the S1 disulphide bond, the 27 C-terminal residues of this subunit are also essential for toxin assembly and secretion [74]. The 47 C-terminal residues of S1 are also involved in the enzyme activity, probably via binding to the Gi/oα protein acceptor substrate. Deletion of this region strongly reduced ADP ribosylation of the G protein, whereas the NAD glycohydrolase activity measured in the absence of the acceptor substrate was not affected [75,76]. Additional amino acid residues important for both ADP ribosyltransferase and NAD glycohydrolase activities were Trp-26 and Glu-129 [77]. While Trp-26 appeared to be important for NAD binding, Glu-129 is directly involved in catalysis. Photoaffinity labelling of S1 resulted in binding of the nicotinamide moiety of NAD to Glu-129 [78]. Furthermore, kinetic studies of an S1 analogue in which Glu-129 was substituted by aspartate showed a reduction in the catalytic rate by more than two orders of magnitude, while the apparent K_m value for NAD was not affected [79]. In addition, fluorescent quenching experiments conducted using increasing concentrations of NAD showed similar dissociation constants between the Glu-129 substitution mutant and wild-type S1, confirming that the substitution did not affect NAD binding in an important way. Importantly, when Glu-129 was replaced by cysteine, a disulphide was formed between Cys-41 and Cys-129, indicating that Glu-129 is located in proximity of the NAD-binding site in the tertiary structure of the protein.

Another residue involved in catalysis is His-35 [80]. Substitutions of this residue resulted in a substantial reduction in k_{cat} values without strongly affecting the K_m value for NAD in an NAD glycohydrolase reaction. Replacement of His-35 by glutamine resulted in a 50-fold decrease in specific activity, while its replacement by asparagine reduced the specific activity by more than 500 fold, indicating that the glutamine side chain can partially mimic the imidazole ring of histidine. Although glutamine can mimic some of the hydrogen-bonding capacity of the ε-N of histidine, it does not have the proton transfer capacity of histidine and can thus not act as a true base. We therefore proposed that His-35 may not be directly involved in proton abstraction but rather in polarization via hydrogen bonding of the acceptor substrate water in the NAD glycohydrolase reaction or the Gi/oα cysteine in the ADP ribosyltransferase reaction. This increases the nucleophilicity of the polarized acceptor substrates, which may then attack the cleavable *N*-glycosidic bond of NAD.

These studies, together with contributions from other laboratories, have allowed us to propose an enzymatic mechanism for PTX S1 [81]. In this model, the C-terminal region of S1 is crucial for Gi/oα binding and positioning of its acceptor amino acid, Cys-347, within the catalytic site of the enzyme. Glu-129 may act on NAD by retrieving the ribose 2'-OH proton, which weakens the nicotinamide–ribosyl bond. His-35 increases the nucleophilicity of the cysteine (or water molecule in the NAD glycohydrolase reaction), which can then attack the weakened *N*-glycosidic bond. Arg-9 within the active site would be involved in productive NAD binding via interaction with the phosphate groups of NAD. Changing its side chain length by replacing Arg-9 with lysine abolishes both NAD glycohydrolase and ADP ribosyltransferase activities.

5.3. Applications: Development of a Live Attenuated Pertussis Vaccine

The most important applications of studies on PTX have been in the field of novel pertussis vaccine development. Genetic detoxification of PTX by replacing Arg-9 and Glu-129 with lysine and glycine, respectively, has been instrumental for the development of a safe and efficacious acellular pertussis vaccine [82]. This vaccine has been used in Italy for several years and is now in use in Thailand. Recent studies have shown that genetically inactivated PTX induces substantially longer-lasting antibody responses than chemically inactivated PTX [83], although the latter is now used in most acellular pertussis vaccines available.

We used the knowledge gained in studies on PTX to develop a live attenuated pertussis vaccine designed for nasal delivery [84]. Considering that *B. pertussis* is a strictly mucosal pathogen and naturally only colonizes the respiratory tract and that its contagiousness is amongst the highest known for a respiratory pathogen [85], we considered that the induction of mucosal immunity may be the optimal strategy to control pertussis at a population level [86]. Mucosal immunity and protection against infection is best achieved by prior infection with *B. pertussis* [87]. We therefore designed a live attenuated *B. pertussis* vaccine strain in order to mimic natural infection as closely as possible without causing disease. This vaccine, named BPZE1, is deficient in the production of dermonecrotic toxin and tracheal cytotoxin and produces genetically inactivated PTX through the replacement of Arg-9 and Glu-129 in S1 with lysine and glycine, respectively [84].

It was shown to protect mice from both lung and nasal colonization by virulent *B. pertussis* and to induce potent systemic and mucosal immune responses [88]. Importantly, the mucosal IgA response induced by BPZE1 appeared to be essential for protection by the vaccine against nasal colonization by *B. pertussis*. Unlike that induced by most current acellular vaccines, protection induced by BPZE1 was long-lasting [89]. It also induced lung and nasal tissue-resident memory T (T_{RM}) cells that produce IL-17 and/or IFN-γ, similar to what has been observed after infection with virulent *B. pertussis*. We have shown that the induction of these T_{RM} cells, especially the IL-17-producing T_{RM} cells, is inhibited by vaccination with acellular pertussis vaccines, which prolongs nasal carriage of *B. pertussis* in acellular pertussis-vaccinated mice [90]. The vaccine was shown to be safe, even in severely immunocompromised mice [91,92]. We also evaluated its safety and protective capacity in non-human primates and showed that at a dose of up to 10^{10} colony-forming units (CFU) it did not induce any sign of disease in baboons yet was highly protective against pertussis disease and colonization by a highly virulent clinical *B. pertussis* isolate [93].

The vaccine has entered clinical development, and a first phase 1, first-in-human study carried out in Sweden has documented that it did not cause any significant adverse events after a single ascending dose from 10^3 over 10^5 to 10^7 CFU delivered as nasal drops, as compared to a placebo administration [94]. It was able to transiently colonize the human respiratory tract in a dose-dependent manner and to induce serum antibodies against PTX, as well as other *B. pertussis* antigens, such as filamentous haemagglutinin, pertactin and fimbriae. These antibody levels did not decline for at least 6 months after vaccination, when the study was terminated. However, even at the highest dose used in that study (10^7 CFU), only 5 out of the 12 participants were colonized by BPZE1 and induced antibodies to *B. pertussis* antigens. However, those that were not colonized had high levels of pre-existing antibodies to *B. pertussis* antigens, although they had never been vaccinated with pertussis vaccines, as they were born at a time when Sweden had stopped pertussis vaccination. This observation suggests that these pre-existing antibodies may have been generated by prior silent *B. pertussis* infection and that this prior infection may have induced immunity, preventing BPZE1 vaccine take, a hypothesis consistent with the notion that *B. pertussis* infection prevents subsequent *B. pertussis* reinfection.

This hypothesis was confirmed by a second phase 1 study also carried out in Sweden [95]. In this phase 1b study, subjects with high pre-existing antibody levels against *B. pertussis* antigens were excluded, and the vaccine was administered at doses of 10^7, 10^8 or 10^9 CFU. In this study, colonization by BPZE1 was detected in roughly 80% of the participants, including in 10 out of 12 subjects who had received 10^7 CFU. This also led to strong seroconversion against the *B. pertussis* antigens, with the highest seroconversion rate found in the 10^9 CFU group. Even in the highest-dose group, the safety profile was comparable to that of the placebo group. Therefore, 10^9 CFU was considered the optimal human dose to be used in the next clinical trials. At this dose, BPZE1 also induced *B. pertussis*-specific T-cell and robust memory B-cell responses [96]. The vaccine was also found to induce opsonizing antibodies, stimulating reactive oxygen species production in neutrophils and exerting bactericidal actions that were superior to those induced by acellular pertussis vaccines.

BPZE1 formulations were optimized as a lyophilized drug product that is stable for at least 2 years at storage temperatures up to room temperature [97]. It can now be delivered by the use of a spray device after reconstitution. The lyophilized and reconstituted formulation applied using the spray device was found to be more immunogenic and equally safe in a phase 2a study compared to the previous formulation administered as nasal drops [98] and was therefore used in a phase 2b trial.

In this phase 2b randomized, double-blind, placebo-controlled trial involving 300 volunteers, the safety of BPZE1 was confirmed at the level of both local tolerability and systemic reactogenicity [99]. The induction of long-lived *B. pertussis*-specific serum antibody production was also confirmed. In addition, this study showed that BPZE1, unlike the acellular vaccine comparator, also induced robust mucosal anti-*B. pertussis* secretory IgA responses, and elevated secretory IgA responses persisted for up to 254 days, when the study was terminated. Most importantly, this phase 2b trial provided proof of concept that nasal BPZE1 vaccination may protect against subsequent *B. pertussis* infection, as for 90% of the participants who had received a first dose of BPZE1, no bacteria were detected when BPZE1 was administered a second time 85 days later as an attenuated challenge dose, compared to only 30% of those who had received the acellular pertussis vaccine instead of BPZE1 as the first vaccine. Furthermore, for the 10% of the BPZE1 recipient for whom BPZE1 bacterial infection could be detected after attenuated challenge, the bacterial loads were substantially lower and cleared much faster than for those that were infected by BPZE1 after acellular pertussis vaccination. BPZE1 also induced PTX-neutralizing serum antibodies and complement-dependent bactericidal antibodies. Interestingly, these antibodies were able to kill pertactin-producing *B. pertussis*, as well as pertactin-deficient *B. pertussis*. In contrast, sera induced by acellular pertussis vaccination were only able to kill pertactin-producing *B. pertussis*. Given that in many high-income countries, pertactin-deficient *B. pertussis* currently largely predominates, this observation may illustrate an important additional asset of BPZE1 in the fight against pertussis.

Further clinical studies are currently underway to test the protective potential of BPZE1 against colonization by virulent *B. pertussis* using a controlled human challenge model (ClinicalTrials.gov: NCT05461131) and to evaluate its performance in school-age children (ClinicalTrials.gov: NCT05116241).

6. Conclusions

Pertussis is essentially a toxin-mediated disease, and in addition to LOS, ACT and PTX are major *B. pertussis* virulence factors involved in the pathogenesis of the disease, often acting in synergy. Scientists at Pasteur Institutes have made major contributions to the understanding of toxin action at the molecular level, their structure–function relationship and their involvement in pathogenesis using murine models. In addition, they have allocated efforts and time to translate this knowledge into novel tools for the prevention or treatment of important human diseases. This is in line with the vision of the founder of these institutes, Louis Pasteur himself, and his immediate disciples, who have devoted their lives to the study of major human health problems using scientific approaches, which have often led to effective solutions to these issues and have helped to improve global health overall.

Funding: This research received no external funding.

Institutional Review Board Statement: Not applicable.

Informed Consent Statement: Not applicable.

Data Availability Statement: This review article does not contain new data.

Conflicts of Interest: The author is listed as inventor on patents related to the live attenuated pertussis vaccine and receives consultancy fees from ILiAD Biotechnologies, the company that acquired the patent portfolio on the live attenuated pertussis vaccine.

References

1. Weigand, M.R.; Peng, Y.; Batra, D.; Burroughs, M.; Davis, J.K.; Knipe, K.; Loparev, V.N.; Johnson, T.; Juieng, P.; Rowe, L.A.; et al. Conserved patterns of symmetric inversion in the genome evolution of *Bordetella* respiratory pathogens. *mSystems* **2019**, *4*, e00702-19. [CrossRef]
2. Bordet, J.; Gengou, O. Le microbe de la coqueluche. *Ann. Inst. Pasteur Paris* **1906**, *20*, 731–741.
3. Bordet, J.; Gengou, O. Le microbe de la coqueluche. *Ann. Inst. Pasteur Paris* **1907**, *21*, 727–732.
4. Bordet, J.; Gengou, O. L'endotoxine coquelucheuse. *Ann. Inst. Pasteur Paris* **1909**, *23*, 415–419.
5. Fukui, A.; Horiguchi, Y. *Bordetella* dermonecrotic toxin exerting toxicity through activation of small GTPase Rho. *J. Biochem.* **2004**, *136*, 415–419. [CrossRef]
6. Caroff, M. Structural variability and originality of the *Bordetella* endotoxins. *J. Endotoxin Res.* **2001**, *7*, 63–68. [CrossRef] [PubMed]
7. Haeffner-Cavaillon, N.; Chaby, R.; Cavaillon, J.M.; Szabo, L. Lipopolysaccharide receptor on rabbit peritoneal macrophages. I. Binding characteristics. *J. Immunol.* **1982**, *128*, 1950–1955. [CrossRef]
8. Haeffner-Cavaillon, N.; Cavaillon, J.M.; Moreau, M.; Szabo, L. Interleukin 1 secretion by human monocytes stimulated by the isolated polysaccharide region of the *Bordetella pertussis* endotoxin. *Mol. Immunol.* **1984**, *21*, 389–395. [CrossRef]
9. Lebbar, S.; Cavaillon, J.M.; Caroff, M.; Ledur, A.; Brade, H.; Sarfati, R.; Haeffner-Cavaillon, N. Molecular requirement for interleukin 1 induction by lipopolysaccharide-stimulated human monocytes: Involvement of the heptosyl-2-keto-3-deoxyoctulosonate region. *Eur. J. Immunol.* **1986**, *16*, 87–91. [CrossRef]
10. Cavaillon, J.M.; Fitting, C.; Caroff, M.; Haeffner-Cavaillon, N. Dissociation of cell-associated interleukin-1 (IL-1) and IL-1 release induced by lipopolysaccharide and lipid A. *Infect. Immun.* **1989**, *57*, 791–797. [CrossRef]
11. Novak, J.; Cerny, O.; Osickova, A.; Linhartova, I.; Masin, J.; Bumba, L.; Sebo, P.; Osicka, R. Structure-function relationships underlying the capacity of *Bordetella* adenylate cyclase toxin to disarm host phagocytes. *Toxins* **2017**, *9*, 300. [CrossRef]
12. Hewlett, E.L.; Urban, M.A.; Manclark, C.R.; Wolff, J. Extracytoplasmic adenylate cyclase of *Bordetella pertussis*. *Proc. Natl. Acad. Sci. USA* **1976**, *73*, 1926–1930. [CrossRef] [PubMed]

13. Glaser, P.; Ladant, D.; Sezer, O.; Pichot, F.; Ullmann, A.; Danchin, A. The calmodulin-sensitive adenylate cyclase of *Bordetella pertussis*: Cloning and expression in *Escherichia coli*. *Mol. Microbiol.* **1988**, *2*, 19–30. [CrossRef]

14. Glaser, P.; Sakamoto, H.; Bellalou, J.; Ullmann, A.; Danchin, A. Secretion of cyclolysin, the calmodulin-sensitive adenylate cyclase-haemolysin bifunctional protein of *Bordetella pertussis*. *EMBO J.* **1988**, *7*, 3997–4004. [CrossRef] [PubMed]

15. Bellalou, J.; Ladant, D.; Sakamoto, H. Synthesis and secretion of *Bordetella pertussis* adenylate cyclase as a 200-kilodalton protein. *Infect. Immun.* **1990**, *58*, 1195–1200. [CrossRef] [PubMed]

16. Chen, Q.; Stibitz, S. The BvgASR virulence regulon of *Bordetella pertussis*. *Curr. Opin. Microbiol.* **2019**, *47*, 74–81. [CrossRef]

17. Laoide, B.M.; Ullmann, A. Virulence dependent and independent regulation of the *Bordetella pertussis cya* operon. *EMBO J.* **1990**, *9*, 999–1005. [CrossRef]

18. Steffen, P.; Goyard, S.; Ullmann, A. Phosphorylated BvgA is sufficient for transcriptional activation of virulence-related genes in *Bordetella pertussis*. *EMBO J.* **1996**, *15*, 102–109. [CrossRef]

19. Ladant, C. Interaction of *Bordetella pertussis* adenylate cyclase with calmodulin. Identification of two separated calmodulin-binding domains. *J. Biol. Chem.* **1988**, *263*, 2612–2618. [CrossRef] [PubMed]

20. Ladant, D.; Michelson, S.; Sarfati, R.; Gilles, A.M.; Predeleanu, R.; Barzu, O. Characterization of the calmodulin-binding and of the catalytic domains of *Bordetella pertussis* adenylate cyclase. *J. Biol. Chem.* **1989**, *264*, 4015–4020. [CrossRef]

21. Sarfati, R.S.; Kansal, V.K.; Munier, H.; Glaser, P.; Gilles, A.M.; Labruyère, E.; Mock, M.; Danchin, A.; Barzu, O. Binding of 3′-anthraniloyl-2′-deoxy-ATP to calmodulin-activated adenylate cyclase from *Bordetella pertussis* and *Bacillus anthracis*. *J. Biol. Chem.* **1990**, *265*, 18902–18906. [CrossRef]

22. Bouhss, A.; Krin, E.; Munier, H.; Gilles, A.M.; Danchin, A.; Glaser, P.; Barzu, O. Cooperative phenomena in binding and activation of *Bordetella pertussis* adenylate cyclase by calmodulin. *J. Biol. Chem.* **1993**, *268*, 1690–1694. [CrossRef]

23. Ladant, D.; Glaser, P.; Ullmann, A. Insertional mutagenesis of *Bordetella pertussis* adenylate cyclase. *J. Biol. Chem.* **1992**, *267*, 2244–2250. [CrossRef]

24. Selwa, E.; Davi, M.; Chenal, A.; Sotomayor-Pérez, A.C.; Ladant, D.; Malliavin, T.E. Allosteric activation of *Bordetella pertussis* adenylate cyclase by calmodulin. Molecular dynamics and mutagenesis studies. *J. Biol. Chem.* **2014**, *289*, 21131–21141. [CrossRef]

25. O'Brian, D.P.; Durand, D.; Voegele, A.; Hourdel, V.; Davi, M.; Chamot-Rooke, J.; Vachette, P.; Brier, S.; Ladant, D.; Chenal, A. Calmodulin fishing with a structurally disordered bait triggers CyaA catalysis. *PLoS Biol.* **2017**, *15*, e2004486. [CrossRef]

26. Glaser, P.; Elmaoglou-Lazaridou, A.; Krin, E.; Ladant, D.; Barzu, O.; Danchin, A. Identification of residues essential for catalysis and binding of calmodulin in *Bordetella pertussis* adenylate cyclase by site-directed mutagenesis. *EMBO J.* **1989**, *8*, 967–972. [CrossRef] [PubMed]

27. Glaser, P.; Munier, H.; Gilles, A.M.; Krin, E.; Porumb, T.; Barzu, O.; Sarfati, R.; Pellecuer, C.; Danchin, A. Functional consequences of single amino acid substitutions in calmodulin-sensitive adenylate cyclase of *Bordetella pertussis*. *EMBO J.* **1991**, *10*, 1683–1688. [CrossRef]

28. Munier, H.; Bouhss, A.; Krin, E.; Danchin, A.; Gilles, A.M.; Glaser, P.; Barzu, O. The role of histidine 63 in the catalytic mechanism of *Bordetella pertussis* adenylate cyclase. *J. Biol. Chem.* **1992**, *267*, 9816–9820. [CrossRef] [PubMed]

29. Sakamoto, H.; Bellalou, J.; Sebo, P.; Ladant, D. *Bordetella pertussis* adenylate cyclase toxin. Structural and functional independence of the catalytic and hemolytic activities. *J. Biol. Chem.* **1992**, *267*, 13598–13602. [CrossRef] [PubMed]

30. Benz, R.; Maier, E.; Ladant, D.; Ullmann, A.; Sebo, P. Adenylate cyclase toxin (CyaA) of *Bordetella pertussis*. Evidence for the formation of small ion-permeable channels and comparison with HlyA of *Escherichia coli*. *J. Biol. Chem.* **1994**, *269*, 27231–27239. [CrossRef]

31. Barry, E.M.; Weiss, A.A.; Ehrmann, I.E.; Gray, M.C.; Hewlett, E.L.; Goodwin, M.S. *Bordetella pertussis* adenylate cyclase toxin and hemolytic activities require a second gene, *cyaC*, for activation. *J. Bacteriol.* **1991**, *173*, 170–736. [CrossRef]

32. Sebo, P.; Glaser, P.; Sakamoto, H.; Ullmann, A. High-level synthesis of active adenylate cyclase toxin of *Bordetella pertussis* in a reconstructed *Escherichia coli* system. *Gene* **1991**, *104*, 19–24. [CrossRef] [PubMed]

33. Heveker, N.; Bonnaffé, D.; Ullmann, A. Chemical fatty acylation confers hemolytic and toxic activities to adenylate cyclase protoxin of *Bordetella pertussis*. *J. Biol. Chem.* **1994**, *269*, 32844–32847. [CrossRef]

34. Hackett, M.; Guo, L.; Shabanowitz, J.; Hunt, D.F.; Hewlett, E.L. Internal lysine palmitoylation in adenylate cyclase toxin from *Bordetella pertussis*. *Science* **1994**, *266*, 433–435. [CrossRef] [PubMed]

35. Hackett, M.; Walker, C.B.; Guo, L.; Gray, M.C.; Van Cuyk, S.; Ullmann, A.; Shabanowitz, J.; Hunt, D.F.; Hewlett, E.L.; Sebo, P. Hemolytic, but not cell-invasive activity, of adenylate cyclase toxin is selectively affected by differential fatty-acylation in *Escherichia coli*. *J. Biol. Chem.* **1995**, *270*, 20250–20253. [CrossRef] [PubMed]

36. Karst, J.C.; Barker, R.; Devi, U.; Swann, M.J.; Davi, M.; Roser, S.J.; Ladant, D.; Chenal, A. Identification of a region that assists membrane insertion and translocation of the catalytic domain of *Bordetella pertussis* CyaA toxin. *J. Biol. Chem.* **2012**, *287*, 9200–9212. [CrossRef]

37. Subrini, O.; Sotomayor-Pérez, A.C.; Hessel, A.; Spiaczka-Karst, J.; Selwa, E.; Sapay, N.; Veneziano, R.; Pansieri, J.; Chopineau, J.; Ladant, A.; et al. Characterization of a membrane-active peptide from the *Bordetella pertussis* CyaA toxin. *J. Biol. Chem.* **2013**, *288*, 32585–32598. [CrossRef]

38. Voegele, A.; Sadi, M.; O'Brian, D.P.; Gehan, P.; Raoux-Barbot, D.; Davi, M.; Hoos, S.; Brûlé, S.; Raynal, B.; Weber, P.; et al. A high-affinity calmodulin-binding site in the CyaA toxin translocation domain is essential for invasion of eukaryotic cells. *Adv. Sci.* **2021**, *8*, 2003630. [CrossRef]

39. Bellalou, J.; Sakamoto, H.; Ladant, D.; Geoffroy, C.; Ullmann, A. Deletions affecting hemolytic and toxin activities of *Bordetella pertussis* adenylate cyclase. *Infect. Immun.* **1990**, *58*, 3242–3247. [CrossRef]

40. Rose, T.; Sebo, P.; Bellalou, J.; Ladant, D. Interaction of calcium with *Bordetella pertussis* adenylate cyclase toxin. Characterization of multiple calcium-binding sites and calcium-induced conformational changes. *J. Biol. Chem.* **1995**, *270*, 26370–26376. [CrossRef]

41. Bauche, C.; Chenal, A.; Knapp, O.; Bodenreider, C.; Benz, R.; Chaffotte, A.; Ladant, D. Structural and functional characterization of an essential RTX subdomain of *Bordetella pertussis* adenylate cyclase toxin. *J. Biol. Chem.* **2006**, *281*, 16914–16926. [CrossRef] [PubMed]

42. Chenal, A.; Guijarro, J.I.; Raynal, B.; Delepierre, M.; Ladant, D. RTX calcium binding motifs are intrinsically disordered in the absence of calcium. Implication for protein secretion. *J. Biol. Chem.* **2009**, *284*, 1781–1789. [CrossRef]

43. Sotomayor-Pérez, A.C.; Ladant, D.; Chenal, A. Calcium-induced folding of intrinsically disordered repeat-in-toxin (RTX) motifs via changes of protein charges and oligomerization states. *J. Biol. Chem.* **2011**, *286*, 16997–17004. [CrossRef]

44. Karst, J.C.; Ntsogo Enguéné, V.Y.; Cannella, S.E.; Subrini, O.; Hessel, A.; Debard, S.; Ladant, D.; Chenal, A. Calcium, acylation, and molecular confinement favor folding of *Bordetella pertussis* adenylate cyclase CyaA toxin into a monomeric and cytotoxic form. *J. Biol. Chem.* **2014**, *289*, 30702–30716. [CrossRef] [PubMed]

45. Guermonprez, P.; Khelef, N.; Blouin, E.; Rieu, P.; Ricciardi-Castagnoli, P.; Guiso, N.; Ladant, D.; Leclerc, C. The adenylate cycloase toxin of *Bordetella pertussis* binds to target cells via the $\alpha_M\beta_2$ integrin (CD11b/CD18). *J. Exp. Med.* **2001**, *193*, 1035–1044. [CrossRef]

46. Khelef, N.; Zychlinsky, A.; Guiso, N. *Bordetella pertussis* induces apoptosis in macrophages: Role of adenyate cyclase-hemolysin. *Infect. Immun.* **1993**, *61*, 4064–4071. [CrossRef] [PubMed]

47. Gueirard, P.; Druilhe, A.; Pretolani, M.; Guiso, N. Role of adenylate cyclase-hemolysin in alveolar macrophage apoptosis during *Bordetella pertussis* infection in vivo. *Infect. Immun.* **1998**, *66*, 1718–1725. [CrossRef]

48. Khelef, N.; Sakamoto, H.; Guiso, N. Both adenylate cyclase and hemolytic activities are required by *Bordetella pertussis* to initiate infection. *Micob. Pathog.* **1992**, *12*, 227–235. [CrossRef]

49. Khelef, N.; Bachelet, C.M.; Vargaftig, B.B.; Guiso, N. Characterization of murine lung inflammation after infection with parental *Bordetella pertussis* and mutants deficient in adhesins and toxins. *Infect. Immun.* **1994**, *62*, 2893–2900. [CrossRef] [PubMed]

50. Njamkepo, E.; Pinot, F.; François, D.; Guiso, N.; Polla, B.S.; Bachelet, M. Adaptive responses of human monocytes infected with *Bordetella pertussis*: The role of adenylate cyclase hemolysin. *J. Cell. Physiol.* **2000**, *183*, 91–99. [CrossRef]

51. Bassinet, L.; Gueirard, P.; Maitre, B.; Housset, B.; Gounon, P.; Guiso, N. Role of adhesins and toxins in invasion of human tracheal epithelial cells by *Bordetella pertussis*. *Infect. Immun.* **2000**, *68*, 1934–1941. [CrossRef]

52. Bassinet, L.; Fitting, C.; Housset, B.; Cavaillon, J.M.; Guiso, N. *Bordetella pertussis* adenylate cyclase-hemolysin induces interleukin-6 secretion by human tracheal epithelial cells. *Infect. Immun.* **2004**, *72*, 5530–5533. [CrossRef]

53. Guiso, N.; Rocancourt, M.; Szatanik, M.; Alonso, J.M. *Bordetella* adenylate cyclase is a virulence associated factor and an immunoprotective antigen. *Microb. Pathog.* **1989**, *7*, 373–380. [CrossRef] [PubMed]

54. Guiso, N.; Szatanik, M.; Rocancourt, M. Protective activity of *Bordetella* adenylate cyclase-hemolysin against bacterial colonization. *Microb. Pathog.* **1991**, *11*, 423–431. [CrossRef] [PubMed]

55. Betsou, F.; Seblo, P.; Guiso, N. The C-terminal domain is essential for protective activity of the *Bordetella pertussis* adenylate cyclase-hemolysin. *Infect. Immun.* **1995**, *63*, 3309–3315. [CrossRef]

56. Betsou, F.; Sebo, P.; Guiso, N. CyaC-mediated activation is important not only for toxic but also for protective activities of *Bordetella pertussis* adenylate cyclase-hemolysin. *Infect. Immun.* **1993**, *61*, 3583–3589. [CrossRef] [PubMed]

57. Sebo, P.; Fayolle, C.; d'Andria, O.; Ladant, D.; Leclerc, C.; Ullmann, A. Cell-invasive activity of epitope-tagged adenylate cyclase of *Bordetella pertussis* allows in vitro presentation of a foreign epitope to CD8$^+$ cytotoxic T cells. *Infect. Immun.* **1995**, *63*, 3851–3857. [CrossRef] [PubMed]

58. Fayolle, C.; Sebo, P.; Ladant, C.; Ullmann, A.; Leclerc, C. In vivo induction of CTL responses by recombinant adenylate cyclase of *Bordetella pertussis* carrying viral CD8$^+$ epitopes. *J. Immunol.* **1996**, *156*, 4697–4706. [CrossRef]

59. Saron, M.F.; Fayolle, C.; Sebo, P.; Ladant, D.; Ullmann, A.; Leclerc, C. Anti-viral protection conferred by recombinant adenylate cyclase toxins from *Bordetella pertussis* carrying a CD8$^+$ T cell epitope from lymphocytic choriomeningitis virus. *Proc. Natl. Acad. Sci. USA* **1997**, *94*, 3314–3319. [CrossRef] [PubMed]

60. Karimova, G.; Fayolle, C.; Gmira, S.; Ullmann, A.; Leclerc, C.; Ladant, D. Charge-dependent translocation of *Bordetella pertussis* adenylate cyclase toxin into eukaryotic cells: Implication for the in vivo delivery of CD8$^+$ T cell epitopes into antigen-presenting cells. *Proc. Natl. Acad. Sci. USA* **1998**, *95*, 12532–12537. [CrossRef]

61. Fayolle, C.; Ladant, D.; Karimova, G.; Ullmann, A.; Leclerc, C. Therapy of murine tumors with recombinant *Bordetella pertussis* adenylate cyclase toxins carrying a cytotoxic T cell epitope. *J. Immunol.* **1999**, *162*, 4157–4162. [CrossRef]

62. Dadaglio, G.; Moukrim, Z.; Lo-Man, R.; Sheshko, V.; Sebo, P.; Leclerc, C. Induction of a polarized Th1 response by insertion of multiple copies of a viral T-cell epitope into adenylate cyclase of *Bordetella pertussis*. *Infect. Immun.* **2000**, *68*, 3867–3872. [CrossRef] [PubMed]

63. Fayolle, C.; Osickova, A.; Osicka, R.; Henry, T.; Rojas, M.J.; Saron, M.F.; Sebo, P.; Leclerc, C. Delivery of multiple epitopes by recombinant detoxified adenylate cyclase of *Bordetella pertussis* induces protective antiviral immunity. *J. Virol.* **2001**, *75*, 7330–7338. [CrossRef]

64. Mascarell, L.; Fayolle, C.; Bauche, C.; Ladant, D.; Leclerc, C. Induction of neutralizing antibodies and Th1-polarized and CD4-independent CD8⁺ T-cell responses following delivery of human deficiency virus type 1 Tat protein by recombinant adenylate cyclase of *Bordetella pertussis. J. Virol.* **2005**, *79*, 9872–9884. [CrossRef]

65. Mascarell, L.; Bauche, C.; Fayolle, C.; Diop, O.M.; Dupuy, M.; Nougarede, N.; Perrault, R.; Ladant, D.; Leclerc, C. Delivery of the HIV-1 Tat protein to dendritic cells by the CyaA vector induces specific Th1 responses and high affinity neutralizing antibodies in non human primates. *Vaccine* **2006**, *24*, 3490–3499. [CrossRef] [PubMed]

66. Préville, X.; Ladant, D.; Timmerman, B.; Leclerc, C. Eradication of established tumors by vaccination with recombinant *Bordetella pertussis* adenylate cyclase carrying the human papillomavirus 16E7 oncoprotein. *Cancer Res.* **2005**, *65*, 641–649. [CrossRef]

67. Van Damme, P.; Bouillette-Marussig, M.; Hens, A.; De Coster, I.; Depuydt, C.; Goubier, A.; Van Tendeloo, V.; Cools, N.; Goossens, H.; Hercend, T.; et al. GTL001, a therapeutic vaccine for women infected with human papillomavirus 16 or 18 and normal cervical cytology: Results of a phase I clinical trial. *Clin. Cancer Res.* **2016**, *22*, 3238–3248. [CrossRef]

68. Karimova, G.; Pidoux, J.; Ullmann, A.; Ladant, D. A bacterial two-hybrid system based on a reconstituted signal transduction pathway. *Proc. Natl. Acad Sci. USA* **1998**, *95*, 5752–5756. [CrossRef]

69. Machitto, K.S.; Smith, S.G.; Locht, C.; Keith, J.M. Nucleotide sequence homology to pertussis toxin gene in *Bordetella bronchiseptica* and *Bordetella parapertussis. Infect. Immun.* **1987**, *55*, 497–501. [CrossRef]

70. Locht, C.; Coutte, L.; Mielcarek, N. The ins and outs of pertussis toxin. *FEBS J.* **2011**, *278*, 4668–4682. [CrossRef] [PubMed]

71. Locht, C.; Keith, J.M. Pertussis toxin gene: Nucleotide sequence and genetic organization. *Science* **1986**, *232*, 1258–1264. [CrossRef]

72. Lobet, Y.; Feron, C.; Dequesne, G.; Simoen, E.; Hauser, P.; Locht, C. Site-specific alterations in the B oligomer that affect receptor-binding activities and mitogenicity of pertussis toxin. *J. Exp. Med.* **1993**, *177*, 79–87. [CrossRef] [PubMed]

73. Raze, D.; Veithen, A.; Sato, H.; Antoine, R.; Menozzi, F.D.; Locht, C. Genetic exchange of the S2 and S3 subunits in pertussis toxin. *Mol. Microbiol.* **2006**, *60*, 1241–1250. [CrossRef]

74. Antoine, R.; Locht, C. Roles of the disulfide bond and the carboxy-terminal region of the S1 subunit in the assembly and biosynthesis of pertussis toxin. *Infect. Immun.* **1990**, *58*, 1518–1526. [CrossRef]

75. Locht, C.; Lobet, Y.; Feron, C.; Cieplak, W.; Keith, J.M. The role of cysteine 41 in the enzymatic activities of the pertussis toxin S1 subunit as investigated by site-directed mutagenesis. *J. Biol. Chem.* **1990**, *265*, 4552–4559. [CrossRef] [PubMed]

76. Locht, C.; Cieplak, W.; Marchitto, K.S.; Sato, H.; Keith, J.M. Activities of complete and truncated forms of pertussis toxin subunits S1 and S2 synthesized by *Escherichia coli. Infect. Immun.* **1987**, *55*, 2146–2553. [CrossRef]

77. Locht, C.; Capiau, C.; Feron, C. Identification of amino acid residues essential for the enzymatic activities of pertussis toxin. *Proc. Natl. Acad. Sci. USA* **1989**, *86*, 3075–3079. [CrossRef] [PubMed]

78. Cieplak, W.; Locht, C.; Mar, V.L.; Burnette, W.N.; Keith, J.M. Photolabelling of mutant forms of the S1 subunit of pertussis toxin with NAD⁺. *Biochem. J.* **1990**, *268*, 547–551. [CrossRef]

79. Antoine, R.; Tallett, A.; van Heyningen, S.; Locht, C. Evidence for a catalytic role of glutamine acid 129 in the NAD-glycohydrolase activity of the pertussis toxin S1 subunit. *J. Biol. Chem.* **1993**, *268*, 24149–24155. [CrossRef] [PubMed]

80. Antoine, R.; Locht, C. The NAD-glycohydrolase activity of the pertussis toxin S1 subunit: Involvement of the catalytic His-35 residue. *J. Biol. Chem.* **1994**, *269*, 6450–6457. [CrossRef]

81. Locht, C.; Antoine, R. A proposed mechanism of ADP-ribosylation catalyzed by the pertussis toxin S1 subunit. *Biochimie* **1995**, *77*, 333–340. [CrossRef] [PubMed]

82. Greco, D.; Salmaso, S.; Mastrantonio, P.; Giuliano, M.; Tozzi, A.E.; Anemona, A.; Ciofi Degli Atti, M.L.; Giammanco, A.; Panei, P.; Blackwelder, W.C.; et al. A controlled trial of two acellular vaccines and one whole-cell vaccine against pertussis. *N. Engl. J. Med.* **1996**, *334*, 341–348. [CrossRef] [PubMed]

83. Pitisuttithum, P.; Dhitavat, J.; Sirivichayakul, C.; Pitisuthitham, A.; Sabmee, Y.; Chinwangso, P.; Kerdsomboon, C.; Fortuna, L.; Spiegel, J.; Chauhan, M.; et al. Antibody persistence 2 and 3 years after booster vaccination of adolescents with recombinant acellular pertussis monovalent aP_gen or combined TdaP_gen vaccines. *EClin. Med.* **2021**, *37*, 100976. [CrossRef] [PubMed]

84. Mielcarek, N.; Debrie, A.S.; Raze, D.; Bertout, J.; Rouanet, C.; Younes, A.B.; Creusy, C.; Engle, J.; Goldman, W.E.; Locht, C. Live attenuated *B. pertussis* as a single-dose nasal vaccine against whooping cough. *PLoS Pathog.* **2006**, *2*, e65. [CrossRef]

85. Anderson, R.M.; May, R.M. Directly transmitted infections diseases: Control by vaccination. *Science* **1982**, *215*, 1053–1060. [CrossRef] [PubMed]

86. Solans, L.; Locht, C. The role of mucosal immunity in pertussis. *Front. Immunol.* **2019**, *9*, 3068. [CrossRef]

87. Warfel, J.M.; Zimmerman, L.I.; Merkel, T.J. Acellular pertussis vaccines protect against disease but fail to prevent infection and transmission in a nonhuman primate model. *Proc. Natl. Acad. Sci. USA* **2014**, *111*, 787–792. [CrossRef]

88. Solans, L.; Debrie, A.S.; Borkner, L.; Aguilo, N.; Thiriard, A.; Coutte, L.; Uranga, S.; Trottein, F.; Martin, C.; Mills, K.H.G.; et al. IL-17-dependent SIgA-mediated protection against nasal *Bordetella pertussis* infection by live attenuated BPZE1 vac-cine. *Mucosal. Immunol.* **2018**, *11*, 1753–1762. [CrossRef]

89. Feunou, P.F.; Kammoun, H.; Debrie, A.S.; Mielcarek, N.; Locht, C. Long-term immunity against pertussis induced by a single nasal administration of live attenuated *B. pertussis* BPZE1. *Vaccine* **2020**, *28*, 7047–7053. [CrossRef]

90. Dubois, V.; Chatagnon, J.; Thiriard, A.; Bauderlique-Le Roy, H.; Debrie, A.S.; Coutte, L.; Locht, C. Suppression of mucosal Th17 memory response by acellular pertussis vaccines enhances nasal *Bordetella pertussis* carriage. *Npj Vaccines* **2021**, *6*, 6. [CrossRef]

91. Skerry, C.M.; Cassidy, J.P.; English, K.; Feunou, P.F.; Locht, C.; Mahon, B.P. A live attenuated *Bordetella pertussis* candi-date vaccine does not cause disseminating infection in gamma interferon receptor knockout mice. *Clin. Vaccine Immunol.* **2009**, *16*, 1344–1351. [CrossRef] [PubMed]
92. Debrie, A.S.; Mielcarek, N.; Lecher, S.; Roux, X.; Sirard, J.C.; Locht, C. Early Protection against pertussis induced by live attenuated *Bordetella pertussis* BPZE1 depends on TLR4. *J. Immunol.* **2019**, *203*, 3293–3300. [CrossRef] [PubMed]
93. Locht, C.; Papin, J.F.; Lecher, S.; Debrie, A.S.; Thalen, M.; Solovay, K.; Rubin, K.; Mielcarek, N. Live attenuated pertussis vaccine BPZE1 protects baboons against *Bordetella pertussis* disease and infection. *J. Infect. Dis.* **2017**, *216*, 117–124. [CrossRef]
94. Thorstensson, R.; Trollfors, B.; Al-Tawil, N.; Jahnmatz, M.; Bergström, J.; Ljungman, M.; Törner, A.; Wehlin, L.; van Broekhoven, A.; Bosman, F.; et al. A phase I clinical study of a live attenuated *Bordetella pertussis* vaccine—BPZE1; A single centre, double-blind, placebo-controlled, dose-escalating study of BPZE1 given intranasally to healthy adult male volunteers. *PLoS ONE* **2014**, *9*, e83449. [CrossRef] [PubMed]
95. Jahnmatz, M.; Richert, L.; Al-Tawil, N.; Storsaeter, J.; Colin, C.; Bauduin, C.; Thalen, M.; Solovay, K.; Rubin, K.; Mielcarek, N.; et al. Safety and immunogenicity of the live attenuated intranasal pertussis vaccine BPZE1: A phase 1b, double-blind, randomised, placebo-controlled dose-escalation study. *Lancet Infect. Dis.* **2020**, *20*, 1290–1301. [CrossRef]
96. Lin, A.; Apostolovic, D.; Jahnmatz, M.; Liang, F.; Ols, S.; Tecleab, T.; Wu, C.; Van Hage, M.; Solovay, K.; Rubin, K.; et al. Live attenuated pertussis vaccine BPZE1 induces a broad antibody response in humans. *J. Clin. Investig.* **2020**, *130*, 2332–2346. [CrossRef]
97. Thalen, M.; Debrie, A.-S.; Coutte, L.; Raze, D.; Solovay, K.; Rubin, K.; Mielcarek, N.; Locht, C. Manufacture of a Stable Lyophilized Formulation of the Live Attenuated Pertussis Vaccine BPZE. *Vaccines* **2020**, *8*, 523. [CrossRef]
98. Creech, C.B.; Jimenez-Truque, N.; Kown, N.; Sokolow, K.; Brady, E.J.; Yoder, S.; Solovay, K.; Rubin, K.; Noviello, S.; Hensel, E.; et al. Safety and immunogenicity of live, attenuated intranasal *Bordetella pertussis* vaccine (BPZE1) in healthy adults. *Vaccine* **2022**, *40*, 6740–6746. [CrossRef]
99. Keech, C.; Miller, V.E.; Rizzardi, B.; Hoyle, C.; Pryor, M.J.; Ferrand, J.; Solovay, K.; Thalen, M.; Noviello, S.; Goldstein, P.; et al. Immunogenicity and safety of BPZE1, an intranasal live attenuated pertussis vaccine, versus tetanus-diphtheria-acellular-pertussis (Tdap) vaccine: A phase 2b randomized, double-blinded, placebo-controlled trial. *Lancet* **2023**; *in press*.

Review

Modulation of Airway Expression of the Host Bactericidal Enzyme, sPLA2-IIA, by Bacterial Toxins

Yongzheng Wu [1], Erwan Pernet [2],* and Lhousseine Touqui [3,4],*

[1] Unité de Biologie Cellulaire de l'Infection Microbionne, CNRS UMR3691, Institut Pasteur, Université de Paris Cité, 75015 Paris, France; yongzheng.wu@pasteur.fr

[2] Groupe de Recherche en Signalisation Cellulaire, Département de Biologie Médicale, Université du Québec à Trois-Rivières, Trois-Rivières, QC G8Z 4M3, Canada

[3] Sorbonne Université, Inserm U938, Centre de Recherche Saint-Antoine (CRSA), 75012 Paris, France

[4] Institut Pasteur, Université de Paris Cité, Mucoviscidose et Bronchopathies Chroniques, 75015 Paris, France

* Correspondence: erwan.pernet@uqtr.ca (E.P.); lhousseine.touqui@pasteur.fr (L.T.)

Abstract: Host molecules with antimicrobial properties belong to a large family of mediators including type-IIA secreted phospholipase A2 (sPLA2-IIA). The latter is a potent bactericidal agent with high selectivity against Gram-positive bacteria, but it may also play a role in modulating the host inflammatory response. However, several pathogen-associated molecular patterns (PAMPs) or toxins produced by pathogenic bacteria can modulate the levels of sPLA2-IIA by either inducing or inhibiting its expression in host cells. Thus, the final sPLA2-IIA concentration during the infection process is determined by the orchestration between the levels of toxins that stimulate and those that downregulate the expression of this enzyme. The stimulation of sPLA2-IIA expression is a process that participates in the clearance of invading bacteria, while inhibition of this expression highlights a mechanism by which certain bacteria can subvert the immune response and invade the host. Here, we will review the major functions of sPLA2-IIA in the airways and the role of bacterial toxins in modulating the expression of this enzyme. We will also summarize the major mechanisms involved in this modulation and the potential consequences for the pulmonary host response to bacterial infection.

Keywords: bacterial toxins; sPLA2; host immunity; inflammation

Key Contribution: This review provides a description of the bacterial toxins that modulate the expression of sPLA2-IIA, the main mechanisms involved in this modulation and their pathophysiological consequences in the airways during infection of the host by pathogenic bacteria.

Citation: Wu, Y.; Pernet, E.; Touqui, L. Modulation of Airway Expression of the Host Bactericidal Enzyme, sPLA2-IIA, by Bacterial Toxins. *Toxins* **2023**, *15*, 440. https://doi.org/10.3390/toxins15070440

Received: 13 April 2023
Revised: 23 June 2023
Accepted: 29 June 2023
Published: 3 July 2023

1. Role of sPLA2-IIA in Infectious and Inflammatory Diseases

1.1. General Biological Functions of sPLA2-IIA

Phospholipase A2 (PLA2) enzymes hydrolyze the sn-2 position of phospholipids, resulting in the production of free fatty acids and lyso-phospholipids [1,2]. These enzymes are classified into two major families: the low molecular weight-secreted PLA2 (sPLA2) and the high molecular weight intracellular PLA2, such as the cytosolic PLA2 (cPLA2) [1,2]. Based on the number and position of their disulfide bridges, sPLA2 can be classified into several different types, one of which is sPLA2-IIA. PLA2 have been shown to release free arachidonic acid (AA), the precursor of proinflammatory eicosanoids, and to bind to specific receptors present on host surface membranes [3,4]. Initially, sPLA2-IIA was suggested to play a role in the development of various inflammatory diseases [5,6]. For example, this enzyme can hydrolyze pulmonary surfactant phospholipids involved in acute respiratory distress syndrome (ARDS). In addition, sPLA2-IIA has been shown to induce neuronal apoptosis in ischemic stroke [7]. Other studies have also shown the involvement of sPLA2-IIA in atherosclerotic lesions [8–10], the hydrolysis of mitochondrial membranes released by platelets [11] and plasma lipoproteins [12]. sPLA2-IIA has also been shown to generate lipid mediators from membrane vesicles of platelets and erythrocytes [13].

Thus, it is clear that sPLA2-IIA can be involved in various pathophysiological processes, but its high bactericidal property (especially against Gram-positive bacteria) is now accepted as its most established biological role [5,6,14]. Therefore, it is important to explore the mechanisms by which bacterial toxins can modulate the expression of this enzyme and the pathophysiological consequences of this modulation.

1.2. PAMPs, Toxins and Innate Immune Response to Bacterial Infections

Pathogen-associated molecular patterns (PAMPs) are microbial motifs that are highly conserved across a wide range of pathogens. They are essential for the survival of these pathogens and their detection by host cells [15]. They include several virulence factors, such as lipopeptides, lipoteichoic acid (LTA) or peptidoglycans (PGN) of Gram-positive bacteria and lipopolysaccharides (LPS), pili or flagellin of Gram-negative bacteria as well as the double-stranded RNA (dsRNA) of certain viruses [16]. Recognition of PAMPs by specialized host cells is the first step in the host immune response, leading to an inflammatory response and elimination of the invading pathogens. This process involves the interactions of PAMPs with cellular receptors called 'pattern recognition receptors', or PRRs.

On the other hand, bacteria also produce a variety of toxins in response to various environmental challenges. These include exotoxins that are actively expressed and secreted into the extracellular media or injected into host cells during the infection process [17,18]. The interaction of pathogens with host cells initiates signaling processes that lead to the production of anti-microbial peptides (AMPs) by these cells. AMPs represent a large family of peptides ranging from 10 to 150 amino acids. In particular, 153 AMPs have been found in humans with net positive charges on the surface of the molecules [19], including defensins, cathelicidin, the type IIA secreted phospholipase A2 (sPLA2-IIA), etc. These antimicrobial molecules interact with bacteria to inhibit the synthesis of bacterial membrane phospholipids, cleave polysaccharides of the bacterial cell wall or increase the permeability of the bacterial membrane [19], which ultimately results in the eradication of the pathogens or the reduction of their proliferation. In particular, sPLA2-IIA has a high net positive charge of +17 [20]. Table 1 shows the reported effects of some PAMPs and toxins on sPLA2-IIA expression by host cells (Table 1).

Table 1. Effects of bacterial PAMPs and toxins on sPLA2-IIA expression by host cells.

PAMPs	Bacterium	Host PRRs	Effect on sPLA2-IIA Expression		References
			gpAMs	BECs	
LPS	G−	TLR4 [a]	Upregulation [b]	No effect [c]	[a] [21], [b] [22], [c] [23]
Peptidoglycan	G+; G−	NOD1 [a], NOD2 [a]	Upregulation [b]	/	[a] [24], [25], [b] [26]
Lipoteichoic acid	G+	TLR2 [a]	No effect [b]	/	[a] [27], [b] [28]
Flagellin	G−	TLR5 [a]	No effect [b]	No effect [c]	[a] [29], [b] our unpublished data, [c] [28]
Pili	G+; G−	CD46, CD48, CD55, etc [a]	Upregulation [b]	No effect [c]	[a] [30], [b] [31], [c] [23]
HSP60	G+; G−	TLR2, TLR4 [a]	/	/	[a] [32]
CpG DNA	G+, G−	TLR9 [a]	/	No effect [b]	[a] [33], [b] [23]
HSL	G−	/	/	No effect [a]	[a] [23]
ExoS	G−	/	/	Upregulation [a]	[a] [23]
Adenosine	G+	Adenosine receptor [a]	downregulation [b]	/	[a] [34], [b] [28]
AC-Hly	G−	CD11b/CD18 integrin [a]	downregulation [b]	/	[a] [35], [b] [28]
Edema toxin	G+	CMG2 [a], TEM8 [a]	downregulation [b]	/	[a] [36], [37], [b] [26]
Lethal toxin	G+	CMG2 [a], TEM8 [a]	downregulation [b]	/	[a] [36], [37], [b] [26]

/: not tested; gAMs: guinea pig alveolar macrophages; BECs: bronchial epithelial cells. a–c: indicate the corresponding references in the table.

1.3. The sPLA2-IIA, an Endogenous Antibiotic-Like Protein of the Host

The bactericidal activity of sPLA2-IIA is related to its ability to efficiently penetrate the cell wall of Gram-positive bacteria. This particular property is due to the high net positive charge of sPLA2-IIA (+17) [20], whereas Gram-positive bacteria have high net anionic charges due to the presence of the D-alanyl moiety covalently linked to lipoteichoic acid (LTA) [38], which is a major membrane component of Gram-positive bacteria. Thus, the highly efficient and rapid binding of sPLA2-IIA to LTA by electrostatic interaction promotes the penetration of sPLA2-IIA into the peptidoglycan layer, another major wall component of Gram-positive bacteria. This leads to efficient hydrolysis of bacterial membrane lipids and subsequent bacterial killing [14]. One study has reported a classification of mouse and human sPLA2 based on their ability to kill the Gram-positive bacterium *Staphylococcus aureus* [39] and showed that sPLA2-IIA is the most bactericidal sPLA2 type. Indeed, the concentration of sPLA2-IIA in human tears of healthy subjects exceeds 30 μg/mL, and only 15–80 ng/mL of this protein is sufficient to kill *S. aureus* [40]. Additionally, concentrations of sPLA2-IIA rapidly increase in host biological fluids as a result of inflammation or bacterial infection [4,41,42], as discussed in Part 4 of this section. These concentrations are virtually sufficient to kill all Gram-positive bacteria that may invade the host. Thus, sPLA2-IIA can be considered a major player in the host's innate immunity.

1.4. Role of the sPLA2-IIA in the Gut Microbiota–Lung Axis

Given its potent and selective antibacterial activity against Gram-positive bacteria, it is tempting to speculate that sPLA2-IIA might be involved in shaping the gut and pulmonary microbiota [43,44]. Indeed, two recent studies have investigated the influence of sPLA2-IIA on the gut microbiota [43,44]. Using gain- or loss-function assays, sPLA2-IIA was shown to play a central role in the composition of the gut microbiota by reducing the proportion of Gram-positive strains [43,44]. Most importantly, sPLA2-IIA-driven changes in the gut microbiota contributed to alterations in local and extra-intestinal immune responses, leading to increased susceptibility to cancer and arthritis [43,44]. This evidence suggests that intestinal sPLA2-IIA has profound effects on the host immune response through modulation of the gut microbiota and it may also have an impact on the pulmonary immune response by influencing the airway microbiota. Interestingly, there is a privileged relationship and communication between the lung and the gut (known as the gut–lung axis) that is mediated by the microbiota [45], and the gut microbiota has been associated with immunity to viral and bacterial infections [45–48]. Therefore, future studies are needed to better address the specific role of sPLA2-IIA in microbiota changes and the associated effects on the gut microbiota–lung axis.

1.5. sPLA2-IIA Levels in Biological Fluids of Infectious and Inflammatory Diseases

sPLA2-IIA was originally identified in synovial fluid from patients with rheumatoid arthritis [49], suggesting its involvement in excessive inflammatory conditions, such as autoimmunity. Subsequent studies have reported elevated levels of sPLA2-IIA in biological fluids from inflammatory diseases, including ARDS [50], pancreatitis, sepsis, cardiovascular disease [4,41,42] and in nasal fluids from patients with allergic rhinitis [51]. ARDS is defined as a life-threatening lung injury characterized by non-cardiogenic pulmonary edema and arterial hypoxemia [50]. The alteration of pulmonary surfactant is a hallmark of ARDS, accounting for increased surface tension at the air–liquid interface, resulting in impaired gas exchange and alveolar collapse. We have shown that sPLA2-IIA hydrolyzes surfactant phospholipids and its expression is inhibited by the surfactant protein A [4].

On the other hand, the pioneering studies of the Weiss group have reported the presence of sPLA2-IIA in rabbit peritoneal exudates [42] and baboon plasma [52] at levels sufficient to kill *S. aureus*. They showed that pre-treatments of these fluids with a specific sPLA2-IIA neutralizing antibody abolished the killing activity of these fluids [42,52]. We have also identified potent bactericidal activity against *B. anthracis* (the causative agent of anthrax) in human bronchoalveolar lavage fluids (BALFs) from patients with ARDS [53].

This bactericidal activity was positively correlated with the levels of sPLA2-IIA in the BALFs, and pretreatment of the BALFs with the sPLA2-IIA inhibitor LY311727 abolished the bactericidal activity [53]. We also showed that alveolar macrophages (AMs), the major source of sPLA2-IIA in a guinea pig model of ARDS [22], released sufficient amounts of sPLA2-IIA to kill *B. anthracis* [53]. This killing activity was inhibited by pretreatment of the AM medium with LY311727, suggesting that sPLA2-IIA is the major anthracidal factor released by AMs [53].

2. Bacterial Toxins That Upregulate sPLA2-IIA Expression

The toxins produced by invading bacteria can modulate sPLA2-IIA levels, either inducing or inhibiting its expression by host cells. Thus, the resulting sPLA2-IIA concentration during the infection process would depend on the balance between the levels of toxins that stimulate and those that downregulate the expression of this enzyme. The synthesis and/or secretion of bacterial toxins will be activated once the bacteria come into contact with host cells. The inhibition of sPLA2-IIA expression by bacterial toxins highlights a mechanism by which certain bacteria can subvert the host immune system. However, the same bacterium can either induce or inhibit sPLA2-IIA production depending on the expression kinetics of the bacterial toxins involved in sPLA2-IIA modulation. In the following sections, we will summarize the most important studies in the literature that have investigated the effects of bacterial toxins on sPLA2-IIA expression in the respiratory system and their pathophysiological consequences (Figure 1).

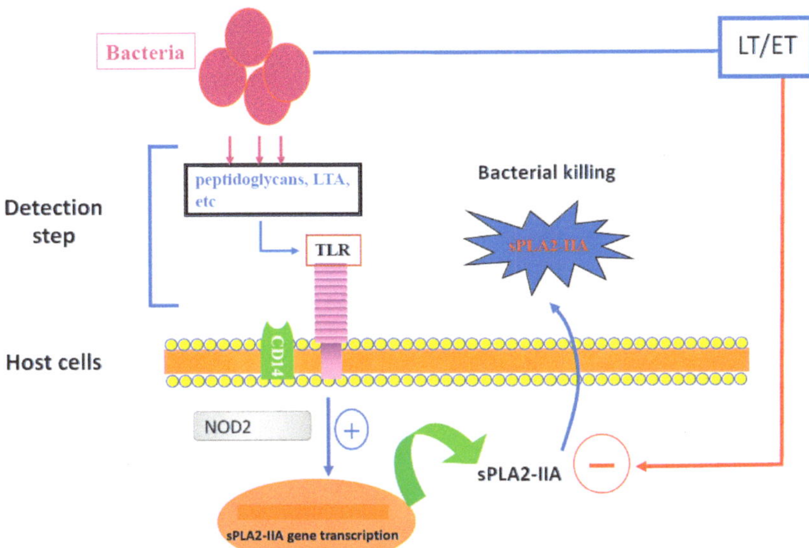

Figure 1. Positive vs. negative modulation of sPLA2-IIA expression in host cells by the *B. anthracis* toxins. At the early step of host cell infection by *B. anthracis*, this bacterium induces sPLA2-IIA expression and secretion by these cells via the action of various PAMPs, including peptidoglycan (PG) or LTA. This expression occurs via a NOD-dependent pathway. The enzyme sPLA2-IIA, once secreted in the medium, interacts with bacteria, leading to their killing. In parallel, *B. anthracis* produces the toxins ET and LT that downregulate sPLA2-IIA expression. The final concentration of this enzyme in the cell medium is the balance between the inducing (by PG) and the inhibiting effects of PLA2-IIA expression (by ET and LT).

2.1. LPS Is the Major Bacterial Toxin Inducing sPLA2-IIA Expression by Host Cells

It is widely accepted that LPS is the major initiator of the inflammatory response that is caused by Gram-negative bacteria [15]. This response involves a range of enzymes and mediators produced by host cells in response to stimulation by LPS. sPLA2-IIA is produced by host cells during the inflammatory response. The stimulation of sPLA2-IIA synthesis in the context of infectious diseases has mostly been attributed to bacterial toxins, such as LPS. LPS is a potent inducer of sPLA2-IIA expression in various rodent models of inflammatory diseases [54]. LPS from *E. Coli*, *P. aeruginosa* and *N. meningitidis* has been shown to induce sPLA2-IIA production from a variety of cells, including AMs and bronchial epithelial cells (BECs) [5,31,55]. However, LPS-induced sPLA2-IIA expression in human BECs is much less potent than in gpAMs [22,55]. Other studies reported that AMs isolated from ARDS patients did not respond to LPS to produce PLA2 [56]. On the contrary, the production of sPLA2-IIA was induced in control patients only after LPS treatment. Thus, PLA2 isoforms may serve as markers of immune status in ARDS [56].

LPS-induced sPLA2-IIA expression has been shown to be cell-specific and species-dependent. We have reported that LPS was able to induce sPLA2-IIA expression in AMs, the major source of this enzyme in a guinea pig model of ARDS [4,57]. In this model, the induction of LPS-induced sPLA2-IIA expression occurs via an autocrine/paracrine process dependent on TNFα. In baboons, TNFα also plays a role as an intermediate in LPS-induced sPLA2-IIA production [52]. These findings contrast with those in rat astrocytes and human hepatoma cells, where LPS-induced sPLA2-IIA expression occurred via a TNFα-independent process [58,59].

sPLA2-IIA has clearly been shown to play a role in the hydrolysis of dipalmitoyl-phosphatidylcholine (DPPC) in an animal model of LPS-induced ARDS [4,57,60]. DPPC is one of the major phospholipids of pulmonary surfactants. Its hydrolysis is associated with the loss of surface tension of the surfactant complex, leading to subsequent alveolar collapse, a key clinical feature of ARDS [4]. Another study reported that pretreatment with the sPLA2-IIA inhibitor S-5920/LY315920Na attenuated lung injury induced by oleic acid [60]. This effect was accompanied by protection against lung surfactant degradation and the associated production of the inflammatory lipid mediators thromboxane A2 and leukotriene B4 [60].

Subsequent studies showed that Azithromycin exerted anti-inflammatory properties on lung epithelial cells by inhibiting LPS-induced sPLA2-IIA expression, but had no effect on AMs [61]. On the other hand, treatment of brain microvascular endothelial cells (BMVECs) with LPS resulted in increased release of sPLA2-IIA and nitrite into the culture medium [33]. This release was decreased by pretreatment of the cells with an NO donor, sodium nitroprusside, suggesting that sPLA2-IIA expression is under the control of the NO–JAK3–STAT1 pathway [62]. Furthermore, sPLA2-IIA stimulates neuronal cell death via apoptosis [7], which appears to be associated with the production of arachidonic acid (AA) metabolites, particularly prostaglandin D2 (PGD2). In addition, sPLA2-IIA is involved in neurodegeneration in the ischemic brain, highlighting the potential therapeutic use of sPLA2-IIA inhibitors in stroke therapy [7].

2.2. Role of Pili in the Induction of sPLA2-IIA Expression in the Airways

In our previous studies, we investigated whether *N. meningitidis* toxins or PAMPs, other than LPS, were able to stimulate sPLA2-IIA expression using gpAMs [31]. This Gram-negative bacterium is the causative agent of meningococcal disease and is exclusively adapted to humans, with the nasopharynx being its natural habitat [63]. The results showed that *N. meningitidis* stimulates sPLA2-IIA synthesis through gpAMs and that pili mediate this stimulation [31]. Indeed, an LPS-deficient mutant of this bacterium was still able to induce sPLA2-IIA synthesis and a pili-deficient mutant showed a significantly lower ability to stimulate sPLA2-IIA expression than the wild-type strain. Moreover, a pili preparation isolated from an LPS-deficient *N. meningitidis* strain stimulated sPLA2-IIA production via a process involving NF-κB activation [31]. These studies demonstrated that pili, in addition to LPS, can induce sPLA2-IIA expression by *N. meningitidis*. However, the receptor(s) and the signaling pathways involved in pili-induced sPLA2-IIA expression were not investigated in this study. However, it should be emphasized that *N. meningitidis* has been shown to exert toxic effects on human epithelial and endothelial cells due to a synergistic effect of LPS and pili [64].

2.3. The Type 3 Secretion System (T3SS) Toxin, Exotoxin S (ExoS), Plays a Key Role in sPLA2-IIA Expression in the Airways

We have also examined the expression and role of sPLA2-IIA in the airways of patients with cystic fibrosis (CF) and showed that the *P. aeruginosa* PAK strain, but not PAMPs, isolated from this bacterium, including LPS, HSL, CpG, flagellin and pili, stimulated sPLA2-IIA synthesis by BECs from these patients [23]. This observation suggests that LPS-induced sPLA2-IIA synthesis is cell type dependent, and also prompted us to identify the *P. aeruginosa* virulence factors involved in sPLA2-IIA production by BECs in the context of CF. Compared to its parental strain, the PAK mutant lacking T3SS induced sPLA2-IIA expression at much lower levels, whereas the T2SS-deficient strain induced it at similar levels [23], suggesting a specific ability of T3SS to induce sPLA2-IIA expression. Moreover, using a pharmacological approach, we showed that the signaling pathways (NF-κB, AP-1 and MAPK), known to induce sPLA2-IIA expression in various cell systems (see ref [5]), did not play a role in ExoS-induced sPLA2-IIA expression in BECs. We established that this expression was under the control of Krüppel-Like Factor 2 (KLF2), a particular transcription factor induced by bacterial toxins [65]. KLF2 belongs to the SP-1 zinc-finger transcription factor [66]. Although the signaling pathways by which ExoS induces KLF2 expression are still unknown, our studies showed that KLF2 induction by ExoS is independent of RhoA, one of the potential targets suggested by others [67]. ExoS is a bifunctional toxin with two enzymatic domains, GAP and ADPRT [17], but only the GAP domain is responsible for RhoA inactivation [68]. We also established that ADPRT, but not the GAP domain, plays a role in ExoS-induced sPLA2-IIA synthesis by BECs from CF patients [23].

These findings led us to investigate whether ExoS was able to induce sPLA2-IIA expression in cells other than epithelial cells. Using the T3SS-deficient PAK strains (ΔpscF mutant), we investigated the role of this toxin in gpAMs. Our results showed that this mutant induced sPLA2-IIA expression at similar levels compared to the parental stain. This suggests that, unlike human BECs, ExoS does not affect sPLA2-IIA expression in gpAMs. We also showed that the mutant lacking T2SS induced sPLA2-IIA expression in gpAMs at similar levels compared to the parental stain. These results clearly indicate that neither T2SS nor T3SS is involved in *P. aeruginosa*-induced sPLA2-IIA expression in gpAMs (Wu et al., manuscript in preparation).

However, the stimulatory effects of ExoS on sPLA2-IIA contrasted with its inhibitory effects on other mediators of the host immune response, such as cytokines and reactive oxygen species (ROS). Indeed, studies have reported that deletion of the exoS gene resulted in a significant increase in interleukin-8 (IL-8) production by Caco-2 cells. This finding suggested that *P. aeruginosa* produces a serine protease capable of degrading IL-8 in the culture medium of infected Caco-2 cells and that the expression of this protease is inhibited in cells infected with the ΔexoS mutant [69]. More recent studies have shown that *P. aeruginosa* can inhibit the ROS burst in neutrophils and that this inhibition occurs, at least in part, through an ExoS-mediated ADP-ribosylation of Ras in neutrophils [70]. However, the pathophysiological consequences of the opposing effects of ExoS on sPLA2-IIA and other mediators of the innate immune response remained to be elucidated.

2.4. Induction of sPLA2-IIA by Porphyromonas Gingivalis and Relevance to Periodontal Disease

P. gingivalis is a Gram-negative bacterium known to play a key role in the development of periodontal disease [71]. This bacterium has been shown to induce dramatic expression of sPLA2-IIA in human oral epithelial cells, suggesting that sPLA2-IIA might play a role in periodontal disease [72]. The induction of sPLA2-IIA expression by *P. gingivalis* occurs via a mechanism dependent on TLR activation or T3SS and was initiated by the production of arginine-gingipains (rgpA and rgpB). The latter belongs to the cysteine proteases group unique to this bacterium [72,73] that target a specific sequence of the Notch-1 extracellular domain. This process stimulates the translocation of the Notch-1 intracellular domain into the nucleus, leading to the transcriptional activation of the sPLA2-IIA gene [72].

2.5. PGN Is the Main Inducer of sPLA2-IIA Expression by Gram-Positive Bacteria

Our studies using gpAMs as a cell model showed that the Gram-positive bacterium *B. anthracis* can stimulate sPLA2-IIA expression. These studies demonstrated that the cell wall PGN purified from *B. anthracis* induced sPLA2-IIA expression through a process depending on NF-κB activation [26]. However, it is still unclear whether PGN stimulated sPLA2-IIA expression via two PGN recognition proteins, TLR2 or NOD-2 [74]. Recent reports showed that NOD may be involved in cell activation by spores of this bacterium [75]. However, we cannot rule out the fact that other bacterial components of the cell wall or secreted by *B. anthracis* are also involved in sPLA2-IIA expression in gpAMs.

In a subsequent study [28], we attempted to identify the *S. aureus* PAMP(s) and associated receptor(s) that induce sPLA2-IIA expression in gpAMs using PGN purified from *S. aureus* (ligand of TLR2 and NOD2) and LTA (ligand of TLR2). Our results revealed that PGN stimulated sPLA2-IIA expression in gpAMs, whereas LTA had only a limited effect on this expression. This suggested the possible involvement of the NOD2 receptor in PGN-induced sPLA2-IIA expression in gpAMPs [28]. To further verify the role of NOD2 in this process, gpAMs were incubated with the minimal active portions of the PGN, motif MDP (ligand of NOD2), MDPc (inactivated MDP), MALP-2 (ligand of TLR2/6), Pam$_3$CSK$_4$ (ligand of TLR2/1) or the cytosine guanosine dinucleotide (CpG, agonist of TLR9). Remarkably, only MDP was shown to induce sPLA2-IIA expression by gpAMs. These results indicate that NOD2 is probably the only receptor involved in PGN-induced sPLA2-IIA expression in gpAMs [28].

3. Bacterial Toxins That Downregulate sPLA2-IIA Expression

3.1. Inhibition of sPLA2-IIA Expression by Bacillus Anthracis Toxins

The exact mechanisms by which *B. anthracis* initiates the anthrax disease are not fully established. However, it is clear that this bacterium multiplicates rapidly in the blood stream, ultimately leading to the death of the host [76]. *B. anthracis* can subvert the host immune response [77,78] via a process involving the action of the *B. anthracis* edema toxin (ET) and lethal toxin (LT). This bacterium releases a binary A-B toxin composed of a single B transporter called a protective antigen (PA) and two alternative A components, lethal factor (LF) or edema factor (EF) [76]. LF and EF act in pairs, with PA leading to LT (=PA + LF) and ET (=PA + EF), respectively. Thus, PA serves as a transporter that delivers LF and EF into the host cell cytosol where they target specific molecular components [76].

3.1.1. Lethal Toxin Downregulates sPLA2-IIA Expression by AMs

Our previous studies showed that the expression of sPLA2-IIA was inhibited by LT during infection with gpAMPs by *B. anthracis* [53]. Next, we investigated the mechanisms by which LT alters sPLA2-IIA synthesis in this cell model. LT exhibits metallo-proteolytic activity toward the N-terminus of the MAPK-kinases. We have shown that this toxin inhibits the phosphorylation of MAPK p38 in gpAMs as well as sPLA2-IIA promoter activity in CHO cells [79]. The inhibition of sPLA2-IIA promoter activity was mimicked by co-transfection with the dominant negative construct of p38 (DN-p38) and reversed by the active form of p38-MAPK (AC-p38) [79]. Both LT and DN-p38 decreased NF-κB luciferase activity. However, neither LT nor a specific p-38 inhibitor interfered with LPS-induced IκBα degradation or NF-κB nuclear translocation in AMs [79]. Therefore, we concluded that sPLA2-IIA expression is induced via sequential MAPK-NF-κB activation and that LT alters this expression by interfering with NF-κB transactivation in the nucleus. This hypothesis is consistent with previous studies suggesting that the regulation of sPLA$_2$-IIA in Jurkat cells involves epigenetic silencing by DNA hypermethylation [80].

3.1.2. Edema Toxin Impairs sPLA2-IIA Expression by AMs

In a subsequent study, we showed that ET inhibited sPLA2-IIA expression in gpAMs at the transcriptional level through a cAMP/protein kinase A-dependent process [26]. Moreover, the ET-deficient strains induced sPLA2-IIA expression, whereas the live *B. anthracis* strains expressing the functional ET inhibited this expression [26]. This suggests that *B. anthracis* can induce sPLA2-IIA expression probably via PGN and that ET injected into gpAMPs reduces this expression. This inhibition seems to occur via a cAMP/protein kinase A-dependent mechanism [26]. Our reports revealed that the cAMP increase was transient, reaching basal values within 24 h, in contrast to previous studies showing that cAMP accumulation increased within 48 h or more following ET addition to NIH/3T3 fibroblasts and RAW 267 macrophages [81].

Regardless of the kinetics of the cAMP increase during infection of gpAMPs by *B. anthracis*, the final concentrations of sPLA2-IIA should be considered as a balance between the inducing effect of PGN and the inhibitory effect of ET. However, it is worth highlighting that the effects of cAMP on sPLA2-IIA expression depend on the cell type considered. For example, in contrast to our results with gpAMs, other studies have shown that cAMP stimulates the transcriptional activity of the sPLA2-IIA gene in rat vascular smooth muscle cells. This stimulation depends on the interplay of the CCAAT/enhancer binding protein (C/EBP), NF-κB and Ets transcription factors [82].

Other studies have examined the effects of *B. anthracis* toxins on human dendritic cells and showed that both ET and LT inhibited the production of proinflammatory chemokines by these cells, with LT exhibiting the higher inhibitory effect. However, only LT impaired neutrophil recruitment in this model [83]. The authors suggested that ET and LT act in concert to suppress chemokine expression by human dendritic cells and that this action leads to an alteration in immune cell recruitment.

Taken together, these reports suggest that ET and LT may interfere with the immune system by altering sPLA2-IIA expression or by impairing chemokine expression and neutrophil recruitment, which may play a role in the innate host response to *B. anthracis* infection.

3.2. Inhibition of sPLA2-IIA Expression by a Bordetella Pertussis AC-Hly Toxin

The ability to inhibit sPLA2-IIA expression in gpAMs seems to be a general mechanism shared by several bacteria. Indeed, we have shown that a toxin secreted by *Bordetella pertussis* named adenylate cyclasehemolysin (AC-Hly) was also able to impair LPS-induced sPLA2-IIA expression by gpAMs. This inhibition is likely due to an increase in cAMP levels by AC-Hly, which is known to display calmodulin-dependent adenylate cyclase activity [84]. It should be noted, however, that the cyclic AMP-elevating capacity of AC-Hly is sufficient to ensure lung infection but not to ensure full virulence of *B. pertussis* [85]. Additionally, AC-Hly did not affect IL-8 production by gpAMs, suggesting that sPLA2-IIA suppression results from a selective effect of this toxin on sPLA2-IIA transcriptional activity. This finding is consistent with our previous studies showing that *B. anthracis* ET can downregulate sPLA2-IIA expression without affecting IL-8 secretion [84].

3.3. S. aureus Adenosine Inhibits sPLA2-IIA Expression and Associated Airway Killing

S. aureus produces several molecules, including adenosine, that dampen the host immunity [86]. Adenosine is produced by the highly conserved cell wall-anchored adenosine synthase A (AdsA) via the degradation of ATP, ADP and AMP [87]. We showed that an adenosine-deficient *S. aureus* mutant (ΔadsA strain) enhanced the pulmonary expression of sPLA2-IIA in a guinea pig model of lung infection and was cleared more efficiently in the airways compared to the wild-type strain [28]. In addition, the ΔadsA strain induced sPLA2-IIA expression by gpAMs after the phagocytosis of *S. aureus* via a NOD2-NF-κB-dependent mechanism. The addition of exogenous adenosine to cultured gpAMs reduced

S. aureus phagocytosis by these cells and impaired sPLA2-IIA synthesis. This occurred through a downregulation of p38 phosphorylation via adenosine receptors A2a-, A2b- and associated protein kinase A activation. In addition to its effects on phagocytosis, adenosine also acts as an anti-inflammatory mediator in macrophages via the cAMP/PKA axis [88]. This effect is mediated by altering NF-κB activity through PKA [89].

Taken together, these studies indicate that in the airway, *S. aureus* can escape sPLA2-IIA-mediated killing via adenosine-mediated alteration of phagocytosis and sPLA2-IIA synthesis. These processes also highlight the contribution of NOD2 in modulating sPLA2-IIA expression and suggest a mechanism by which *S. aureus* subverts host immunity based on the inhibition of the production of this antimicrobial enzyme.

4. Conclusions

As a mediator of the inflammatory response, sPLA2-IIA appears to act as a double-edged sword for the host (Figure 2). Indeed, in addition to the bactericidal functions and their regulation summarized in this review, sPLA2-IIA has been shown to play a role as a proinflammatory mediator involved in pathogenic processes. This includes alteration of pulmonary surfactant, hydrolysis of mitochondrial membrane released by platelets and interaction with plasma lipoproteins. sPLA2-IIA has also been shown to generate lipid mediators from membrane vesicles of platelets and erythrocytes and to play a role in neurotoxicity and cell apoptosis. The bactericidal activity of sPLA2-IIA has emerged as a central function of this enzyme, leading to the elimination of invading pathogenic bacteria and the modulation of microbiota composition. However, further work is required to identify the precise mechanisms involved in the crosstalk between sPLA2-IIA-mediated effects on gut and airway microbiota as well as on lung immune responses.

In general, Gram-negative bacteria, such as *P. aeruginosa*, are insensitive to the bactericidal effects of sPLA2-IIA but can induce its expression by host cells. This results in the selective elimination of Gram-positive bacteria, such as *S. aureus*, which are highly sensitive to sPLA2-IIA killing, a process that is advantageous for Gram-negative bacteria. However, the Gram-positive bacteria can also inhibit sPLA2-IIA expression, highlighting an evolutionary adaptive mechanism by which these bacteria evade the host's innate immunity.

As shown in this review, several bacterial toxins can regulate sPLA2-IIA levels by either inducing or inhibiting their expression in host cells. These effects depend on the toxins and the host cell type considered. Thus, the final sPLA2-IIA concentration in host biological fluids can be considered as a balance between the stimulatory and inhibitory effects of these toxins. Our review was focused on bacterial toxins, but it is necessary to investigate whether other respiratory pathogens (viruses and fungi) can also modulate sPLA2-IIA expression in the airways and its potential role in the outcome of the infection. Nevertheless, the beneficial bactericidal functions of sPLA2-IIA highlight a potential therapeutic value of this enzyme in respiratory infectious diseases, especially in the context of the antibiotic resistance crisis.

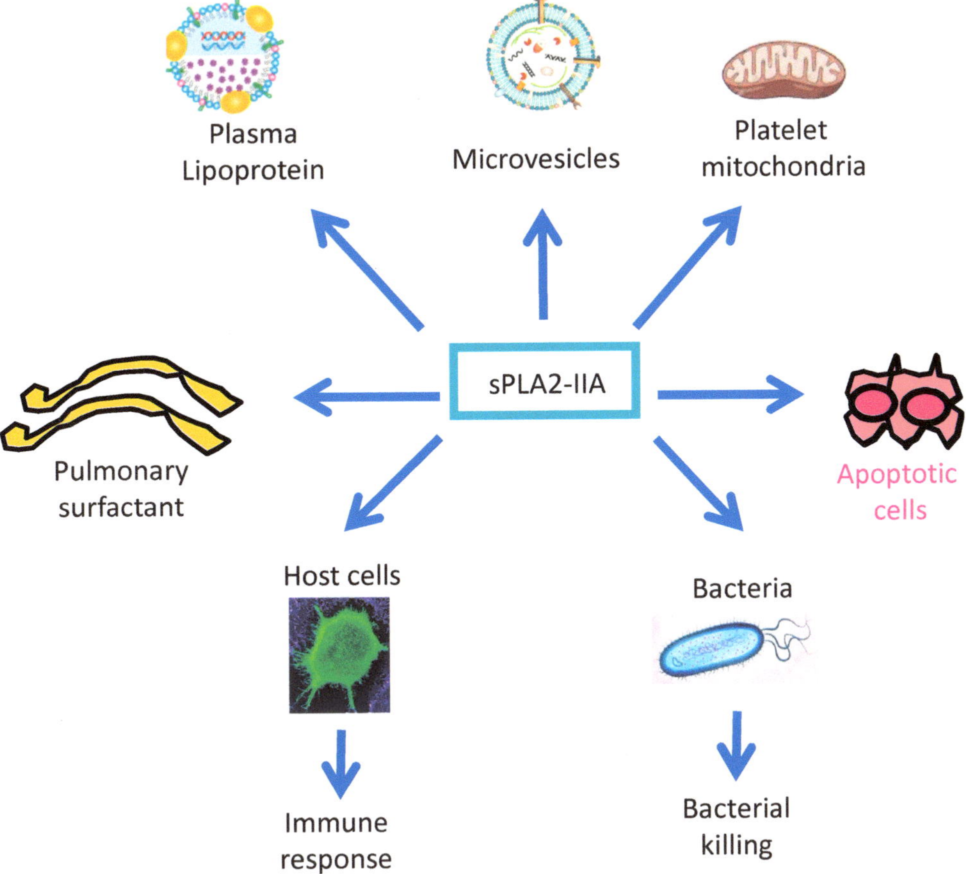

Figure 2. Examples of established biological functions of sPLA2-IIA. sPLA2-IIA targets phospholipids of plasma lipoproteins, microvesicles or mitochondria released by platelets or erythrocytes, thus contributing to the inflammatory processes associated with diseases such as atherosclerosis and sepsis. This enzyme can also hydrolyze pulmonary surfactant phospholipids, an important feature of ARDS pathogenesis. sPLA2-IIA has been also shown to promote neurodegeneration through activation of neuronal cell apoptosis. But, the most studied functions of sPLA2-IIA are related to its bactericidal action and ability to release lipid mediators through the hydrolysis of host cell phospholipids.

Author Contributions: This manuscript was written by Y.W., E.P. and L.T. All authors have read and agreed to the published version of the manuscript.

Funding: This research received no external funding.

Institutional Review Board Statement: Not applicable.

Informed Consent Statement: Not applicable.

Data Availability Statement: Not applicable.

Acknowledgments: The authors warmly thank Nora Touqui for her precious and meticulous proof-reading and English improvement of the present article.

Conflicts of Interest: The authors declare no conflict of interest.

Abbreviations

PAMP	pathogen-associated molecular pattern
PRRs	pattern recognition receptors
PLA2	phospholipase A2
sPLA2-IIA	type-IIA secreted phospholipase A2
cPLA2	cytosolic PLA2
G+ bacterium	Gram-positive bacterium
G- bacterium	Gram-negative bacterium
LTA	lipoteichoic acid
PGN	peptidoglycan
LPS	lipopolysaccharides
dsDNA	double-stranded DNA
ARDS	acute respiratory distress syndrome
BALF	bronchoalveolar lavage fluid
AMs	alveolar macrophages
gpAMs	guinea pig alveolar macrophages
BECs	bronchial epithelial cells
DPPC	dipalmitoyl-phosphatidylcholine
BMVECs	brain microvascular endothelial cells
AA	arachidonic acid
PGD2	prostaglandin D2
LOS	lipooligosaccharides
T3SS	type 3 secretion system
ExoS	exotoxin S
CF	cystic fibrosis
KLF2	Krüppel-Like Factor 2
GAP	GTPase activating protein
ADPRT	ADP ribosyltransferase
MDP	muramyl dipeptide
MALP-2	macrophage-activating lipopeptide 2 kDa
CpG	cytosine guanosine dinucleotide
ET	edema toxin
LT	lethal toxin
EF	edema factor
PA	protective antigen
DN-p38	dominant negative construct of p38
AC-Hly	adenylate cyclasehemolysin
C/EBP	CCAAT/enhancer binding protein
cAMP	cyclic adenosine monophosphate
AdsA	adenosine synthase A

References

1. Dennis, E.A. The growing phospholipase A2 superfamily of signal transduction enzymes. *Trends Biochem. Sci.* **1997**, *22*, 1–2. [CrossRef] [PubMed]
2. Ghosh, M.; Tucker, D.E.; Burchett, S.A.; Leslie, C.C. Properties of the Group IV phospholipase A2 family. *Prog. Lipid Res.* **2006**, *45*, 487–510. [CrossRef] [PubMed]
3. Murakami, M.; Lambeau, G. Emerging roles of secreted phospholipase A(2) enzymes: An update. *Biochimie* **2013**, *95*, 43–50. [CrossRef] [PubMed]
4. Touqui, L.; Arbibe, L. A role for phospholipase A2 in ARDS pathogenesis. *Mol. Med. Today* **1999**, *5*, 244–249. [CrossRef]
5. Touqui, L.; Alaoui-El-Azher, M. Mammalian secreted phospholipases A2 and their pathophysiological significance in inflammatory diseases. *Curr. Mol. Med.* **2001**, *1*, 739–754. [CrossRef] [PubMed]
6. Nevalainen, T.J.; Graham, G.G.; Scott, K.F. Antibacterial actions of secreted phospholipases A2. Review. *Biochim. Biophys. Acta* **2008**, *1781*, 1–9. [CrossRef]
7. Yagami, T.; Ueda, K.; Asakura, K.; Hata, S.; Kuroda, T.; Sakaeda, T.; Takasu, N.; Tanaka, K.; Gemba, T.; Hori, Y. Human group IIA secretory phospholipase A2 induces neuronal cell death via apoptosis. *Mol. Pharm.* **2002**, *61*, 114–126. [CrossRef]

8. Kugiyama, K.; Ota, Y.; Takazoe, K.; Moriyama, Y.; Kawano, H.; Miyao, Y.; Sakamoto, T.; Soejima, H.; Ogawa, H.; Doi, H.; et al. Circulating levels of secretory type II phospholipase A(2) predict coronary events in patients with coronary artery disease. *Circulation* **1999**, *100*, 1280–1284. [CrossRef]

9. Mattsson, N.; Magnussen, C.G.; Ronnemaa, T.; Mallat, Z.; Benessiano, J.; Jula, A.; Taittonen, L.; Kahonen, M.; Juonala, M.; Viikari, J.S.; et al. Metabolic syndrome and carotid intima-media thickness in young adults: Roles of apolipoprotein B, apolipoprotein A-I, C-reactive protein, and secretory phospholipase A2: The cardiovascular risk in young Finns study. *Arter. Thromb. Vasc. Biol.* **2010**, *30*, 1861–1866. [CrossRef]

10. Rosenson, R.S.; Gelb, M.H. Secretory phospholipase A2: A multifaceted family of proatherogenic enzymes. *Curr. Cardiol. Rep.* **2009**, *11*, 445–451. [CrossRef]

11. Boudreau, L.H.; Duchez, A.C.; Cloutier, N.; Soulet, D.; Martin, N.; Bollinger, J.; Pare, A.; Rousseau, M.; Naika, G.S.; Levesque, T.; et al. Platelets release mitochondria serving as substrate for bactericidal group IIA-secreted phospholipase A2 to promote inflammation. *Blood* **2014**, *124*, 2173–2183. [CrossRef]

12. Pruzanski, W.; Lambeau, G.; Lazdunski, M.; Cho, W.; Kopilov, J.; Kuksis, A. Hydrolysis of minor glycerophospholipids of plasma lipoproteins by human group IIA, V and X secretory phospholipases A2. *Biochim. Biophys. Acta* **2007**, *1771*, 5–19. [CrossRef]

13. Fourcade, O.; Simon, M.F.; Viode, C.; Rugani, N.; Leballe, F.; Ragab, A.; Fournie, B.; Sarda, L.; Chap, H. Secretory phospholipase A2 generates the novel lipid mediator lysophosphatidic acid in membrane microvesicles shed from activated cells. *Cell* **1995**, *80*, 919–927. [CrossRef]

14. van Hensbergen, V.P.; Wu, Y.; van Sorge, N.M.; Touqui, L. Type IIA Secreted Phospholipase A2 in Host Defense against Bacterial Infections. *Trends Immunol.* **2020**, *41*, 313–326. [CrossRef]

15. Janeway, C.A., Jr. Approaching the asymptote? Evolution and revolution in immunology. *Cold. Spring Harb. Symp. Quant. Biol.* **1989**, *54 Pt 1*, 1–13. [CrossRef]

16. Kumar, V. Toll-like receptors in sepsis-associated cytokine storm and their endogenous negative regulators as future immunomodulatory targets. *Int. Immunopharmacol.* **2020**, *89*, 107087. [CrossRef] [PubMed]

17. Hauser, A.R. The type III secretion system of *Pseudomonas aeruginosa*: Infection by injection. *Nat. Rev. Microbiol.* **2009**, *7*, 654–665. [CrossRef] [PubMed]

18. Shenoy, A.R.; Furniss, R.C.D.; Goddard, P.J.; Clements, A. Modulation of Host Cell Processes by T3SS Effectors. *Curr. Top Microbiol. Immunol.* **2018**, *416*, 73–115. [CrossRef] [PubMed]

19. Wang, G. Human antimicrobial peptides and proteins. *Pharmaceuticals* **2014**, *7*, 545–594. [CrossRef]

20. Baker, S.F.; Othman, R.; Wilton, D.C. Tryptophan-containing mutant of human (group IIa) secreted phospholipase A2 has a dramatically increased ability to hydrolyze phosphatidylcholine vesicles and cell membranes. *Biochemistry* **1998**, *37*, 13203–13211. [CrossRef]

21. Poltorak, A.; He, X.; Smirnova, I.; Liu, M.Y.; Van Huffel, C.; Du, X.; Birdwell, D.; Alejos, E.; Silva, M.; Galanos, C.; et al. Defective LPS signaling in C3H/HeJ and C57BL/10ScCr mice: Mutations in Tlr4 gene. *Science* **1998**, *282*, 2085–2088. [CrossRef]

22. Arbibe, L.; Vial, D.; Rosinski-Chupin, I.; Havet, N.; Huerre, M.; Vargaftig, B.B.; Touqui, L. Endotoxin induces expression of type II phospholipase A2 in macrophages during acute lung injury in guinea pigs: Involvement of TNF-alpha in lipopolysaccharide-induced type II phospholipase A2 synthesis. *J. Immunol.* **1997**, *159*, 391–400. [CrossRef] [PubMed]

23. Pernet, E.; Guillemot, L.; Burgel, P.R.; Martin, C.; Lambeau, G.; Sermet-Gaudelus, I.; Sands, D.; Leduc, D.; Morand, P.C.; Jeammet, L.; et al. *Pseudomonas aeruginosa* eradicates Staphylococcus aureus by manipulating the host immunity. *Nat. Commun.* **2014**, *5*, 5105. [CrossRef]

24. Chaput, C.; Boneca, I.G. Peptidoglycan detection by mammals and flies. *Microbes Infect.* **2007**, *9*, 637–647. [CrossRef] [PubMed]

25. Inohara, N.; Ogura, Y.; Fontalba, A.; Gutierrez, O.; Pons, F.; Crespo, J.; Fukase, K.; Inamura, S.; Kusumoto, S.; Hashimoto, M.; et al. Host recognition of bacterial muramyl dipeptide mediated through NOD2. Implications for Crohn's disease. *J. Biol. Chem.* **2003**, *278*, 5509–5512. [CrossRef]

26. Raymond, B.; Leduc, D.; Ravaux, L.; Le Goffic, R.; Candela, T.; Raymondjean, M.; Goossens, P.L.; Touqui, L. Edema toxin impairs anthracidal phospholipase A2 expression by alveolar macrophages. *PLoS Pathog.* **2007**, *3*, e187. [CrossRef] [PubMed]

27. Schwandner, R.; Dziarski, R.; Wesche, H.; Rothe, M.; Kirschning, C.J. Peptidoglycan- and lipoteichoic acid-induced cell activation is mediated by toll-like receptor 2. *J. Biol. Chem.* **1999**, *274*, 17406–17409. [CrossRef] [PubMed]

28. Pernet, E.; Brunet, J.; Guillemot, L.; Chignard, M.; Touqui, L.; Wu, Y. Staphylococcus aureus Adenosine Inhibits sPLA2-IIA-Mediated Host Killing in the Airways. *J. Immunol.* **2015**, *194*, 5312–5319. [CrossRef]

29. Hayashi, F.; Smith, K.D.; Ozinsky, A.; Hawn, T.R.; Yi, E.C.; Goodlett, D.R.; Eng, J.K.; Akira, S.; Underhill, D.M.; Aderem, A. The innate immune response to bacterial flagellin is mediated by Toll-like receptor 5. *Nature* **2001**, *410*, 1099–1103. [CrossRef] [PubMed]

30. Proft, T.; Baker, E.N. Pili in Gram-negative and Gram-positive bacteria—Structure, assembly and their role in disease. *Cell Mol. Life Sci.* **2009**, *66*, 613–635. [CrossRef]

31. Touqui, L.; Paya, M.; Thouron, F.; Guiyoule, A.; Zarantonelli, M.L.; Leduc, D.; Wu, Y.; Taha, M.K.; Alonso, J.M. Neisseria meningitidis pili induce type-IIA phospholipase A2 expression in alveolar macrophages. *FEBS Lett.* **2005**, *579*, 4923–4927. [CrossRef]

32. Vabulas, R.M.; Ahmad-Nejad, P.; da Costa, C.; Miethke, T.; Kirschning, C.J.; Hacker, H.; Wagner, H. Endocytosed HSP60s use toll-like receptor 2 (TLR2) and TLR4 to activate the toll/interleukin-1 receptor signaling pathway in innate immune cells. *J. Biol. Chem.* **2001**, *276*, 31332–31339. [CrossRef]

33. Bauer, S.; Kirschning, C.J.; Hacker, H.; Redecke, V.; Hausmann, S.; Akira, S.; Wagner, H.; Lipford, G.B. Human TLR9 confers responsiveness to bacterial DNA via species-specific CpG motif recognition. *Proc. Natl. Acad. Sci. USA* **2001**, *98*, 9237–9242. [CrossRef] [PubMed]

34. Hasko, G.; Linden, J.; Cronstein, B.; Pacher, P. Adenosine receptors: Therapeutic aspects for inflammatory and immune diseases. *Nat. Rev. Drug. Discov.* **2008**, *7*, 759–770. [CrossRef] [PubMed]

35. Guermonprez, P.; Khelef, N.; Blouin, E.; Rieu, P.; Ricciardi-Castagnoli, P.; Guiso, N.; Ladant, D.; Leclerc, C. The adenylate cyclase toxin of *Bordetella pertussis* binds to target cells via the alpha(M)beta(2) integrin (CD11b/CD18). *J. Exp. Med.* **2001**, *193*, 1035–1044. [CrossRef] [PubMed]

36. Bradley, K.A.; Mogridge, J.; Mourez, M.; Collier, R.J.; Young, J.A. Identification of the cellular receptor for anthrax toxin. *Nature* **2001**, *414*, 225–229. [CrossRef]

37. Scobie, H.M.; Rainey, G.J.; Bradley, K.A.; Young, J.A. Human capillary morphogenesis protein 2 functions as an anthrax toxin receptor. *Proc. Natl. Acad. Sci. USA* **2003**, *100*, 5170–5174. [CrossRef]

38. Neuhaus, F.C.; Baddiley, J. A continuum of anionic charge: Structures and functions of D-alanyl-teichoic acids in gram-positive bacteria. *Microbiol. Mol. Biol. Rev.* **2003**, *67*, 686–723. [CrossRef]

39. Koduri, R.S.; Gronroos, J.O.; Laine, V.J.; Le Calvez, C.; Lambeau, G.; Nevalainen, T.J.; Gelb, M.H. Bactericidal properties of human and murine groups I, II, V, X, and XII secreted phospholipases A(2). *J. Biol. Chem.* **2002**, *277*, 5849–5857. [CrossRef]

40. Qu, X.D.; Lehrer, R.I. Secretory phospholipase A2 is the principal bactericide for staphylococci and other gram-positive bacteria in human tears. *Infect. Immun.* **1998**, *66*, 2791–2797. [CrossRef]

41. Murakami, M.; Nakatani, Y.; Atsumi, G.; Inoue, K.; Kudo, I. Regulatory functions of phospholipase A2. *Crit. Rev. Immunol.* **1997**, *17*, 225–283. [CrossRef]

42. Weinrauch, Y.; Elsbach, P.; Madsen, L.M.; Foreman, A.; Weiss, J. The potent anti-Staphylococcus aureus activity of a sterile rabbit inflammatory fluid is due to a 14-kD phospholipase A2. *J. Clin. Investig.* **1996**, *97*, 250–257. [CrossRef]

43. Miki, Y.; Taketomi, Y.; Kidoguchi, Y.; Yamamoto, K.; Muramatsu, K.; Nishito, Y.; Park, J.; Hosomi, K.; Mizuguchi, K.; Kunisawa, J.; et al. Group IIA secreted phospholipase A2 controls skin carcinogenesis and psoriasis by shaping the gut microbiota. *JCI Insight* **2022**, *7*, e152611. [CrossRef] [PubMed]

44. Dore, E.; Joly-Beauparlant, C.; Morozumi, S.; Mathieu, A.; Levesque, T.; Allaeys, I.; Duchez, A.C.; Cloutier, N.; Leclercq, M.; Bodein, A.; et al. The interaction of secreted phospholipase A2-IIA with the microbiota alters its lipidome and promotes inflammation. *JCI Insight* **2022**, *7*, e152638. [CrossRef]

45. Enaud, R.; Prevel, R.; Ciarlo, E.; Beaufils, F.; Wieers, G.; Guery, B.; Delhaes, L. The Gut-Lung Axis in Health and Respiratory Diseases: A Place for Inter-Organ and Inter-Kingdom Crosstalks. *Front. Cell. Infect. Microbiol.* **2020**, *10*, 9. [CrossRef] [PubMed]

46. Ichinohe, T.; Pang, I.K.; Kumamoto, Y.; Peaper, D.R.; Ho, J.H.; Murray, T.S.; Iwasaki, A. Microbiota regulates immune defense against respiratory tract influenza A virus infection. *Proc. Natl. Acad. Sci. USA* **2011**, *108*, 5354–5359. [CrossRef]

47. Khan, N.; Mendonca, L.; Dhariwal, A.; Fontes, G.; Menzies, D.; Xia, J.; Divangahi, M.; King, I.L. Intestinal dysbiosis compromises alveolar macrophage immunity to Mycobacterium tuberculosis. *Mucosal Immunol.* **2019**, *12*, 772–783. [CrossRef] [PubMed]

48. Brown, R.L.; Sequeira, R.P.; Clarke, T.B. The microbiota protects against respiratory infection via GM-CSF signaling. *Nat. Commun.* **2017**, *8*, 1512. [CrossRef]

49. Pruzanski, W.; Vadas, P. Secretory synovial fluid phospholipase A2 and its role in the pathogenesis of inflammation in arthritis. *J. Rheumatol.* **1988**, *15*, 1601–1603. [PubMed]

50. Nakos, G.; Kitsiouli, E.; Hatzidaki, E.; Koulouras, V.; Touqui, L.; Lekka, M.E. Phospholipases A2 and platelet-activating-factor acetylhydrolase in patients with acute respiratory distress syndrome. *Crit. Care Med.* **2005**, *33*, 772–779. [CrossRef]

51. Touqui, L.; Herpin-Richard, N.; Gene, R.M.; Jullian, E.; Aljabi, D.; Hamberger, C.; Vargaftig, B.B.; Dessange, J.F. Excretion of platelet activating factor-acetylhydrolase and phospholipase A2 into nasal fluids after allergenic challenge: Possible role in the regulation of platelet activating factor release. *J. Allergy Clin. Immunol.* **1994**, *94*, 109–119. [CrossRef]

52. Weinrauch, Y.; Abad, C.; Liang, N.S.; Lowry, S.F.; Weiss, J. Mobilization of potent plasma bactericidal activity during systemic bacterial challenge. Role of group IIA phospholipase A2. *J. Clin. Investig.* **1998**, *102*, 633–638. [CrossRef]

53. Gimenez, A.P.; Wu, Y.Z.; Paya, M.; Delclaux, C.; Touqui, L.; Goossens, P.L. High bactericidal efficiency of type iia phospholipase A2 against Bacillus anthracis and inhibition of its secretion by the lethal toxin. *J. Immunol.* **2004**, *173*, 521–530. [CrossRef]

54. Hamaguchi, K.; Kuwata, H.; Yoshihara, K.; Masuda, S.; Shimbara, S.; Oh-ishi, S.; Murakami, M.; Kudo, I. Induction of distinct sets of secretory phospholipase A(2) in rodents during inflammation. *Biochim. Biophys. Acta* **2003**, *1635*, 37–47. [CrossRef]

55. Medjane, S.; Raymond, B.; Wu, Y.; Touqui, L. Impact of CFTR DeltaF508 mutation on prostaglandin E2 production and type IIA phospholipase A2 expression by pulmonary epithelial cells. *Am. J. Physiol. Lung Cell. Mol. Physiol.* **2005**, *289*, L816–L824. [CrossRef] [PubMed]

56. Hatzidaki, E.; Nakos, G.; Galiatsou, E.; Lekka, M.E. Impaired phospholipases A2 production by stimulated macrophages from patients with acute respiratory distress syndrome. *Biochim. Biophys. Acta* **2010**, *1802*, 986–994. [CrossRef] [PubMed]

57. Arbibe, L.; Koumanov, K.; Vial, D.; Rougeot, C.; Faure, G.; Havet, N.; Longacre, S.; Vargaftig, B.B.; Bereziat, G.; Voelker, D.R.; et al. Generation of lyso-phospholipids from surfactant in acute lung injury is mediated by type-II phospholipase A2 and inhibited by a direct surfactant protein A-phospholipase A2 protein interaction. *J. Clin. Investig.* **1998**, *102*, 1152–1160. [CrossRef] [PubMed]
58. Oka, S.; Arita, H. Inflammatory factors stimulate expression of group II phospholipase A2 in rat cultured astrocytes. Two distinct pathways of the gene expression. *J. Biol. Chem.* **1991**, *266*, 9956–9960. [CrossRef]
59. Crowl, R.M.; Stoller, T.J.; Conroy, R.R.; Stoner, C.R. Induction of phospholipase A2 gene expression in human hepatoma cells by mediators of the acute phase response. *J. Biol. Chem.* **1991**, *266*, 2647–2651. [CrossRef]
60. Furue, S.; Kuwabara, K.; Mikawa, K.; Nishina, K.; Shiga, M.; Maekawa, N.; Ueno, M.; Chikazawa, Y.; Ono, T.; Hori, Y.; et al. Crucial role of group IIA phospholipase A(2) in oleic acid-induced acute lung injury in rabbits. *Am. J. Respir. Crit. Care Med.* **1999**, *160*, 1292–1302. [CrossRef]
61. Kitsiouli, E.; Antoniou, G.; Gotzou, H.; Karagiannopoulos, M.; Basagiannis, D.; Christoforidis, S.; Nakos, G.; Lekka, M.E. Effect of azithromycin on the LPS-induced production and secretion of phospholipase A2 in lung cells. *Biochim. Biophys. Acta* **2015**, *1852*, 1288–1297. [CrossRef] [PubMed]
62. Wang, G.; Qian, P.; Xu, Z.; Zhang, J.; Wang, Y.; Cheng, S.; Cai, W.; Qian, G.; Wang, C.; Decoster, M.A. Regulatory effects of the JAK3/STAT1 pathway on the release of secreted phospholipase A$_2$-IIA in microvascular endothelial cells of the injured brain. *J. Neuroinflamm.* **2012**, *9*, 170. [CrossRef]
63. Taha, M.K.; Alonso, J.M. Molecular epidemiology of infectious diseases: The example of meningococcal disease. *Res. Microbiol.* **2008**, *159*, 62–66. [CrossRef]
64. Dunn, K.L.; Virji, M.; Moxon, E.R. Investigations into the molecular basis of meningococcal toxicity for human endothelial and epithelial cells: The synergistic effect of LPS and pili. *Microb. Pathog.* **1995**, *18*, 81–96. [CrossRef]
65. O'Grady, E.P.; Mulcahy, H.; O'Callaghan, J.; Adams, C.; O'Gara, F. *Pseudomonas aeruginosa* infection of airway epithelial cells modulates expression of Kruppel-like factors 2 and 6 via RsmA-mediated regulation of type III exoenzymes S and Y. *Infect. Immun.* **2006**, *74*, 5893–5902. [CrossRef] [PubMed]
66. McConnell, B.B.; Yang, V.W. Mammalian Kruppel-like factors in health and diseases. *Physiol. Rev.* **2010**, *90*, 1337–1381. [CrossRef] [PubMed]
67. Dach, K.; Zovko, J.; Hogardt, M.; Koch, I.; van Erp, K.; Heesemann, J.; Hoffmann, R. Bacterial toxins induce sustained mRNA expression of the silencing transcription factor klf2 via inactivation of RhoA and Rhophilin 1. *Infect. Immun.* **2009**, *77*, 5583–5592. [CrossRef]
68. Goehring, U.M.; Schmidt, G.; Pederson, K.J.; Aktories, K.; Barbieri, J.T. The N-terminal domain of *Pseudomonas aeruginosa* exoenzyme S is a GTPase-activating protein for Rho GTPases. *J. Biol. Chem.* **1999**, *274*, 36369–36372. [CrossRef]
69. Okuda, J.; Hayashi, N.; Tanabe, S.; Minagawa, S.; Gotoh, N. Degradation of interleukin 8 by the serine protease MucD of *Pseudomonas aeruginosa*. *J. Infect. Chemother.* **2011**, *17*, 782–792. [CrossRef]
70. Vareechon, C.; Zmina, S.E.; Karmakar, M.; Pearlman, E.; Rietsch, A. *Pseudomonas aeruginosa* Effector ExoS Inhibits ROS Production in Human Neutrophils. *Cell Host Microbe* **2017**, *21*, 611–618. [CrossRef]
71. Hajishengallis, G.; Liang, S.; Payne, M.A.; Hashim, A.; Jotwani, R.; Eskan, M.A.; McIntosh, M.L.; Alsam, A.; Kirkwood, K.L.; Lambris, J.D.; et al. Low-abundance biofilm species orchestrates inflammatory periodontal disease through the commensal microbiota and complement. *Cell Host Microbe* **2011**, *10*, 497–506. [CrossRef] [PubMed]
72. Al-Attar, A.; Alimova, Y.; Kirakodu, S.; Kozal, A.; Novak, M.J.; Stromberg, A.J.; Orraca, L.; Gonzalez-Martinez, J.; Martinez, M.; Ebersole, J.L.; et al. Activation of Notch-1 in oral epithelial cells by *P. gingivalis* triggers the expression of the antimicrobial protein PLA(2)-IIA. *Mucosal. Immunol.* **2018**, *11*, 1047–1059. [CrossRef]
73. Potempa, J.; Sroka, A.; Imamura, T.; Travis, J. Gingipains, the major cysteine proteinases and virulence factors of Porphyromonas gingivalis: Structure, function and assembly of multidomain protein complexes. *Curr. Protein Pept. Sci.* **2003**, *4*, 397–407. [CrossRef] [PubMed]
74. Dziarski, R.; Gupta, D. Peptidoglycan recognition in innate immunity. *J. Endotoxin Res.* **2005**, *11*, 304–310. [CrossRef]
75. Glomski, I.J.; Fritz, J.H.; Keppler, S.J.; Balloy, V.; Chignard, M.; Mock, M.; Goossens, P.L. Murine splenocytes produce inflammatory cytokines in a MyD88-dependent response to Bacillus anthracis spores. *Cell. Microbiol.* **2007**, *9*, 502–513. [CrossRef]
76. Mock, M.; Fouet, A. Anthrax. *Annu. Rev. Microbiol.* **2001**, *55*, 647–671. [CrossRef]
77. Lincoln, R.E.; Hodges, D.R.; Klein, F.; Mahlandt, B.G.; Jones, W.I., Jr.; Haines, B.W.; Rhian, M.A.; Walker, J.S. Role of the lymphatics in the pathogenesis of anthrax. *J. Infect. Dis.* **1965**, *115*, 481–494. [CrossRef] [PubMed]
78. Turnbull, P.C. Anthrax vaccines: Past, present and future. *Vaccine* **1991**, *9*, 533–539. [CrossRef]
79. Raymond, B.; Ravaux, L.; Mémet, S.; Wu, Y.; Sturny-Leclère, A.; Leduc, D.; Denoyelle, C.; Goossens, P.L.; Payá, M.; Raymondjean, M.; et al. Anthrax lethal toxin down-regulates type-IIA secreted phospholipase A(2) expression through MAPK/NF-kappaB inactivation. *Biochem. Pharm.* **2010**, *79*, 1149–1155. [CrossRef]
80. Menschikowski, M.; Hagelgans, A.; Kostka, H.; Eisenhofer, G.; Siegert, G. Involvement of epigenetic mechanisms in the regulation of secreted phospholipase A2 expressions in Jurkat leukemia cells. *Neoplasia* **2008**, *10*, 1195–1203. [CrossRef]
81. Voth, D.E.; Hamm, E.E.; Nguyen, L.G.; Tucker, A.E.; Salles, I.I.; Ortiz-Leduc, W.; Ballard, J.D. Bacillus anthracis oedema toxin as a cause of tissue necrosis and cell type-specific cytotoxicity. *Cell. Microbiol.* **2005**, *7*, 1139–1149. [CrossRef] [PubMed]

82. Antonio, V.; Brouillet, A.; Janvier, B.; Monne, C.; Bereziat, G.; Andreani, M.; Raymondjean, M. Transcriptional regulation of the rat type IIA phospholipase A2 gene by cAMP and interleukin-1beta in vascular smooth muscle cells: Interplay of the CCAAT/enhancer binding protein (C/EBP), nuclear factor-kappaB and Ets transcription factors. *Biochem. J.* **2002**, *368*, 415–424. [CrossRef] [PubMed]

83. Cleret-Buhot, A.; Mathieu, J.; Tournier, J.N.; Quesnel-Hellmann, A. Both lethal and edema toxins of Bacillus anthracis disrupt the human dendritic cell chemokine network. *PLoS ONE* **2012**, *7*, e43266. [CrossRef] [PubMed]

84. Wu, Y.; Raymond, B.; Goossens, P.L.; Njamkepo, E.; Guiso, N.; Paya, M.; Touqui, L. Type-IIA secreted phospholipase A2 is an endogenous antibiotic-like protein of the host. *Biochimie* **2010**, *92*, 583–587. [CrossRef]

85. Skopova, K.; Tomalova, B.; Kanchev, I.; Rossmann, P.; Svedova, M.; Adkins, I.; Bibova, I.; Tomala, J.; Masin, J.; Guiso, N.; et al. Cyclic AMP-Elevating Capacity of Adenylate Cyclase Toxin-Hemolysin Is Sufficient for Lung Infection but Not for Full Virulence of *Bordetella pertussis*. *Infect. Immun.* **2017**, *85*, e00937-16. [CrossRef]

86. Foster, T.J. Immune evasion by staphylococci. *Nat Rev Microbiol* **2005**, *3*, 948–958. [CrossRef]

87. McCarthy, A.J.; Lindsay, J.A. Genetic variation in Staphylococcus aureus surface and immune evasion genes is lineage associated: Implications for vaccine design and host-pathogen interactions. *BMC Microbiol.* **2010**, *10*, 173. [CrossRef] [PubMed]

88. Gonzales, J.N.; Gorshkov, B.; Varn, M.N.; Zemskova, M.A.; Zemskov, E.A.; Sridhar, S.; Lucas, R.; Verin, A.D. Protective effect of adenosine receptors against lipopolysaccharide-induced acute lung injury. *Am. J. Physiol. Lung Cell. Mol. Physiol.* **2014**, *306*, L497–L507. [CrossRef]

89. Takahashi, N.; Tetsuka, T.; Uranishi, H.; Okamoto, T. Inhibition of the NF-kappaB transcriptional activity by protein kinase A. *Eur. J. Biochem.* **2002**, *269*, 4559–4565. [CrossRef]

Review

From Bacterial Toxin to Therapeutic Agent: The Unexpected Fate of Mycolactone

Daniela Ricci [†] **and Caroline Demangel** [*,†]

Institut Pasteur, Université Paris Cité, Inserm U1224, Immunobiology and Therapy Unit, 75015 Paris, France; daniela.ricci@pasteur.fr
* Correspondence: caroline.demangel@pasteur.fr
† These authors contributed equally to this work.

Abstract: "Recognizing a surprising fact is the first step towards discovery." This famous quote from Louis Pasteur is particularly appropriate to describe what led us to study mycolactone, a lipid toxin produced by the human pathogen *Mycobacterium ulcerans*. *M. ulcerans* is the causative agent of Buruli ulcer, a neglected tropical disease manifesting as chronic, necrotic skin lesions with a "surprising" lack of inflammation and pain. Decades after its first description, mycolactone has become much more than a mycobacterial toxin. This uniquely potent inhibitor of the mammalian translocon (Sec61) helped reveal the central importance of Sec61 activity for immune cell functions, the spread of viral particles and, unexpectedly, the viability of certain cancer cells. We report in this review the main discoveries that marked our research into mycolactone, and the medical perspectives they opened up. The story of mycolactone is not over and the applications of Sec61 inhibition may go well beyond immunomodulation, viral infections, and oncology.

Keywords: mycolactone; *Mycobacterium ulcerans*; Sec61; immunomodulation; viral infection; cancer

Key Contribution: In this work we review the history of mycolactone, from its discovery as a bacterial toxin from *M. ulcerans* to its key role in determining the therapeutic potential of Sec61 inhibitors as immunomodulators, anti-viral and anti-cancer molecules.

Citation: Ricci, D.; Demangel, C. From Bacterial Toxin to Therapeutic Agent: The Unexpected Fate of Mycolactone. *Toxins* **2023**, *15*, 369. https://doi.org/10.3390/toxins15060369

Received: 13 May 2023
Revised: 27 May 2023
Accepted: 28 May 2023
Published: 30 May 2023

1. Introduction

Buruli ulcer (BU) is a neglected tropical disease caused by skin infection with the environmental pathogen *Mycobacterium ulcerans*. Following inoculation into subcutaneous tissues, the bacteria establish a local infection that escapes the control of the immune system and causes chronic ulcers characterized by a relative lack of inflammation and pain [1]. In 1999, pioneering studies conducted by George et al. revealed that *M. ulcerans* contains a plasmid coding for mycolactone, a polyketide synthase product underpinning bacterial pathogenicity and virulence [2,3]. Indeed, bacteria genetically modified to be mycolactone-deficient were unable to survive in experimentally infected guinea pigs, and the subcutaneous injection of purified mycolactone in these animal models was sufficient to induce BU-like lesions [2,3]. Why significant amounts of bacteria and dead cells in Buruli ulcers do not trigger inflammatory responses and pain, and how mycolactone contributes to these abnormalities, were intriguing questions that we decided to investigate in the laboratory [3].

2. The Immunomodulatory Effects of Mycolactone

We and other research groups started to investigate the dual, immunomodulatory and cytotoxic effects of mycolactone in models of immune cells, using mycolactone purified from bacterial cultures with a protocol adapted from the initial methodology developed by George et al. [2,4]. From these cellular studies, mycolactone was found to be a double-trigger weapon of *M. ulcerans* capable of rapidly paralyzing the effector functions of immune cells in conditions not affecting their viability, while inducing their apoptosis in the longer term [3]. Notably, mycolactone impacted both the innate and adaptive components of immunity, and in multiple ways [1]. In particular, it suppressed the ability of multiple immune cells (macrophages, dendritic cells (DCs), neutrophils, lymphocytes) to produce cytokines and chemokines upon activation [5–9]. Intriguingly, mycolactone did not affect the production of cytokines and chemokines at the mRNA level, and it did not affect protein synthesis in a global way. Cytokine production was interrupted at the intracellular level, pointing to a defect upstream of the secretion pathway.

Cytokines and chemokines were not the only proteins affected by mycolactone, and membrane receptors whose expression by immune cells was impacted by mycolactone were subsequently discovered. A striking example is L-selectin (also called CD62-L), a membrane protein that is expressed at high levels by naive T lymphocytes and mediates their trafficking to the draining lymph nodes through molecular interaction with high endothelial venules [10]. We found that mycolactone treatment downregulates the basal expression of CD62-L by human T cells in vitro, thereby altering their capacity to traffic to peripheral lymph nodes in vivo. Other examples of membrane proteins that were found highly susceptible to mycolactone treatment include cytokine receptors, such as the interferon (IFN) receptors. In T lymphocytes exposed to mycolactone in vitro, we observed within 6 h a significant decrease in cell surface expression of the receptors of IFN-α and IFN-γ. Similarly in macrophages, mycolactone rapidly decreased the cell surface expression of the IFN-γ receptor, reducing their ability to express nitric oxide synthase upon stimulation with IFN-γ, and thereby their antimicrobial functions [11].

Given their central role in initiating adaptive immune responses and immunity to mycobacteria [12], we were particularly interested to characterize the effects of mycolactone on the viability and functional biology of DCs. Both the phenotypic and functional maturations of DCs were inhibited by treatment with non-cytotoxic concentrations of mycolactone, leading to a defective emigration capacity of DCs upon adoptive transfer in mice, a decreased ability to activate allogeneic T cell priming and to produce inflammatory cytokines and chemokines upon in vitro stimulation [5]. A unique feature of DCs is their ability to capture antigens released by the surrounding cells and present them in the context of major histocompatibility complex class I (MHC-I) molecules, a process referred to as cross-presentation. In collaboration with the group of Sebastian Amigorena (Institut Curie, Paris, France), we showed that mycolactone dramatically affects DC production of MHC-I molecules, and therefore their ability to cross-present antigens [13].

Collectively, these data revealed mycolactone as a novel type of natural immunosuppressor, with the ability to prevent the generation of anti-mycobacterial responses at multiple stages [3]. Because mycolactone shared structural and functional features with the macrocyclic triene rapamycin and the macrolide lactone FK506, both immunosuppressive compounds operating through interaction with the intracellular receptor FKBP12 to inhibit the mammalian target of rapamycin C1 (mTORC1), we hypothesized that it could target the same signaling pathway [5]. While an inhibitory effect of mycolactone on mTOR signaling was reported by Bieri et al. [14], mycolactone did not alter the lipopolysaccharide-driven activation of mTORC1 in monocytes in an independent study by Simmonds et al. [6]. In our hands, mycolactone did not alter the constitutive activation of mTORC1 in the Jurkat T cell line, or the TCR-induced activation of mTORC1 in peripheral blood-derived lymphocytes [7]. Having excluded the known mechanisms of immune suppression and highlighted the unique immunosuppressive features of mycolactone, we concluded that mycolactone operates via a novel, selective and post-transcriptional mechanism of protein biogenesis inhibition.

3. Mycolactone Targets the Sec61 Translocon

With these discoveries, identifying the molecular target of mycolactone took a new dimension. Indeed, in addition to providing an effective treatment of BU, it could lead to the generation of novel immunomodulatory molecules. In 2014, Hall et al. made a breakthrough by demonstrating that mycolactone prevents the translocation of model secretory proteins into the endoplasmic reticulum (ER), leading to their degradation in the cytosol by the ubiquitin–proteasome system [15]. Using cell-free assays, McKenna et al. then showed that mycolactone selectively affects the step of cotranslational translocation of secreted and integral transmembrane proteins (TMPs) into the ER [16]. In eukaryotes, the cotranslational protein translocation pathway is initiated by recognition of signal peptides or transmembrane domains by the signal recognition particle (SRP). The SRP then targets the ribosome-nascent polypeptide complex to the Sec61 translocon. Sec61 is a heterotrimeric complex embedded in the ER membrane that ensures the transport of most secreted proteins (with a signal peptide but no transmembrane domain) and single-spanning TMPs into the ER. TMPs exclusively relying on Sec61 for insertion into the ER membrane include the Type I (with a signal peptide) and the Type II (without a signal peptide and a cytosolic N terminus) TMPs. Instead, the rare subsets of Type III TMPs (without a signal peptide and the opposite N terminal topology) and the C-terminal tail-anchored proteins can use alternative pathways for membrane integration at the ER [17,18], while mitochondrial membrane proteins depend on independent TIM/TOM complexes for mitochondrial membrane insertion.

In collaboration with Ville Paavilainen (University of Helsinki, Helsinki, Finland), we demonstrated that mycolactone prevents the cotranslational translocation of proteins into the ER by directly targeting the Sec61 translocon. Indeed, a single amino acid mutation (R66G) in the pore-forming (alpha) subunit of Sec61 conferred full resistance to mycolactone activity in bioassays [11]. Mycolactone was not the first reported inhibitor of Sec61. In 2005, Garrison et al. had discovered cotransin, a fungal product derivative with the capacity to inhibit Sec61 in a selective, substrate-specific manner [19,20]. Competition assays between cotransin and mycolactone, and mutant Sec61 studies, indicated that the binding sites of the two compounds overlap [11]. However, our in vitro translocation assays and global profiling of mycolactone-susceptible proteins in T cells, DCs and sensory neurons showed that, contrary to cotransin, mycolactone is not substrate-selective [11,13,21]. In addition, these proteomic analyses made it possible to characterize for the first time the signature of the Sec61 blockade at the cellular level [21].

Consistent with its mechanism of action, most of the detected Sec61 clients (secreted proteins, Type I and Type II TMPs) were massively downregulated by mycolactone in the three tested cell types. In contrast, mycolactone did not affect the cellular levels of Type III TMPs, C-tail anchored and mitochondrial membrane proteins [21]. Recent cryo-EM studies suggest that mycolactone and cotransin both operate by maintaining Sec61 in a close, inactive conformation through interactions with its sealing plug and lateral gate. However, the structural determinants of their differential selectivity for Sec61 substrates remain to be elucidated [22].

4. Structure–Activity Relationships: The Input of Synthetic Chemistry

Identifying the minimal structural determinants of biological activity is key for the development of mycolactone-inspired derivatives that are compatible with large-scale synthesis. Since the initial description by George et al. [2], other structurally divergent mycolactones from genetically related mycobacterial species have been discovered and functionally characterized [23–25]. The most active in cellular assays is mycolactone A/B, produced by a subset of *M. ulcerans* clinical isolates (referred to as mycolactone in this review and depicted in Figure 1) [26]. The structure of mycolactone A/B is defined by a macrolactone ring with 12 atoms substituted at C11 by a C12–C20 carbon chain with hydroxyl functions at its end, and at C5 by a pentaenic chain C1′–C16′. In the natural state, mycolactone A/B is a mixture of geometric isomers bearing C4′–C5′ unsaturation with a Z-/E- ratio of 3:2. The Z-isomer defined as mycolactone A is in the majority due to the allylic constraints A of the E-isomer (mycolactone B).

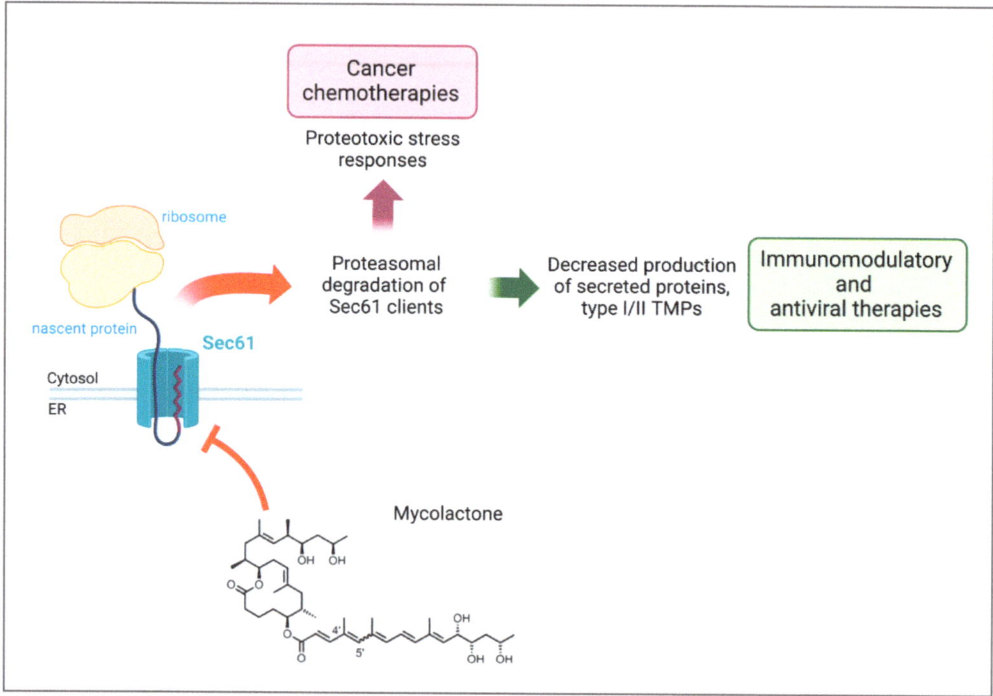

Figure 1. Sec61 blockade, its consequences at the cellular level and potential therapeutic applications. Mycolactone binds to Sec61 and blocks the translocation of its protein clients (secreted proteins and type I/II TMPs) in the ER, which in turn induces their degradation by the proteasome in the cytosol. This leads on the one hand to decreased levels of Sec61 clients, making it a potential immunomodulatory and antiviral agent, and on the other hand to proteotoxic stress responses and eventually to apoptosis, a useful addition to the cancer chemotherapies portfolio.

Because *M. ulcerans* slowly multiplies and must be grown in safety laboratories of Level > 2, significant efforts were made to generate synthetic mycolactone and variants. In 2002, Song et al. reported the first total synthesis of mycolactone [27]. Additional synthetic routes engineered by different groups followed (Yin et al. [28], Feyen et al. [29] and Chany et al. [30]), which altogether provided researchers with a synthetic compound that was equivalent to the *M. ulcerans* factor in cellular assays. These stereo-selective, time-consuming syntheses involving at least 30 steps used different synthetical concepts, making a strong contribution to organic synthesis and paving the way for structure–activity relationship (SAR) studies [24].

One major finding of these SAR studies was that truncating mycolactone's structure in any way is detrimental to its biological activity. This suggested that mycolactone is the result of an evolution of the pathogen for maximal inhibition of the host Sec61. In particular, an intact and full-length lower side chain of mycolactone was critical for Sec61 inhibition, and the core structure devoid of polyketide chains was biologically inert [31,32]. Using a diverse set of C8-desmethylmycolactone analogues generated by our collaborator Nicolas Blanchard (University of Haute Alsace, Mulhouse, France), we attempted to identify the simplest synthetic version of mycolactone presenting an optimal ratio between immunomodulatory and cytotoxic properties [8,30]. All synthetic compounds were compared to natural mycolactone for their capacity to block the activation-induced production of interleukin (IL)-2 by the human Jurkat T cell line (as a read-out of their immunomodulatory properties), and for their cytopathic activity in the human epithelial cell line HeLa (an epithelial cell model that is highly susceptible to mycolactone-mediated anoikis). A structural module of natural mycolactone lacking the upper side chain and core C8-methyl (hereafter named mini-mycolactone) was selected. In further assays using human primary cells, mini-mycolactone significantly retained natural mycolactone's ability to inhibit the production of key inflammatory cytokines such as the tumor necrosis factor (TNF)-α, while displaying a relatively lower cytotoxicity in primary dermal fibroblasts modeling ulcerative activity [8]. Therefore, in addition to being synthetically simpler than the natural product, mini-mycolactone was expected to display a broader therapeutic window in vivo.

5. Sec61 Activity, Immunity and Apoptosis

The identification of mycolactone-resistant Sec61 mutants enabled us to examine the role of the translocon in its immunomodulatory and cytotoxic effects. Production of IFN-γ by T cells and activation of the IFN-γ receptor signaling pathway in infected macrophages are key components of anti-mycobacterial immunity [33]; both of which were markedly impaired by mycolactone. Expression of the Sec61α-R66G mutant in mycolactone-treated T cells rescued their effector functions in vitro and in vivo. When expressed in macrophages, the mycolactone-resistant mutant restored IFN-γ receptor-mediated anti-microbial responses [11]. These experiments demonstrated that Sec61 is the host receptor mediating the diverse immunomodulatory effects of mycolactone. Further, they revealed a novel mechanism of immune evasion evolved by *M. ulcerans*. Beyond the control of IFN-γ and IFN-γ receptor production, our proteomic studies showed that inhibiting protein translocation has the potential to suppress inflammatory responses. In mouse models, the systemic administration of mycolactone could be used therapeutically to limit chronic skin inflammation, rheumatoid arthritis and inflammatory pain [8] (Figure 1). While less potent than natural mycolactone in the conditions used, mini-mycolactone also conferred significant protection in these disease models, confirming the potential of mycolactone-derived structures as prospective immunosuppressants.

Notably, the Sec61α-R66G mutant protected mycolactone-treated cells from undergoing apoptosis [11], showing that mycolactone cytotoxicity is a late consequence of Sec61 blockade. Looking at the proteins that were up-regulated by mycolactone in treated cells, we identified the hallmarks of cytosolic and ER stress responses. Thapsigargin, tunicamycin, and MG132 are canonical ER stressors targeting Ca^{2+} ATPases, protein glycosylation or the proteasome, respectively, which trigger an unfolded protein response (UPR) to restore protein homeostasis. Like them, mycolactone upregulated the UPR in treated cells, leading to the expression of the pro-apoptotic factor C/EBP homologous protein (Chop). This provided an explanation for how a prolonged exposure to saturating amounts of mycolactone can lead to cell apoptosis (Figure 1). However, unlike canonical ER stressors, mycolactone did not augment in parallel the expression of the ER chaperone GRP78/BiP [21,34]. Because BiP increases the cell's ability to resolve ER stress and prevents the transition from protective to terminal UPR, this suggested that ER stress caused by mycolactone is more prone to evolve towards apoptosis.

6. Sec61 Blockers for Oncology

Our observation that mycolactone causes terminal UPR led us to investigate whether the proteotoxic impact of Sec61 blockade could also be exploited therapeutically. Despite the fact that considerable advances have been made over the years, the plasma cell malignancy multiple myeloma (MM) still remains an incurable disease. The current first line of treatment consists of proteasome inhibitors (bortezomib and derivatives) and immunomodulators (lenalidomide and derivatives), but eventually the majority of MM patients will relapse over time because of the generation of drug-resistant cancerous cells [35]. Therefore, the development of novel drugs with different mechanisms of action is vital to turn the tide of the battle against MM. We reasoned that Sec61 blockade may represent a novel therapeutic approach of interest in MM, by inducing proteotoxic stress responses while preventing the expression of membrane receptors that are key to MM cell division and dissemination.

With regard to this, recent work in our laboratory established protein translocation inhibition as a novel, useful tool against MM. By broadly inhibiting Sec61 with mycolactone, we showed that the translocon's activity has a central role in determining MM cells' fate. Indeed, mycolactone efficiently reduced MM cell line production of immunoglobulins and multiple type I/II TMP receptors such as CD138, a hallmark of MM that allows the survival of cancerous cells in the bone marrow by promoting growth factor signaling. Mycolactone treatment also decreased MM cell expression of the pro-survival IL-6 receptor and CD40, whose activation stimulates IL-6 production. As a later effect, mycolactone induced a pro-apoptotic ER stress-response in MM cell lines and tumors isolated from MM patients. This was used as proof of concept to show that Sec61 inhibitors have the potential to be used as an anti-cancer treatment against MM, alone or in combination with currently used chemotherapies [36]. Strikingly, mycolactone combined with bortezomib significantly delayed MM tumor growth in mice without significant toxicity. Equally importantly, mycolactone showed a synergistic action with lenalidomide, and was even effective at inducing cell death in bortezomib- and or lenalinomide-resistant cells [36,37], giving hope for the treatment of relapsed/refractory MM (Figure 1).

As described in the previous sections of this review, type III TMPs translocate into the ER in a Sec61-independent manner, and their levels are therefore not blunted by Sec61 inhibitors. The plasma cell-specific B cell maturation antigen (BCMA) belongs to the class of type III TMPs and ever since its overexpression and activation have been associated with MM, it has attracted great attention from the medical research community. As a result, BCMA is currently used as a biomarker for MM diagnosis and tracking. Furthermore, BCMA is used as a target to treat MM through specific antibodies and chimeric antigen receptor (CAR)-T cell therapy, both of which are giving promising results in clinics for an efficient and durable MM cancer treatment [38]. As expected, we found that mycolactone does not decrease BCMA expression via MM cell lines, but surprisingly its expression was greatly increased after treatment in a dose-dependent manner [36]. Even though we do not have mechanistic insights as to why mycolactone increases BCMA levels, we hypothesize that this is a secondary effect of the stress response induced by the Sec61 blocker. Nonetheless, it stands to reason that the effect of mycolactone on BCMA may potentiate the anti-BCMA therapies that are currently emerging against MM, giving mycolactone an additional key role in MM treatment.

Mycolactone is not the only Sec61 inhibitor that showed promising results by targeting clinically relevant proteins in the Oncology field. In fact, the cyclodepsipeptide apratoxins, a class of molecules isolated from marine cyanobacteria, showed a significantly strong antiproliferative activity in different cancer cells and only later was its mechanism of action revealed. Like cotransin and mycolactone, apratoxin A inhibits the secretion pathway through the direct inhibition of Sec61 [39]. The ability of apratoxin A to inhibit vascular endothelial growth factor (VEGF)-A expression inspired Cai et al. to test its anti-angiogenic effect, a hallmark of solid tumors. The group showed that a synthetic analog of apratoxin A, which was improved through structure–activity relationship studies and named apratoxin S10, inhibits angiogenesis and cancer cell growth in vitro. The antiangiogenic effect was mainly due to the ability of apratoxin S10 to downregulate VEGF receptor on endothelial cells while the anti-tumor effect was shown to result from a reduction in secretion of VEGF-A and IL-6, which have a role in both promoting tumor cell growth and the formation of new blood vessels [40]. The dual anti-tumor and anti-angiogenic properties of apratoxin A are extremely promising for solid tumor treatments, but the extensive pancreatic toxicity of the molecule in vivo [39] has blocked it from further clinical studies. How it compares to mycolactone and cotransin, in terms of Sec61 substrate selectivity and mechanism of inhibition, remains largely unknown.

As mentioned above, a different molecule belonging to the same family, the cyclodep-sipeptide natural product cotransin, which was first discovered through a Sec61-unrelated screen of molecules that aimed at discovering inhibitors of the expression of cell adhesion molecules, proved to be a selective Sec61 inhibitor. Unlike mycolactone, cotransin only inhibits a subset of Sec61 clients, including the vascular cell adhesion molecule 1 (VCAM-1) and human epidermal growth factor receptor 3 (HER3), by preventing the recognition of specific signal sequences at the N-terminus end of target proteins by the translocon [20]. HER3 is a therapeutically interesting Sec61 client because it has a key role in tumorigenesis in several types of tumors, induces cell proliferation and has been linked to chemotherapy resistance. Much effort has been devoted to targeting HER3 through diverse methods against different types of cancers [41]. Ruiz-Saenz et al. showed that CT8, which belongs to the class of cotransins, selectively inhibits the expression of HER3 by blocking its transloca-tion in the ER and subsequently inducing its proteasomal degradation in the cytosol, with no effect on the expression of other protein members of the HER family. This selective effect was in fact due to the specific sequence of the N-terminal signal peptide of HER3, as part of the mechanism of action of cotransins, and proved to enhance the pro-apoptotic effects of chemotherapeutic drugs currently used against breast cancer and other solid tumors [42].

What we learned from the action of these natural products shed light on the great therapeutic potential of Sec61 blockers: their anti-tumor, pro-apoptotic properties, together with the possibility of developing specific inhibitors that target selective clinically relevant Sec61 proteins to avoid/minimize normal cell toxicity, show that Sec61 inhibitors may make great contributions to the field of Oncology in the near future.

7. Indications beyond Inflammation and Oncology

In recent years, the potential of Sec61 inhibitors has been explored as a weapon against other threats to global public health such as viral infections (Figure 1). Viruses take advantage of many cellular machineries to enter the cell and to replicate themselves and the translocon is one of those, which viruses use to form mature viral particles ready to perpetuate the infection. It stands to reason that many groups worldwide have assayed the ability of inhibitors of protein translocation to prevent viral particle formation. Through an elegant influenza virus–host interactome analysis during infection, Heaton et al. proved for the first time in 2016 that Sec61 is one of the most effective interactors of several influenza virus proteins. The group's analysis was not limited to influenza, and by pharmacologic inhibition and genetic knockdown of Sec61 they showed that an active translocon is required for the replication of influenza, HIV and dengue viruses [43]. Subsequently, our laboratory used the influenza virus model to prove that mycolactone can target virus envelope proteins just as well as endogenous targets, with specificity for type I/II but not type III TMPs [21]. Expanding the scope of its anti-viral activity, mycolactone was employed by Monel et al. to help understand the mechanisms of the cytopathic effects of the Zika virus. In this context, it was revealed that Sec61 activity drives the Zika virus-induced cell death program. Translocon inhibition through mycolactone prevented viral replication, the formation of the characteristic ER-derived vacuoles, and the ensuing cell death [44]. This study highlighted novel host–virus interaction mechanisms and possible pharmaceutical targets to prevent Zika virus-induced cytopathy, which can cause serious problems when the infection occurs during pregnancy, in some cases leading to birth defects and even fetal lethality. Along this same line, following a sequential affinity purification-mass spectrometry (AP-MS) analysis and RNAi screening aimed at generating a dengue–human protein–protein interaction map, Shah et al. revealed that the translocon machinery is a key interactor of dengue and Zika virus transmembrane proteins. Interestingly, Sec61 inhibition through cotransins not only completely abolished viral protein production in mammalian cells, but also inhibited dengue and Zika viral particle production and replication in cells derived from mosquitoes, the transmission vectors of these viruses [45]. These results identified Sec61 as a targetable host factor against dengue and Zika infections in two different hosts, humans and mosquitoes, amplifying their potential and interest.

Sec61 blockers even came under the spotlight during the recent COVID-19 pandemic. Driven again by a screen aimed at identifying protein–protein interactions between SARS-

CoV-2 and human proteins, Sec61 was identified as one of the leading druggable human protein candidates to prevent such interactions. Multiple SARS-CoV-2 proteins were predicted to access the ER through Sec61 and in fact, by chemically blocking the translocon activity, the viral replication was inhibited in a virus plaque assay [46]. Ipomoeassin-F is a synthetic derivative of a glycoresin isolated from the leaves of Ipomoea squamosa [47], subsequently found to be another potent Sec61 inhibitor [48]. O'Keefe et al. [49] recently demonstrated through in vitro assays that ipomoeassin-F is a potent anti-SARS-CoV-2 agent. As a matter of fact, through the blockade of Sec61, ipomoeassin-F inhibited the expression of the viral proteins spike and ORF8 while at the same time inducing the degradation of the angiotensin-converting enzyme 2, a key host cell receptor that allows SARS-CoV-2 viral entry [49]. Therefore, Sec61 inhibitors show a dual action by decreasing both SARS-CoV-2 virion production and the cell entry receptor, promising efficacy against the early and late stages of viral infection. The discovery of novel anti-viral drugs with new modes of action is essential, given the emergence of drug resistance in existing viral infections and the threat of future pandemics. With regard to this, Sec61 inhibitors showed broad spectrum anti-viral activities, by acting both on the formation of essential viral proteins and by inhibiting viral entry in the host through different mechanisms. We believe that Sec61 blockers have the potential to be novel and powerful anti-viral tools.

8. Conclusions and Perspectives

Mycolactone may have started its journey through research as a detrimental mycobacterial toxin, but it subsequently turned out to be a powerful tool: not only has it provided us with many important lessons on Sec61 biology, but it is also paving the way for clinical applications of translocon inhibitors (Figure 1). Comparisons between mycolactone and other, less potent natural inhibitors of Sec61 such as cotransins suggest that it may be possible to achieve selective inhibition of protein targets. This may be used against clinically relevant Sec61 clients while bypassing the cytotoxicity issues of broad inhibitors. Importantly, natural inhibitors also provided structural information on the inactive state of Sec61 [22], which will be key in the development of small-molecule blockers of the translocon activity.

Mycolactone and the other natural inhibitors of Sec61 may not be suitable for employment in clinics due to their complexity, laborious synthesis and toxicity, but their activity and that of their derivatives have captured the attention of many researchers worldwide, who are already working on overcoming these issues. The race for the development of drug-like Sec61 inhibitors has already started [50,51] and, to date, Oncology is the primary indication for the usage of these small molecules, which have the potential to be added to the treatment portfolio of many cancer types to target multiple tumor drivers at once. Furthermore, Sec61 blockers proved to possess potent anti-inflammatory activities by targeting the expression of several cytokines, chemokines and cytokines receptors, paving the way for the development of novel treatments for inflammatory pain and inflammatory diseases. In Virology, the development of translocon blockers to decrease viral titers is particularly relevant because Sec61 is a host factor, and that can limit the likelihood of viral resistance. However, the potential of Sec61 inhibitors as antivirals will have to be confirmed in animal models of infection.

Author Contributions: Conceptualization, D.R. and C.D.; writing—original draft preparation, D.R. and C.D.; writing—review and editing, D.R. and C.D.; funding acquisition, C.D. All authors have read and agreed to the published version of the manuscript.

Funding: Besides the core funding provided by the Institut Pasteur and INSERM U1224, the research work described in this review benefited from financial support allocated by the Fondation Raoul Follereau, the Fondation pour la Recherche Médicale (Equipe FRM 2012), the European Community's Seventh Framework Programme (FP7 N° 241500), the Cancéropôle Ile de France (Emergence 2019), the Fondation Française pour la Recherche contre le Myélome et les Gammapathies (FFRMG) and the Sanofi iAwards program 2020.

Institutional Review Board Statement: Not applicable.

Informed Consent Statement: Not applicable.

Data Availability Statement: No new data were created or analyzed in this study.

Acknowledgments: We thank Serge Mignani for his valuable input in the section relating to the structure and chemistry of mycolactones.

Conflicts of Interest: The authors declare no conflict of interest.

References

1. Demangel, C.; Stinear, T.P.; Cole, S.T. Buruli ulcer: Reductive evolution enhances pathogenicity of *Mycobacterium ulcerans*. *Nat. Rev. MicroBiol.* **2009**, *7*, 50–60. [CrossRef] [PubMed]
2. George, K.M.; Chatterjee, D.; Gunawardana, G.; Welty, D.; Hayman, J.; Lee, R.; Small, P.L. Mycolactone: A polyketide toxin from *Mycobacterium ulcerans* required for virulence. *Science* **1999**, *283*, 854–857. [CrossRef]
3. Demangel, C. Immunity against *Mycobacterium ulcerans*: The subversive role of mycolactone. *Immunol. Rev.* **2021**, *301*, 209–221. [CrossRef] [PubMed]
4. Rifflet, A.; Demangel, C.; Guenin-Macé, L. Mycolactone Purification from *M. ulcerans* Cultures and HPLC-Based Approaches for Mycolactone Quantification in Biological Samples. In *Mycobacterium ulcerans*; Pluschke, G., Röltgen, K., Eds.; Methods in Molecular Biology; Springer: New York, NY, USA, 2022; Volume 2387, pp. 117–130. ISBN 978-1-07-161778-6.
5. Coutanceau, E.; Decalf, J.; Martino, A.; Babon, A.; Winter, N.; Cole, S.T.; Albert, M.L.; Demangel, C. Selective suppression of dendritic cell functions by *Mycobacterium ulcerans* toxin mycolactone. *J. Exp. Med.* **2007**, *204*, 1395–1403. [CrossRef] [PubMed]
6. Simmonds, R.E.; Lali, F.V.; Smallie, T.; Small, P.L.; Foxwell, B.M. Mycolactone inhibits monocyte cytokine production by a posttranscriptional mechanism. *J. Immunol.* **2009**, *182*, 2194–2202. [CrossRef] [PubMed]
7. Boulkroun, S.; Guenin-Mace, L.; Thoulouze, M.I.; Monot, M.; Merckx, A.; Langsley, G.; Bismuth, G.; Di Bartolo, V.; Demangel, C. Mycolactone suppresses T cell responsiveness by altering both early signaling and posttranslational events. *J. Immunol.* **2010**, *184*, 1436–1444. [CrossRef]
8. Guenin-Mace, L.; Baron, L.; Chany, A.C.; Tresse, C.; Saint-Auret, S.; Jonsson, F.; Le Chevalier, F.; Bruhns, P.; Bismuth, G.; Hidalgo-Lucas, S.; et al. Shaping mycolactone for therapeutic use against inflammatory disorders. *Sci. Transl. Med.* **2015**, *7*, 289ra85. [CrossRef]
9. Pahlevan, A.A.; Wright, D.J.; Andrews, C.; George, K.M.; Small, P.L.; Foxwell, B.M. The inhibitory action of *Mycobacterium ulcerans* soluble factor on monocyte/T cell cytokine production and NF-kappa B function. *J. Immunol.* **1999**, *163*, 3928–3935. [CrossRef] [PubMed]
10. Guenin-Macé, L.; Carrette, F.; Asperti-Boursin, F.; Le Bon, A.; Caleechurn, L.; Di Bartolo, V.; Fontanet, A.; Bismuth, G.; Demangel, C. Mycolactone impairs T cell homing by suppressing microRNA control of L-selectin expression. *Proc. Natl. Acad. Sci. USA* **2011**, *108*, 12833–12838. [CrossRef]
11. Baron, L.; Paatero, A.O.; Morel, J.-D.; Impens, F.; Guenin-Macé, L.; Saint-Auret, S.; Blanchard, N.; Dillmann, R.; Niang, F.; Pellegrini, S.; et al. Mycolactone subverts immunity by selectively blocking the Sec61 translocon. *J. Exp. Med.* **2016**, *213*, 2885–2896. [CrossRef]
12. Demangel, C.; Britton, W.J. Interaction of dendritic cells with mycobacteria: Where the action starts. *Immunol. Cell Biol.* **2000**, *78*, 318–324. [CrossRef] [PubMed]
13. Grotzke, J.E.; Kozik, P.; Morel, J.-D.; Impens, F.; Pietrosemoli, N.; Cresswell, P.; Amigorena, S.; Demangel, C. Sec61 blockade by mycolactone inhibits antigen cross-presentation independently of endosome-to-cytosol export. *Proc. Natl. Acad. Sci. USA* **2017**, *114*, E5910–E5919. [CrossRef]
14. Bieri, R.; Scherr, N.; Ruf, M.-T.; Dangy, J.-P.; Gersbach, P.; Gehringer, M.; Altmann, K.-H.; Pluschke, G. The Macrolide Toxin Mycolactone Promotes Bim-Dependent Apoptosis in Buruli Ulcer through Inhibition of mTOR. *ACS Chem. Biol.* **2017**, *12*, 1297–1307. [CrossRef]
15. Hall, B.S.; Hill, K.; McKenna, M.; Ogbechi, J.; High, S.; Willis, A.E.; Simmonds, R.E. The pathogenic mechanism of the *Mycobacterium ulcerans* virulence factor, mycolactone, depends on blockade of protein translocation into the ER. *PLoS Pathog.* **2014**, *10*, e1004061. [CrossRef]
16. McKenna, M.; Simmonds, R.E.; High, S. Mechanistic insights into the inhibition of Sec61-dependent co- and post-translational translocation by mycolactone. *J. Cell Sci.* **2016**, *129*, 1404–1415. [CrossRef] [PubMed]
17. Casson, J.; McKenna, M.; Haßdenteufel, S.; Aviram, N.; Zimmerman, R.; High, S. Multiple pathways facilitate the biogenesis of mammalian tail-anchored proteins. *J. Cell Sci.* **2017**, *130*, 3851–3861. [CrossRef]
18. O'Keefe, S.; Zong, G.; Duah, K.B.; Andrews, L.E.; Shi, W.Q.; High, S. An alternative pathway for membrane protein biogenesis at the endoplasmic reticulum. *Commun. Biol.* **2021**, *4*, 1–15. [CrossRef]
19. Mackinnon, A.L.; Paavilainen, V.O.; Sharma, A.; Hegde, R.S.; Taunton, J. An allosteric Sec61 inhibitor traps nascent transmembrane helices at the lateral gate. *Elife* **2014**, *3*, e01483. [CrossRef]
20. Garrison, J.L.; Kunkel, E.J.; Hegde, R.S.; Taunton, J. A substrate-specific inhibitor of protein translocation into the endoplasmic reticulum. *Nature* **2005**, *436*, 285–289. [CrossRef]

21. Morel, J.-D.; Paatero, A.O.; Wei, J.; Yewdell, J.W.; Guenin-Macé, L.; Van Haver, D.; Impens, F.; Pietrosemoli, N.; Paavilainen, V.O.; Demangel, C. Proteomics Reveals Scope of Mycolactone-mediated Sec61 Blockade and Distinctive Stress Signature. *Mol. Cell. Proteom.* **2018**, *17*, 1750–1765. [CrossRef]
22. Itskanov, S.; Wang, L.; Junne, T.; Sherriff, R.; Xiao, L.; Blanchard, N.; Shi, W.Q.; Forsyth, C.; Hoepfner, D.; Spiess, M.; et al. A Common Mechanism of Sec61 Translocon Inhibition by Small Molecules. *Nat. Chem. Biol.* **2022**, 1–9. [CrossRef] [PubMed]
23. Hong, H.; Demangel, C.; Pidot, S.J.; Leadlay, P.F.; Stinear, T. Mycolactones: Immunosuppressive and cytotoxic polyketides produced by aquatic mycobacteria. *Nat. Prod. Rep.* **2008**, *25*, 447. [CrossRef] [PubMed]
24. Gehringer, M.; Altmann, K.-H. The chemistry and biology of mycolactones. *Beilstein J. Org. Chem.* **2017**, *13*, 1596–1660. [CrossRef] [PubMed]
25. Guenin-Macé, L.; Ruf, M.-T.; Pluschke, G.; Demangel, C. Mycolactone: More than Just a Cytotoxin. In *Buruli Ulcer*; Pluschke, G., Röltgen, K., Eds.; Springer International Publishing: Cham, Switzerland, 2019; pp. 117–134. ISBN 978-3-030-11113-7.
26. Hong, H.; Stinear, T.; Porter, J.; Demangel, C.; Leadlay, P.F. A Novel Mycolactone Toxin Obtained by Biosynthetic Engineering. *ChemBioChem* **2007**, *8*, 2043–2047. [CrossRef]
27. Song, F.; Fidanze, S.; Benowitz, A.B.; Kishi, Y. Total Synthesis of the Mycolactones. *Org. Lett.* **2002**, *4*, 647–650. [CrossRef]
28. Yin, N.; Wang, G.; Qian, M.; Negishi, E. Stereoselective Synthesis of the Side Chains of Mycolactones A and B Featuring Stepwise Double Substitutions of 1,1-Dibromo-1-alkenes. *Angew. Chem. Int. Ed.* **2006**, *45*, 2916–2920. [CrossRef]
29. Feyen, F.; Jantsch, A.; Altmann, K.-H. Synthetic Studies on Mycolactones: Synthesis of the Mycolactone Core Structure through Ring-Closing Olefin Metathesis. *Synlett* **2007**, *2007*, 0415–0418. [CrossRef]
30. Chany, A.-C.; Casarotto, V.; Schmitt, M.; Tarnus, C.; Guenin-Macé, L.; Demangel, C.; Mirguet, O.; Eustache, J.; Blanchard, N. A Diverted Total Synthesis of Mycolactone Analogues: An Insight into Buruli Ulcer Toxins. *Chem. Eur. J.* **2011**, *17*, 14413–14419. [CrossRef]
31. Scherr, N.; Gersbach, P.; Dangy, J.-P.; Bomio, C.; Li, J.; Altmann, K.-H.; Pluschke, G. Structure-Activity Relationship Studies on the Macrolide Exotoxin Mycolactone of *Mycobacterium ulcerans*. *PLoS Negl. Trop. Dis.* **2013**, *7*, e2143. [CrossRef]
32. Gersbach, P.; Jantsch, A.; Feyen, F.; Scherr, N.; Dangy, J.-P.; Pluschke, G.; Altmann, K.-H. A Ring-Closing Metathesis (RCM)-Based Approach to Mycolactones A/B. *Chem. Eur. J.* **2011**, *17*, 13017–13031. [CrossRef]
33. Flynn, J.L.; Chan, J. Immunology of tuberculosis. *Annu. Rev. Immunol.* **2001**, *19*, 93–129. [CrossRef] [PubMed]
34. Demangel, C.; High, S. Sec61 blockade by mycolactone: A central mechanism in Buruli ulcer disease. *Biol. Cell* **2018**, *110*, 237–248. [CrossRef] [PubMed]
35. Kumar, S.; Baizer, L.; Callander, N.S.; Giralt, S.A.; Hillengass, J.; Freidlin, B.; Hoering, A.; Richardson, P.G.; Schwartz, E.I.; Reiman, A.; et al. Gaps and opportunities in the treatment of relapsed-refractory multiple myeloma: Consensus recommendations of the NCI Multiple Myeloma Steering Committee. *Blood Cancer J.* **2022**, *12*, 98. [CrossRef] [PubMed]
36. Domenger, A.; Choisy, C.; Baron, L.; Mayau, V.; Perthame, E.; Deriano, L.; Arnulf, B.; Bories, J.; Dadaglio, G.; Demangel, C. The Sec61 translocon is a therapeutic vulnerability in multiple myeloma. *EMBO Mol. Med.* **2022**, *14*, e14740. [CrossRef]
37. Domenger, A.; Ricci, D.; Mayau, V.; Majlessi, L.; Marcireau, C.; Dadaglio, G.; Demangel, C. Sec61 blockade therapy overrides resistance to proteasome inhibitors and immunomodulatory drugs in multiple myeloma. *Front. Oncol.* **2023**, *13*, 1110916. [CrossRef]
38. Shah, N.; Chari, A.; Scott, E.; Mezzi, K.; Usmani, S.Z. B-cell maturation antigen (BCMA) in multiple myeloma: Rationale for targeting and current therapeutic approaches. *Leukemia* **2020**, *34*, 985–1005. [CrossRef]
39. Huang, K.-C.; Chen, Z.; Jiang, Y.; Akare, S.; Kolber-Simonds, D.; Condon, K.; Agoulnik, S.; Tendyke, K.; Shen, Y.; Wu, K.-M.; et al. Apratoxin A Shows Novel Pancreas-Targeting Activity through the Binding of Sec 61. *Mol. Cancer Ther.* **2016**, *15*, 1208–1216. [CrossRef]
40. Cai, W.; Chen, Q.-Y.; Dang, L.H.; Luesch, H. Apratoxin S10, a Dual Inhibitor of Angiogenesis and Cancer Cell Growth To Treat Highly Vascularized Tumors. *ACS Med. Chem. Lett.* **2017**, *8*, 1007–1012. [CrossRef]
41. Mishra, R.; Patel, H.; Alanazi, S.; Yuan, L.; Garrett, J.T. HER3 signaling and targeted therapy in cancer. *Oncol. Rev.* **2018**, *12*, 355. [CrossRef]
42. Ruiz-Saenz, A.; Sandhu, M.; Carrasco, Y.; Maglathlin, R.L.; Taunton, J.; Moasser, M.M. Targeting HER3 by interfering with its Sec61-mediated cotranslational insertion into the endoplasmic reticulum. *Oncogene* **2015**, *34*, 5288–5294. [CrossRef]
43. Heaton, N.S.; Moshkina, N.; Fenouil, R.; Gardner, T.J.; Aguirre, S.; Shah, P.S.; Zhao, N.; Manganaro, L.; Hultquist, J.F.; Noel, J.; et al. Targeting Viral Proteostasis Limits Influenza Virus, HIV, and Dengue Virus Infection. *Immunity* **2016**, *44*, 46–58. [CrossRef] [PubMed]
44. Monel, B.; Compton, A.A.; Bruel, T.; Amraoui, S.; Burlaud-Gaillard, J.; Roy, N.; Guivel-Benhassine, F.; Porrot, F.; Génin, P.; Meertens, L.; et al. Zika virus induces massive cytoplasmic vacuolization and paraptosis-like death in infected cells. *EMBO J.* **2017**, *36*, 1653–1668. [CrossRef] [PubMed]
45. Shah, P.S.; Link, N.; Jang, G.M.; Sharp, P.P.; Zhu, T.; Swaney, D.L.; Johnson, J.R.; Von Dollen, J.; Ramage, H.R.; Satkamp, L.; et al. Comparative Flavivirus-Host Protein Interaction Mapping Reveals Mechanisms of Dengue and Zika Virus Pathogenesis. *Cell* **2018**, *175*, 1931–1945.e18. [CrossRef] [PubMed]
46. Gordon, D.E.; Jang, G.M.; Bouhaddou, M.; Xu, J.; Obernier, K.; White, K.M.; O'Meara, M.J.; Rezelj, V.V.; Guo, J.Z.; Swaney, D.L.; et al. A SARS-CoV-2 protein interaction map reveals targets for drug repurposing. *Nature* **2020**, *583*, 459–468. [CrossRef] [PubMed]

47. Cao, S.; Guza, R.C.; Wisse, J.H.; Miller, J.S.; Evans, R.; Kingston, D.G.I. Ipomoeassins A−E, Cytotoxic Macrocyclic Glycoresins from the Leaves of *Ipomoea s quamosa* from the Suriname Rainforest. *J. Nat. Prod.* **2005**, *68*, 487–492. [CrossRef]
48. Zong, G.; Hu, Z.; O'Keefe, S.; Tranter, D.; Iannotti, M.J.; Baron, L.; Hall, B.; Corfield, K.; Paatero, A.O.; Henderson, M.J.; et al. Ipomoeassin F Binds Sec61α to Inhibit Protein Translocation. *J. Am. Chem. Soc.* **2019**, *141*, 8450–8461. [CrossRef] [PubMed]
49. O'Keefe, S.; Roboti, P.; Duah, K.B.; Zong, G.; Schneider, H.; Shi, W.Q.; High, S. Ipomoeassin-F inhibits the in vitro biogenesis of the SARS-CoV-2 spike protein and its host cell membrane receptor. *J. Cell Sci.* **2021**, *134*, jcs257758. [CrossRef]
50. Klein, W.; Rutz, C.; Eckhard, J.; Provinciael, B.; Specker, E.; Neuenschwander, M.; Kleinau, G.; Scheerer, P.; von Kries, J.-P.; Nazaré, M.; et al. Use of a sequential high throughput screening assay to identify novel inhibitors of the eukaryotic SRP-Sec61 targeting/translocation pathway. *PLoS ONE* **2018**, *13*, e0208641. [CrossRef]
51. Rehan, S.; Tranter, D.; Sharp, P.P.; Lowe, E.; Anderl, J.L.; Muchamuel, T.; Abrishami, V.; Kuivanen, S.; Wenzell, N.; Jennings, A.; et al. Signal peptide mimicry primes Sec61 for client-selective inhibition. *Nat. Chem. Biol.* **2022**, 1–9. [CrossRef]

Review

Cyanotoxins and Other Bioactive Compounds from the Pasteur Cultures of Cyanobacteria (PCC)

Muriel Gugger *, Anne Boullié and Thierry Laurent

Institut Pasteur, Université Paris Cité, Collection of Cyanobacteria, 75015 Paris, France
* Correspondence: muriel.gugger@pasteur.fr

Abstract: In tribute to the bicentenary of the birth of Louis Pasteur, this report focuses on cyanotoxins, other natural products and bioactive compounds of cyanobacteria, a phylum of Gram-negative bacteria capable of carrying out oxygenic photosynthesis. These microbes have contributed to changes in the geochemistry and the biology of Earth as we know it today. Furthermore, some bloom-forming cyanobacterial species are also well known for their capacity to produce cyanotoxins. This phylum is preserved in live cultures of pure, monoclonal strains in the Pasteur Cultures of Cyanobacteria (PCC) collection. The collection has been used to classify organisms within the Cyanobacteria of the bacterial kingdom and to investigate several characteristics of these bacteria, such as their ultrastructure, gas vacuoles and complementary chromatic adaptation. Thanks to the ease of obtaining genetic and further genomic sequences, the diversity of the PCC strains has made it possible to reveal some main cyanotoxins and to highlight several genetic loci dedicated to completely unknown natural products. It is the multidisciplinary collaboration of microbiologists, biochemists and chemists and the use of the pure strains of this collection that has allowed the study of several biosynthetic pathways from genetic origins to the structures of natural products and, eventually, their bioactivity.

Keywords: cyanobacteria; the Pasteur Cultures of Cyanobacteria; cyanotoxins; natural products

Key Contribution: The Pasteur Cultures of Cyanobacteria collection has helped scientists for decades to reveal the potential of cyanotoxins and natural products.

Citation: Gugger, M.; Boullié, A.; Laurent, T. Cyanotoxins and Other Bioactive Compounds from the Pasteur Cultures of Cyanobacteria (PCC). *Toxins* **2023**, *15*, 388. https://doi.org/10.3390/toxins15060388

Received: 12 May 2023
Revised: 5 June 2023
Accepted: 7 June 2023
Published: 9 June 2023

1. Introduction

Cyanobacteria represent a monophyletic lineage of Gram-negative oxygenic photosynthetic bacteria [1]. This phylum has inhabited the Earth for 2.8 billion years, contributing to changes in the geochemistry and the biology of the globe [2]. They are ubiquitous and found in diverse ecological niches, from aquatic ecosystems such as lakes, rivers and oceans to deserts, Polar Regions, caves and even in symbiosis with other organisms, such as fungi, to form lichens, for example. Cyanobacteria are also the ancestors of chloroplasts that are found in most plants and algae, while recurrent examples of various cyanobacterial morphotypes are found associated with the leaves or in the roots of plants. Although cyanobacteria are famous for the oxygenation of the Earth [2,3], they are also sadly famous for the toxic blooms they can massively develop in marine and fresh waters all around the globe [4]. The field of cyanotoxins and other cyanobacterial bioactive compounds has greatly extended over the last century, notably through chemical analysis and compound structure elucidation, and due to the availability of enough material for toxicological studies. Despite 1100 cyanobacterial natural products (NPs) discovered by these approaches [5], we still know little about them. In this review, we will first introduce the diversity of the cyanobacterial phylum; secondly, we will present the repository dedicated to this phylum at the Institut Pasteur, and finally, we will discuss how to exploit the cyanobacterial biobank to reveal novel compounds, corresponding pathways, new enzymes and even intriguing chemistry. We will end this overview with examples of bioactivity studies on cyanotoxins and NPs.

2. The Phylum Cyanobacteria, Diversity in Terms of Morphology and Genome

The phylum of Cyanobacteria containing all bacteria capable of performing oxygenic photosynthesis [1] presents a wide breadth of habitats and, thus, of ecology, physiology and morphology. Their most simple morphotypes can be unicellular, as single cells, e.g., *Prochlorococcus*, Chisolm et al. (1992), that colonize the oceans [6], or as colonies of single cells embedded in mucilage, e.g., *Microcystis*, Kützing ex Lemmermann (1907), that form toxic blooms in lakes [7]. The colonies of unicellular cyanobacteria can be tightly organized in one layer of arranged cells of *Merispomedia*, Meyen (1839), or can appear as cell aggregations of various sizes surrounded by mucilaginous envelopes such as the tiny *Chroococus*, Nägeli (1849), and the large *Gloeocapsa*, Kützing (1843). More complex morphotypes of single cells in a colony are the *Pleurocapsa*, Thuret in Hauck (1885), and other baeocytous cyanobacteria that proliferate in desert environments. These cyanobacteria are extremely resistant to desiccation; containing baeocytes that will revive a novel colony when the environmental conditions are more favorable for the growth of these bacteria. Filamentous cyanobacteria also have a more organized cell division which is perpendicular to the growing axis of the filament, such as the solitary trichomes found in toxic freshwater blooms such as *Planktothrix*, Gaget et al. (2015) or the several trichomes embedded in a common sheath such as *Hydrocoleum*, Kützing ex Gomont (1892). The filamentous heterocystous cyanobacteria, exemplified by *Byssus flos-aquae*, Linnaeus (1753), are even more complex; this taxa is currently invalid and has been replaced by *Aphanizomenon flos-aquae*, Ralfs ex Bornet and Flahault (1886). This morphotype bears differentiated cells such as akinetes, used to revive a novel filament after harsh conditions, and heterocytes, used to fix atmospheric nitrogen. *Aphanizomenon* is also a toxic bloom-forming cyanobacterium found in lakes and rivers. Finally, the most complex morphotypes, and often larger than other cyanobacterial morphologies, are the filamentous cyanobacteria with differentiated cells (heterocytes and akinetes) and true ramifications, surrounded or not by mucilage and sheaths. The best ubiquitous representative of this cyanobacterial morphotype is *Fischerella* (Bornet and Flahault) Gomont, 1895. This summarizes at a glance the morphological diversity of Cyanobacteria that can be seen with the naked eye or using a light microscope. A glimpse into the morphologies encountered in the cyanobacterial phylum can be found in Figure 1; a more detailed view of the morphological diversity, through a botanical approach with camera lucida drawings, can be found in the book series Süsswasserflora von Mitteleuropa on Cyanoprokaryota [8–10]. Moreover, a detailed view of the morphological diversity through a bacteriological approach with photography of the various morphologies and electron microscopy photographs of the ultrastructural arrangement within the cyanobacterial cells can be found in the Bergey's Manual of Systematic Bacteriology [11].

In response to the morphological diversity within the phylum of Cyanobacteria, their genomes are also extremely diverse. The publicly available genomes present sizes ranging from 1.5 to 15 Mb and a GC content from 30 to 68 %. The smallest genome size corresponds to the picocyanobacterial genus of *Prochlorococcus*, while the largest ones were found in filamentous cyanobacteria with differentiated cells and ramifications. On the contrary, the highest and lowest GC contents were reported in the genus *Synechococcus*, Nägeli (1849), while the filamentous cyanobacteria with differentiated cells and with or without branching presented a GC content centered on approximately 42%. The comparison of the morphologies and ultrastructural data with genomic sequence data did not show any solid groupings corresponding to the different morphotypes; however, a correlation between the ultrastructures and genes coding for cellular inclusions was identified [12].

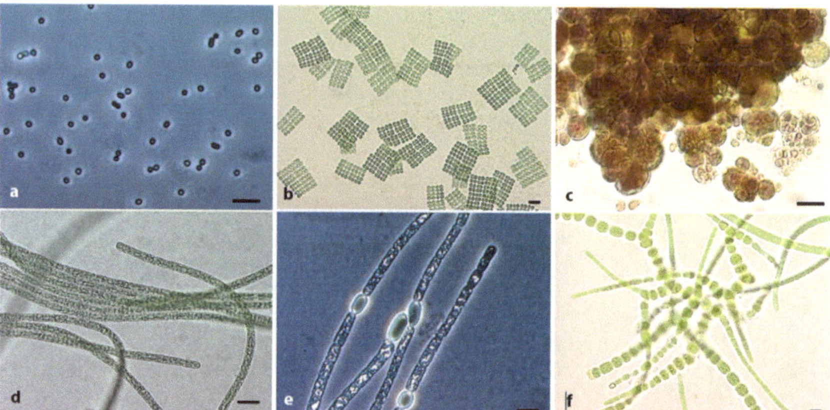

Figure 1. Morphologies of cyanobacteria. Unicellular morphotypes are on the top row: from tiny single cells of *Cyanobium* sp. PCC 7001 in dark field (**a**), to flat square colonies of *Merismopedia*, from an environmental sample (**b**) to the baeocytous former *Stanieria* sp. PCC 7301 (**c**). Filamentous morphotypes on the bottom row: from *Planktothrix agardhii* PCC 10110 with bright gas vacuoles (**d**) to *Aphanizomenon flos aquae* PCC 7905 in dark field with gas vacuoles and barrel-shaped heterocyte (**e**), to *Fischerella* sp. in the late stage of purification (**f**). All scale bars represent 5 μm.

3. Collection of Cyanobacteria and the Pasteur Cultures of Cyanobacteria (PCC)

Several collections of cyanobacteria have been constituted in universities and institutions around the world, as cyanobacterial blooms occur everywhere on Earth. In addition, a few major collections often conserved these bacteria along with algal isolates, protists and other bacteria (for example, NIES-MCC in Japan; UTEX and ATCC in the USA; GCC, NFMC and NCCS in India; CCAP in Scotland, CCALA in Czech Republic; BCCM-ULC in Belgium; NIVA in Norway; UHCC in Finland; SAG in Germany; PCC, PMC, TCC and RCC in France). The collection of cyanobacteria currently present at the Institut Pasteur arrived from the University of California, Berkeley, with Pr. Roger Stanier (1916–1982), who worked on the classification of the so-called blue-green algae at that time and included them among the bacterial classification as the Cyanobacteria [13]. Pr. Stanier led a research team named Microbial Physiology Unit, which focused on various subjects such as pigments, photosynthesis, gas vacuoles, fatty acids and taxonomy, upon of these 150 pure cyanobacterial strains. From the same laboratory, the isolation of a purple cyanobacterium without thylakoids, internal membranes on which the light-harvesting complexes sit, led to the description of the first representative of the basal clade of the phylum Cyanobacteria, *Gloeobacter violaceaus*, Rippka, Waterbury and Cohen-Bazire (1974), strain PCC 7421 [14]. After a short period (1982–1988) during which Dr. Germaine Cohen-Bazire Stanier (1920–2001) was heading this team and working on the ultrastructure of cyanobacteria [15,16], Pr. Nicole Tandeau de Marsac (1944–2020) was named head of a novel team called the Unit of Cyanobacteria from 1988–2009. This team maintained the cyanobacterial collection of the Institut Pasteur, which had greatly expanded since its arrival in France through the effort of several researchers and visitors. She and collaborators worked extensively on a few model PCC strains to thoroughly investigate the phycobilisome and the photoregulation, the cellular differentiation of hormogonium, the gas vesicle genes, and the phosphorylation of the signal transducer PII [17–21]. Pr. Tandeau de Marsac also revealed the complementary chromatic adaptation [22–24]. Towards the end of her research career, she majorly focused on the toxic cyanobacterial isolate *Microcystis aeruginosa*, PCC 7806 [7,25–27]. In July 2009, Dr. Muriel Gugger was appointed head of the Collection of Cyanobacteria with the mission to maintain, distribute and valorize the Pasteur Cultures of Cyanobacteria collection. The PCC collection is also used as a reference for this phylum and to exemplify the bacterial classification in Bergey's Manual [11].

Since Stanier's time, several collaborators and visitors cooperated with Rosemarie Rippka (1944–today), the curator of the PCC collection until June 2009, to carry out research [28–30] and, in the meantime, bred up to 800 axenic living PCC strains isolated from all over the world (Figure 2). From the 150 strains that originally arrived with Stanier, 102 are still maintained at the PCC, with the emblematic strain *Synechococcus elongatus*, PCC 6301, isolated in 1956 from the USA, rendered monoclonal and axenic by 1963 and maintained alive since then as well as entered in cryopreservation stocks several times as all other PCC strains [31]. In 1973 and 1989, strong efforts generated about 100 cyanobacterial monoclonal purified isolates incorporated in the collection, but over time, several isolates were lost or stopped growing. The purification of cyanobacteria is a long delicate process that can take several years, as exemplified by the two years required to obtain an axenic monoclonal culture of *Prochlorococcus marinus*, PCC 9511 [32], which vanished in 2013 due to the replacement of the incubator in which it was maintained. The last strains incorporated in the collection also belong to the genus *Synechococcus*; two come from India and are closely related to the PCC 6301 strain but represent another species, and one comes from Singapore described as a transformable strain with a fast-growing capacity under high light [31,33–35]. Since 2006, the PCC collection was integrated into the Biological Resource Center of the Institut Pasteur (CRBIP) along with the Collection of Bacteria of the Institut Pasteur (CIP) and the fungal collection (former CMIP) and joined later under the same umbrella of biobank and quality management by the Collection Nationale de Cultures de Microorganismes (CNCM) and the Integrated Collections for Adaptive Research in Biomedicine (ICAReB-Biobank). Today, the PCC collection contains around 800 monoclonal axenic cultures, 600 maintained alive in liquid or solid cultures, and all of them cryopreserved. The maintenance of the collection takes up to 4500 transfers, along with 4500 purity tests per year. While the PCC collection represents a breadth of the cyanobacterial phylum, the study of their genomes of diverse PCC strains revealed a plethora of gene clusters for NPs [3,36].

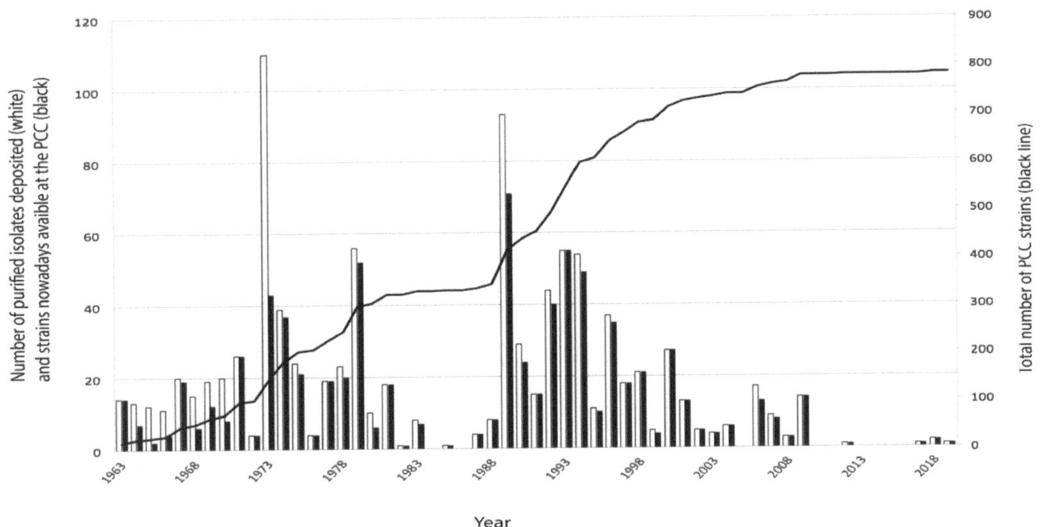

Figure 2. Number of axenic monoclonal cyanobacterial strains of the PCC obtained over the years 1963–2019 (in white), cultures still preserved today (2023, in black), and total number of PCC strains (black line and second vertical axis).

4. Natural Products of Cyanobacteria—From Toxins to Novel Compounds

In the 1870s, a report documented the occurrence of cattle poisonings from Australian lakes [37]. Over 100 years later, dog deaths became a vivid subject connected to cyanotoxins from cyanobacterial developments on the surface or benthos of water ecosystems in North America and Europe, as well as in South Africa and New Zealand [38–44]. More recently, a human fatality during a renal dialysis treatment in Brazil and a fatality in bald eagles in the southeastern USA were demonstrated to be due to freshwater cyanobacterial occurrences [45,46]. Several wildlife intoxications or deaths have also been associated with cyanotoxins, for example, with flamingos and most probably with elephants in Africa and with fishes in Canada [47–49]. The above examples clearly demonstrate that for the last 145 years, the recurrent problem of cyanobacterial blooms in fresh and marine water bodies has presented a threat to animals and humans. Moreover, the cyanotoxins released by bloom-forming cyanobacteria into water bodies used for drinking water create a global public health issue. The World Health Organization has developed guidance values for the most common cyanotoxins present in recreational water and drinking water [50]. In the USA, the simultaneous incident of animal and human disease around one lake with blooms of toxic cyanobacteria has been documented and has called for the development of a proactive relationship between the healthcare system and veterinarians to protect human health [51,52]. In the field of cyanotoxins, the monoclonal and axenic PCC strains have been useful since 1988. For example, the *Microcystis aeruginosa* PCC 7820 was used to monitor the hepatotoxic effect of microcystin-LR on mice and rat liver damages and pulmonary emboli leading to acute toxicities and death [53]. In particular, two strains have made it possible to discover the genetic bases of three cyanotoxins: *Microcystis aeruginosa* PCC 7806 for the discovery of the microcystin biosynthetic gene cluster, and *Kamptonema* sp. PCC 6506 to reveal anatoxin-a and related compounds as well as cylindrospermopsins [54–56]. Indeed, a recent bibliographic survey (27th April 2023, in Pubmed: Cyanobacteria AND PCC AND Toxin) reveals about 150 publications dedicated to cyanotoxin discovery, cyanotoxin effects (larvicidal, antifungal, . . .), toxin-antitoxin and treatment against toxic cyanobacteria based on PCC strains.

Cyanobacteria do not only produce toxins; they contain a real diversity in terms of natural substances. In order to have greater visibility of this diversity, we undertook the sequencing of the genomes of the living axenic strains preserved in the PCC collection. First, we obtained the genomes of 54 PCC strains selected on their morphology, their ecology and their physiology to better represent the breadth of cyanobacterial phylum. Combined with genomic data from 72 publicly available strains, a phylogenetic tree based on 31 genes conserved in Bacteria was constructed to reflect the evolution and relationship between these organisms. In parallel, a systematic analysis of gene clusters coding for NPs and toxins was undertaken on this dataset with the search for ribosomally synthesized and post-translationally modified peptides (RiPPs), non-ribosomal peptides synthetases (NRPS) and polyketide synthases (PKSs) [3]. This analysis showed the presence of these three classes of metabolites, with *cis* and *trans* AT-PKS, peptides from both ribosomal and non-ribosomal pathways, and terpenes throughout the phylum represented by these 126 genomes. In line with this work, a focus on the NRPS and PKS in the same dataset made it possible to highlight 452 biosynthetic gene clusters (BGCs) distributed into 286 cluster families based on the similarity of the modules of the NRPS and PKS and the length of the regions compared [36]. Interestingly, one-fourth of the BGCs were hybrids of NRPS and PKS, mostly distributed in the late branches of the cyanobacterial phylogeny, whereas the early branches contained mainly PKS. In addition, 80% of these cluster families did not correspond to any known NPs, giving an idea of the scope of the investigative work, which could be devolved by chemists and biochemists.

Based on the above findings and the availability of cyanobacterial strains potentially producing unknown NPs in the PCC collection, collaborations with various chemist and biochemist colleagues helped reveal the NPs derived from these unattributed BGCs. For the BGCs smaller than 15 kb, a cloning strategy performed in a heterologous host was more straightforward to discover compounds of interest and to find them further in the cyanobacterium of interest, such as the schizokinen-like siderophore of *Leptolyngbya* sp., PCC 7376 [57]. However, most of the NRPS-PKS BGCs were larger than 20 kb, and they could not be produced with this approach. For the investigations of these large BGCs, we sometimes cultivated liters of biomass of pure cyanobacterium, potentially producing the desired metabolites in our laboratory conditions. Through numerous collaborations, we have discovered more than 20 novel NPs and/or their BGC, thanks to the cyanobacteria in the PCC collection and the collaborators, as well as other researchers with PCC strains (Table 1).

Finally, in the investigation course of these NPs by a multidisciplinary consortium of chemists, biochemists and microbiologists, we were fascinated to find novel enzymes and unknown chemistry from the metabolisms of the cyanobacterial strains of the PCC, examples of which are discussed hereafter. First, the search for the proteusins of the cyanobacteria and the radical *S*-adenosyl methionine epimerase (rSAM) of this pathway revealed regioselective *D*-configured amino acids into peptidic NPs. For this, an ingenious methodology was developed to understand how these rSAMs work to irreversibly insert multiple *D*-amino acids in the peptides from the strains *Kamptonema* sp. PCC 6506, *Pleurocapsa* sp. PCC 7319, and *Anabaena variabilis* ATCC 29413 [58,59]. Secondly, the same talented chemists working with the genomic data of *Pleurocapsa* sp. PCC 7319 discovered non-canonical protein splicing via a post-translational excision of a tyramine equivalent, leading to an α-keto-β-amino amide [60]. Third, in the RiPPs family of chemically diverse cyanobactins, the BGC is highly conserved and thus, the genes coding the enzymes of these pathways are named consistently from A to F. Nevertheless, several F enzymes enlarged the prenyltransferase family, with two of these enzymes in the muscoride pathway acting differently by introducing a regioselective prenylation on the amino acid termini of the produced linear cyanobactin in *Nostoc* sp., PCC 7906 [61], or with another prenyltransferase which places a forward-prenyl on a threonine residue in the cyclic cyanobactin tolypamide of *Tolypothrix* sp. PCC 7601 [62]. Finally, the strain *Lyngbya* sp. PCC 8601 and two other cyanobacterial strains were used to uncover a suite of post-translational modifying bacterial enzymes that install single or multiple strained cyclophane macrocycles. As the cyclophane natural products are found in fungi, plants and bacteria; this enzyme family is widely distributed in nature [63].

5. From Molecules to Bioactivity

From the discovery of new molecules or their genetic heritage to the knowledge of their activity, the path is not straightforward. Initially, the origin of a toxic event was sought before finding the cyanotoxin responsible for it. Several reviews described the potential of cyanobacterial compounds to become drug products such as anticancer agents and antibiotics, for example [64–67]. Chemists have found more than 1100 NPs from cyanobacteria, but less than 20% of them are associated with a biosynthesis pathway [5]. In these chemical studies, it has often been attempted to find bioactivity associated with it by means of conventional screening techniques, in particular, to find a therapeutic potential. However, almost half of the compounds were not tested or detected in any bioactivity assay. More recently, a review on the bioactivity of NPs of cyanobacteria found 1630 unique molecules, classified into 260 families of metabolites [68]. Importantly, most of the compounds were not tested for their bioactivities. This is because bioactivity testing requires different knowledge and specialty than that needed to discover the compounds or the genetic data that encodes them. In addition, as the characterisation of a compound will often require the extraction and collection of grams of it from the biomass of the producing organism, it is often for lack of material that the bioactivity test cannot be carried out or confirmed.

Table 1. Natural products (NPs) and their biosynthetic gene clusters (BGC) discovered based on the monoclonal and axenic strains of the collection PCC.

NPs or Their BGCs	Type of NPs	Gene Cluster	Strain	Reference
Microcyclamide	RiPPs, cyanobactins	*mca*, 13 kb	*Microcystis aeruginosa* PCC 7806	[26]
Viridisamide Aeruginosamide B and C3	RiPPs, 1st linear cyanobactins	Variation from *pat* *	*Oscillatoria viridis* PCC 7112 *Microcystis aeruginosa* PCC 9432	[69]
Muscoride	RiPPs, linear cyanobactin	*mus*, 12.7 kb	12 *Nostoc* strains, in which 6 PCC strains	[61]
Tolypamide	RiPPs, cyanobactins	*tol*, 10.4 kb	*Tolypothrix* sp. PCC 7601	[62]
Geosmin	RiPPs, sesquiterpene	*geosmin synthase*	*Nostoc* spp. PCC 7310 and PCC 7120, *Kamptonema* sp. PCC 6506	[70,71]
Merosterol A and B + isomer	RiPPs, meroterpene	*mst*, 29 kb	*Scytonema* sp. PCC 10023	[72]
Cyclophanes	RiPPs, cyclopetide alkaloids	*lsc*, 2.6 kb	*Lyngbya* sp. PCC 8106 and various other strains	[63]
Landornamides	RiPPs, proteusins	*osp*, 12 kb	*Kamptonema* sp. PCC 6506 and 6 other *Kamptonema* PCC strains	[73]
Kamptornamide	RiPPs, 1st ribosomal fatty-acylated lipo-petides, selidamides	*ksp*, 6.3 kb	*Kamptonema* sp. PCC 6506 and *Nostoc punctiforme* PCC 73102	[74]
Microguanidine amide Aeruginoguanidine BGC	NRPS	*agd*, 34 kb	11 *Microcystis*, in which 7 PCC strains	[75]
Hassallidin E	NRPS	*has*, 48 kb	*Planktothrix serta* PCC 8927	[76]
Cyanopeptolin	NRPS	*oci* *, 31.5 kb	*Microcystis aeruginosa* PCC 7806 *Scytonema hofmanni* PCC 7110	[77,78]
Scyptolins	NRPS		*Scytonema hofmanni* PCC 7110	[79]
Anatoxin-a and dihydroanatoxin	PKS, alkaloid	*ana*, 20 kb	*Kamptonema* sp. PCC 6506, *Cylindrospermum* sp. PCC 7417 and 13 other PCC strains	[36,55]
Cylindrospermopsins	PKS, alkaloid	*cyr*, 42 kb	*Kamptonema* sp. PCC 6506	[56]
Luminaolide B Tolytoxin BGC Tolytoxin, Scytophycin	Trans AT-PKS	*lum*, 99 kb *tto*, >100 kb *tto*, 92.8 kb	*Planktothrix paucivesiculata* PCC 9631 *Scytonema* sp. PCC 10023 *Planktothrix* sp. PCC 11201	[80,81]
Leptolyngbyalide	Trans AT-PKS	*lept*, 96.7 kb	*Leptolyngbya* sp. PCC 7375	[82]
Alkene and alkanes	PKS, Hydrocarbon	*ols* 10 kb	*Synechococcus* sp. PCC 7002 16 unicellular PCC strains	[83–85]
Heterocyte glycolipids	PKS, polyunsaturated fatty acid	CF1 *	*Nostoc* sp. PCC 7120, 18 Nostocales strains and *Microchaete* sp. PCC 7126	[86–88]
Microcystin	NRPS-PKS Hybrid	*mcy*, 55 kb	*Microcystis aeruginosa* PCC 7806, *Fischerella* sp. PCC 9339	[36,54]
Nostopeptolide	NRPS-PKS Hybrid	*pks2*, 62.7 kb	*Nostoc puctiforme* ATCC29133/PCC 73102	[89,90]
Aranazoles	NRPS-PKS Hybrid	*arz*, 43 kb	*Fischerella* sp. PCC 9339	[91]

* Indicates: *pat* from *Prochoron*, a cyanobacterial symbiont of a tunicate [92]; *oci* from diverse cyanobacteria [93]; and CF1 correspond to the cluster in *Nostoc* sp. PCC 7120 [36,88].

The study of biological activity can be directed to help human interest, for a pharmacological application or a biotechnological development. Tolytoxin has the potential to be used for therapeutic application, as this cyanobacterial macrolide inhibits actin filament dynamics and was proposed as a potential anti-cancer drug [94–96]. However, this molecule was also proved to be extremely toxic at nM concentrations and to induce cell death [94,97]. As we recently described tolytoxin producers and the tolytoxin biosynthetic gene cluster from PCC pure strains [80], we found several other PCC strains capable of producing this molecule [81,98]. We revisited the activity of tolytoxin in human cells from neuronal and epithelial origins with the goal of reducing disease transmission by tunneling nanotubes mainly constituting of actin [98]. In this experiment, with the two strains we used, we noticed a strong decrease in tolytoxin dose needed (3 and 15 nM) to obtain an inhibitory effect without setting off the toxic side effects previously observed in cells. During the isolation of pure tolytoxin from *Planktothrix serta* PCC 8926 and *Scytonema* sp. PCC 10023, we noticed that a fatty acid was extracted along with the tolytoxin almost until the end of our extraction procedure. To perform this experiment and extract enough purified tolytoxin, we worked in a chemist's laboratory, thanks to J. Piel's team and R. Ueoka in particular, without whom we would have missed this trace of contamination. For the activity of tolytoxin, we collaborated with specialists in the cells and nanotubes to be tested, thanks to C Zurzolo's team and A. Dilsizoglu-Senol in particular [98]. In addition, the odorous volatile compound geosmin is also of concern for human health with its biological activity. Using the geosmin-producing strain *Kamptonema* sp. PCC 6506, and the non-producing strain *Leptolyngbya* sp. PCC 8913 isolated from a lake colonized by mosquitoes in the south of France [99], we collaborated with researchers working on insect olfaction at Lund University (Sweden) to reveal the attraction of the *Aedes aegypti* mosquito for this compound and check if this odour is an indicator of egg-laying site for this insect [70].

The study of biological activity can also be investigated to learn more about its need for producing cyanobacteria. Two clear examples of useful NPs for producing cyanobacteria have been reported. The first one is the production of heterocyte glycolipids by a PKS cluster [88]. When a vegetative cell differentiates into a future heterocyte, the nascent cell only becomes active in fixing nitrogen when a layer of glycolipids covers it. This layer prevents oxygen, produced by adjacent cells, to enter into the heterocyte and to inhibit the nitrogenase. This mechanism must be tightly regulated and programmed because heterocytes can only survive 3 to 4 days before being replaced by another heterocyte resulting from the differentiation of a vegetative cell. The second example is from the product of a PKS cluster, the nostopeptolide that governs the cellular differentiation of a symbiotic *Nostoc* [89,90]. This example also indicates a clear scheduling of the production of the natural product at the time needed by the producing organism. Finally, the last example of the need of the producer to produce certain molecules at a certain time can be seen through the study of the toxic-bloom forming *Microcystis aeruginosa* PCC 7806. With an ingenious culture system consisting of two compartments separated by a filter, which allows compounds but not cells to pass, Briand and his collaborators demonstrated that this strain produced certain NPs in the medium only by sharing it with another strain of *Microcystis* such as PCC 9432 or *Planktothrix agardhii* PCC 7805 [100,101]. Thus, *Microcystis aeruginosa* PCC 7806 produces these NPs when it detects another cyanobacterium in its environment. This allelopathic research deserves further study because it illustrates a very controlled production of these metabolites beyond the genetic potential of the producer. It can also lead to the discovery of so-called cryptic NPs.

6. Concluding Remarks

In conclusion, the Pasteur Cultures of Cyanobacteria collection has been a living biobank and a research tool since its creation at the Institut Pasteur. The status of these strains has allowed research in the global scientific community. Within the framework of cyanobacterial toxins, the strains of the PCC collection led to the discovery of cyanotoxins and NPs. While several cyanotoxins were already structurally known, the pure strains maintained at the Institut Pasteur for 52 years have made it possible to discover the genetic origins of these toxins, intriguing enzymes, even unprecedented chemistry, and certain bioactivities. The genomics of the strains of the PCC collection highlights the wide diversity of NPs that we are still fully investigating.

Author Contributions: Writing—original draft preparation, M.G., A.B. and T.L.; writing—review and editing, M.G. All authors have read and agreed to the published version of the manuscript.

Funding: This research received no external funding.

Institutional Review Board Statement: Not applicable.

Informed Consent Statement: Not applicable.

Data Availability Statement: Not applicable.

Acknowledgments: We thank Aniket Saraf, Collection of Cyanobacteria at the Institut Pasteur, for helpful comments. We also are grateful to our genomic and chemical collaborators for helping to valorize the collection. We acknowledge former researchers from the laboratories Unité de Physiologie Microbienne and Unité des Cyanobacteries who have preserved and developed the PCC. The Institut Pasteur funds the PCC collection and our research activities.

Conflicts of Interest: The authors declare no conflict of interest.

References

1. Garcia-Pichel, F.; Zehr, J.P.; Bhattacharya, D.; Pakrasi, H.B. What's in a name? The case of cyanobacteria. *J. Phycol.* **2020**, *56*, 1–5. [CrossRef] [PubMed]
2. Schirrmeister, B.E.; Gugger, M.; Donoghue, P.C. Cyanobacteria and the Great Oxidation Event: Evidence from genes and fossils. *Palaeontology* **2015**, *58*, 769–785. [CrossRef] [PubMed]
3. Shih, P.M.; Wu, D.; Latifi, A.; Axen, S.D.; Fewer, D.P.; Talla, E.; Calteau, A.; Cai, F.; Tandeau de Marsac, N.; Rippka, R.; et al. Improving the coverage of the cyanobacterial phylum using diversity-driven genome sequencing. *Proc. Natl. Acad. Sci. USA* **2013**, *110*, 1053–1058. [CrossRef]
4. Huisman, J.; Codd, G.A.; Paerl, H.W.; Ibelings, B.W.; Verspagen, J.M.H.; Visser, P.M. Cyanobacterial blooms. *Nat. Rev. Microbiol.* **2018**, *16*, 471–483. [CrossRef]
5. Dittmann, E.; Gugger, M.; Sivonen, K.; Fewer, D.P. Natural product biosynthetic diversity and comparative genomics of the Cyanobacteria. *Trends Microbiol.* **2015**, *23*, 642–652. [CrossRef]
6. Partensky, F.; Hess, W.R.; Vaulot, D. *Prochlorococcus*, a marine photosynthetic prokaryote of global significance. *Microbiol. Mol. Biol. Rev.* **1999**, *63*, 106–127. [CrossRef]
7. Humbert, J.F.; Barbe, V.; Latifi, A.; Gugger, M.; Calteau, A.; Coursin, T.; Lajus, A.; Castelli, V.; Oztas, S.; Samson, G.; et al. A tribute to disorder in the genome of the bloom-forming freshwater cyanobacterium *Microcystis aeruginosa. PLoS ONE* **2013**, *8*, e70747. [CrossRef] [PubMed]
8. Komarek, J.; Anagnostidis, K. *Cyanoprokaryota I (Chroococcales)*; Pascher, A., Ettl, H., Gärtner, G., Heyning, H., Mollenhauer, D., Eds.; Gustav Fischer: Jena, Germany, 1998; Volume 19/1.
9. Komarek, J.; Anagnostidis, K. *Cyanoprokaryota II (Oscillatoriales)*; Büdel, B., Gärtner, G., Krietniz, L., Schagerl, M., Eds.; Spektrum Akademischer Verlag: Jena, Germany, 2005; Volume 19/2.
10. Komarek, J. *Cyanoprokaryota III (Heterocytous Genera)*; Springer Spektrum: Berlin/Heidelberg, Germany, 2013; Volume 19/3.
11. Castenholz, R.W.; Phylum, B.X. Cyanobacteria, Oxygenic Photosynthetic Bacteria. In *Bergey's Manual of Systematic Bacteriology*, 2nd ed.; Boone, D.R., Castenholz, R.W., Garrity, G.M., Eds.; Volume 1—The Archaea and the Deeply Branching and Phototrophic Bacteria; Springer: New York, NY, USA, 2001; pp. 473–487.
12. Gonzalez-Esquer, C.R.; Smarda, J.; Rippka, R.; Axen, S.D.; Guglielmi, G.; Gugger, M.; Kerfeld, C.A. Cyanobacterial ultrastructure in light of genomic sequence data. *Photosynth. Res.* **2016**, *129*, 147–157. [CrossRef]
13. Stanier, R.Y.; Sistrom, W.R.; Hansen, T.A.; Whitton, B.A.; Castenholz, R.W.; Pfennig, N.; Gorlenko, V.N.; Kondratieva, E.N.; Eimhjellen, K.E.; Whittenbury, R.; et al. Proposal to place the nomemclature of the cyanobacteria (blue-green algae) under the rules of the International Code of Nomenclature of Bacteria. *Int. J. Syst. Bacteriol.* **1978**, *28*, 335–336. [CrossRef]
14. Rippka, R.; Waterbury, J.; Cohen-Bazire, G. A cyanobacterium which lacks thylakoids. *Arch. Microbiol.* **1974**, *100*, 419–436. [CrossRef]
15. Guglielmi, G.; Cohen-Bazire, G. Structure et distribution des pores et des perforations de l'enveloppe de peptidoglycane chez quelques cyanobactéries. *Protistologica* **1982**, *18*, 151–165.
16. Guglielmi, G.; Cohen-Bazire, G. Etude taxonomique d'un genre de cyanobactérie oscillatoriacée: Le genre *Pseudanabaena* Lauterborn. I. Etude ultrastructurale. *Protistologica* **1984**, *20*, 377–391.

17. Campbell, D.; Houmard, J.; Tandeau de Marsac, N. Electron transport regulates cellular differentiation in the filamentous cyanobacterium *Calothrix*. *Plant Cell* **1993**, *5*, 451–463. [CrossRef] [PubMed]
18. Damerval, T.; Castets, A.M.; Guglielmi, G.; Houmard, J.; Tandeau de Marsac, N. Occurrence and distribution of gas vesicle genes among cyanobacteria. *J. Bacteriol.* **1989**, *171*, 1445–1452. [CrossRef]
19. Forchhammer, K.; Tandeau de Marsac, N. Phosphorylation of the PII protein (glnB gene product) in the cyanobacterium *Synechococcus* sp. strain PCC 7942: Analysis of in vitro kinase activity. *J. Bacteriol.* **1995**, *177*, 5812–5817. [CrossRef]
20. Herdman, M.; Coursin, T.; Rippka, R.; Houmard, J.; Tandeau de Marsac, N. A new appraisal of the prokaryotic origin of eukaryotic phytochromes. *J. Mol. Evol.* **2000**, *51*, 205–213. [CrossRef]
21. Tandeau de Marsac, N.; Mazel, D.; Damerval, T.; Guglielmi, G.; Capuano, V.; Houmard, J. Photoregulation of gene expression in the filamentous cyanobacterium *Calothrix* sp. PCC 7601: Light-harvesting complexes and cell differentiation. *Photosynth. Res.* **1988**, *18*, 99–132. [CrossRef]
22. Mazel, D.; Houmard, J.; Tandeau de Marsac, N. A multigene family in *Calothrix* sp. PCC 7601 encodes phycocyanin, the major component of the cyanobacterial light-harvesting antenna. *Mol. Gen. Genet.* **1988**, *211*, 296–304. [CrossRef]
23. Damerval, T.; Guglielmi, G.; Houmard, J.; Tandeau de Marsac, N. Hormogonium differentiation in the cyanobacterium Calothrix: A photoregulated developmental process. *Plant Cell* **1991**, *3*, 191–201. [CrossRef]
24. Liotenberg, S.; Campbell, D.; Rippka, R.; Houmard, J.; Tandeau de Marsac, N. Effect of the nitrogen source on phycobiliprotein synthesis and cell reserves in a chromatically adapting filamentous cyanobacterium. *Microbiology* **1996**, *142*, 611–622. [CrossRef]
25. Frangeul, L.; Quillardet, P.; Castets, A.M.; Humbert, J.F.; Matthijs, H.C.; Cortez, D.; Tolonen, A.; Zhang, C.C.; Gribaldo, S.; Kehr, J.C.; et al. Highly plastic genome of *Microcystis aeruginosa* PCC 7806, a ubiquitous toxic freshwater cyanobacterium. *BMC Genom.* **2008**, *9*, 274. [CrossRef] [PubMed]
26. Ziemert, N.; Ishida, K.; Quillardet, P.; Bouchier, C.; Hertweck, C.; Tandeau de Marsac, N.; Dittmann, E. Microcyclamide biosynthesis in two strains of *Microcystis aeruginosa*: From structure to genes and *vice versa*. *Appl. Environ. Microbiol.* **2008**, *74*, 1791–1797. [CrossRef] [PubMed]
27. Zilliges, Y.; Kehr, J.C.; Mikkat, S.; Bouchier, C.; Tandeau de Marsac, N.; Börner, T.; Dittmann, E. An extracellular glycoprotein is implicated in cell-cell contacts in the toxic cyanobacterium *Microcystis aeruginosa* PCC 7806. *J. Bacteriol.* **2008**, *190*, 2871–2879. [CrossRef] [PubMed]
28. Rippka, R.; Deruelles, J.; Waterbury, J.B.; Herdman, M.; Stanier, R.Y. Generic assignments, strain histories and properties of pure cultures of cyanobacteria. *J. Gen. Microbiol.* **1979**, *111*, 1–61. [CrossRef]
29. Rippka, R.; Herdman, M. Pasteur Culture Collection of Cyanobacterial Strains in Axenic Culture. In *Catalogue & Taxonomic Handbook*; Institut Pasteur: Paris, France, 1992; pp. 1–103.
30. Palinska, K.A.; Jahns, T.; Rippka, R.; Tandeau de Marsac, N. *Prochlorococcus marinus* strain PCC 9511, a picoplanktonic cyanobacterium, synthesizes the smallest urease. *Microbiology* **2000**, *146*, 3099–3107. [CrossRef]
31. Adomako, M.; Ernst, D.; Simkovsky, R.; Chao, Y.Y.; Wang, J.; Fang, M.; Bouchier, C.; Lopez-Igual, R.; Mazel, D.; Gugger, M.; et al. Comparative genomics of *Synechococcus elongatus* explains the phenotypic diversity of the strains. *mBio* **2022**, *13*, e0086222. [CrossRef]
32. Rippka, R.; Coursin, T.; Hess, W.; Lichtlé, C.; Scanlan, D.J.; Palinska, K.A.; Iteman, I.; Partensky, F.; Houmard, J.; Herdman, M. *Prochlorococcus marinus* Chisholm et al. 1992, subsp. nov. pastoris, strain PCC 9511, the first axenic chlorophyll a2/b2-containing cyanobacterium (Oxyphotobacteria). *Int. J. Syst. Evol. Microbiol* **2000**, *50*, 1833–1847. [CrossRef]
33. Włodarczyk, A.; Selão, T.T.; Norling, B.; Nixon, P.J. Newly discovered *Synechococcus* sp. PCC 11901 is a robust cyanobacterial strain for high biomass production. *Commun. Biol.* **2020**, *3*, 215. [CrossRef]
34. Jaiswal, D.; Sengupta, A.; Sohoni, S.; Sengupta, S.; Phadnavis, A.G.; Pakrasi, H.B.; Wangikar, P.P. Genome features and biochemical characteristics of a robust, fast growing and naturally transformable cyanobacterium *Synechococcus elongatus* PCC 11801 isolated from India. *Sci. Rep.* **2018**, *8*, 16632. [CrossRef]
35. Jaiswal, D.; Sengupta, A.; Sengupta, S.; Madhu, S.; Pakrasi, H.B.; Wangikar, P.P. A novel cyanobacterium *Synechococcus elongatus* PCC 11802 has distinct genomic and metabolomic characteristics compared to its neighbor PCC 11801. *Sci. Rep.* **2020**, *10*, 191. [CrossRef]
36. Calteau, A.; Fewer, D.P.; Latifi, A.; Coursin, T.; Laurent, T.; Jokela, J.; Kerfeld, C.A.; Sivonen, K.; Piel, J.; Gugger, M. Phylum-wide comparative genomics unravel the diversity of secondary metabolism in Cyanobacteria. *BMC Genom.* **2014**, *15*, 977. [CrossRef] [PubMed]
37. Francis, G. Poisonous Australian Lake. *Nature* **1878**, *18*, 11–12. [CrossRef]
38. Edwards, C.; Beattie, K.A.; Scrimgeour, C.M.; Codd, G.A. Identification of anatoxin-A in benthic cyanobacteria (blue-green algae) and in associated dog poisonings at Loch Insh, Scotland. *Toxicon* **1992**, *30*, 1165–1175. [CrossRef] [PubMed]
39. Gugger, M.; Lenoir, S.; Berger, C.; Ledreux, A.; Druart, J.C.; Humbert, J.F.; Guette, C.; Bernard, C. First report in a river in France of the benthic cyanobacterium *Phormidium favosum* producing anatoxin-a associated with dog neurotoxicosis. *Toxicon* **2005**, *45*, 919–928. [CrossRef]
40. Harding, W.R.; Rowe, N.; Wessels, J.C.; Beattie, K.A.; Codd, G.A. Death of a dog attributed to the cyanobacterial (blue-green algal) hepatotoxin nodularin in South Africa. *J. S. Afr. Vet. Assoc.* **1995**, *66*, 256–259.
41. Kelly, L.T.; Bouma-Gregson, K.; Puddick, J.; Fadness, R.; Ryan, K.G.; Davis, T.W.; Wood, S.A. Multiple cyanotoxin congeners produced by sub-dominant cyanobacterial taxa in riverine cyanobacterial and algal mats. *PLoS ONE* **2019**, *14*, e0220422. [CrossRef]

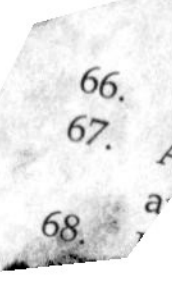

42. Mahmood, N.A.; Carmichael, W.W.; Pfahler, D. Anticholinesterase poisonings in dogs from a cyanobacterial (blue-green algae) bloom dominated by *Anabaena flos-aquae*. *Am. J. Vet. Res.* **1988**, *49*, 500–503.

43. Turner, A.D.; Turner, F.R.I.; White, M.; Hartnell, D.; Crompton, C.G.; Bates, N.; Egginton, J.; Branscombe, L.; Lewis, A.M.; Maskrey, B.H. Confirmation Using Triple Quadrupole and High-Resolution Mass Spectrometry of a Fatal Canine Neurotoxicosis following Exposure to Anatoxins at an Inland Reservoir. *Toxins* **2022**, *14*, 804. [CrossRef]

44. Wood, S.A.; Heath, M.W.; Holland, P.T.; Munday, R.; McGregor, G.B.; Ryan, K.G. Identification of a benthic microcystin-producing filamentous cyanobacterium (Oscillatoriales) associated with a dog poisoning in New Zealand. *Toxicon* **2010**, *55*, 897–903. [CrossRef]

45. Azevedo, S.M.; Carmichael, W.W.; Jochimsen, E.M.; Rinehart, K.L.; Lau, S.; Shaw, G.R.; Eaglesham, G.K. Human intoxication by microcystins during renal dialysis treatment in Caruaru-Brazil. *Toxicology* **2002**, *181–182*, 441–446. [CrossRef]

46. Breinlinger, S.; Phillips, T.J.; Haram, B.N.; Mareš, J.; Martínez Yerena, J.A.; Hrouzek, P.; Sobotka, R.; Henderson, W.M.; Schmieder, P.; Williams, S.M.; et al. Hunting the eagle killer: A cyanobacterial neurotoxin causes vacuolar myelinopathy. *Science* **2021**, *371*, eaax9050. [CrossRef] [PubMed]

47. Krienitz, L.; Ballot, A.; Kotut, K.; Wiegand, C.; Pütz, S.; Metcalf, J.S.; Codd, G.A.; Pflugmacher, S. Contribution of hot spring cyanobacteria to the mysterious deaths of Lesser Flamingos at Lake Bogoria, Kenya. *FEMS Microbiol. Ecol.* **2003**, *43*, 141–148. [CrossRef] [PubMed]

48. Skafi, M.; Duy, S.V.; Munoz, G.; Dinh, Q.T.; Simon, D.F.; Juneau, P.; Sauvé, S. Occurrence of microcystins, anabaenopeptins and other cyanotoxins in sh from a freshwater wildlife reserve impacted by harmful cyanobacterial blooms. *Toxicon* **2021**, *194*, 44–52. [CrossRef] [PubMed]

49. Veerman, J.; Kumar, A.; Mishra, D.R. Exceptional landscape-wide cyanobacteria bloom in Okavango Delta, Botswana in 2020 coincided with a mass elephant die-off event. *Harmful Algae* **2022**, *111*, 102145. [CrossRef]

50. Chorus, I.; Welker, M. *Toxic Cyanobacteria in Water: A Guide to Their Pubic Health Consequences, Monitoring and Management*, 2nd ed.; Welker, I.C.M., Ed.; CRC Press: Boca Raton, FL, USA; on behalf of the World Health Organization: Geneva, Switzerland, 2021.

51. Trevino-Garrison, I.; DeMent, J.; Ahmed, F.S.; Haines-Lieber, P.; Langer, T.; Ménager, H.; Neff, J.; van der Merwe, D.; Carney, E. Human illnesses and animal deaths associated with freshwater harmful algal blooms-Kansas. *Toxins* **2015**, *7*, 353–366. [CrossRef]

52. Hilborn, E.D.; Beasley, V.R. One health and cyanobacteria in freshwater systems: Animal illnesses and deaths are sentinel events for human health risks. *Toxins* **2015**, *7*, 1374–1395. [CrossRef]

53. Theiss, W.C.; Carmichael, W.W.; Wyman, J.; Bruner, R. Blood pressure and hepatocellular effects of the cyclic heptapeptide toxin produced by the freshwater cyanobacterium (blue-green alga) *Microcystis aeruginosa* strain PCC-7820. *Toxicon* **1988**, *26*, 603–613. [CrossRef]

54. Tillett, D.; Dittmann, E.; Erhard, M.; von Döhren, H.; Börner, T.; Neilan, B.A. Structural organization of microcystin biosynthesis in *Microcystis aeruginosa* PCC7806: An integrated peptide-polyketide synthetase system. *Chem. Biol.* **2000**, *7*, 753–764. [CrossRef]

55. Méjean, A.; Mann, S.; Maldiney, T.; Vassiliadis, G.; Lequin, O.; Ploux, O. Evidence that biosynthesis of the neurotoxic alkaloids anatoxin-a and homoanatoxin-a in the cyanobacterium *Oscillatoria* PCC 6506 occurs on a modular polyketide synthase initiated by L-proline. *J. Am. Chem. Soc.* **2009**, *131*, 7512–7513. [CrossRef]

56. Mazmouz, R.; Chapuis-Hugon, F.; Mann, S.; Pichon, V.; Méjean, A.; Ploux, O. Biosynthesis of cylindrospermopsin and 7-epicylindrospermopsin in *Oscillatoria* sp. strain PCC 6506: Identification of the cyr gene cluster and toxin analysis. *Appl. Environ. Microbiol.* **2010**, *76*, 4943–4949. [CrossRef]

57. Wang, S.; Pearson, L.A.; Mazmouz, R.; Liu, T.; Neilan, B.A. Heterologous expression and biochemical analysis reveal a schizokinen-based siderophore pathway in *Leptolyngbya* (Cyanobacteria). *Appl. Environ. Microbiol.* **2022**, *88*, e0237321. [CrossRef] [PubMed]

58. Morinaka, B.I.; Verest, M.; Freeman, M.F.; Gugger, M.; Piel, J. An orthogonal D2 O-based induction system that provides insights into d-amino acid pattern formation by radical S-adenosylmethionine peptide epimerases. *Angew. Chem. Int. Ed.* **2017**, *56*, 762–766. [CrossRef]

59. Morinaka, B.I.; Vagstad, A.L.; Helf, M.J.; Gugger, M.; Kegler, C.; Freeman, M.F.; Bode, H.B.; Piel, J. Radical S-adenosyl methionine epimerases: Regioselective introduction of diverse D-amino acid patterns into peptide natural products. *Angew. Chem. Int. Ed.* **2014**, *53*, 8503–8507. [CrossRef] [PubMed]

60. Morinaka, B.I.; Lakis, E.; Verest, M.; Helf, M.J.; Scalvenzi, T.; Vagstad, A.L.; Sims, J.; Sunagawa, S.; Gugger, M.; Piel, J. Natural noncanonical protein splicing yields products with diverse β-amino acid residues. *Science* **2018**, *359*, 779–782. [CrossRef] [PubMed]

61. Mattila, A.; Andsten, R.M.; Jumppanen, M.; Assante, M.; Jokela, J.; Wahlsten, M.; Mikula, K.M.; Sigindere, C.; Kwak, D.H.; Gugger, M.; et al. Biosynthesis of the bis-prenylated alkaloids muscoride A and B. *ACS Chem. Biol.* **2019**, *14*, 2683–2690. [CrossRef] [PubMed]

62. Purushothaman, M.; Sarkar, S.; Morita, M.; Gugger, M.; Schmidt, E.W.; Morinaka, B.I. Genome-mining-based discovery of the cyclic peptide tolypamide and tolF, a Ser/Thr forward O-prenyltransferase. *Angew. Chem. Int. Ed.* **2021**, *60*, 8460–8465. [CrossRef]

63. Nguyen, T.Q.N.; Tooh, Y.W.; Sugiyama, R.; Nguyen, T.P.D.; Purushothaman, M.; Leow, L.C.; Hanif, K.; Yong, R.H.S.; Agatha, I.; Winnerdy, F.R.; et al. Post-translational formation of strained cyclophanes in bacteria. *Nat. Chem.* **2020**, *12*, 1042–1053. [CrossRef]

64. Chlipala, G.E.; Mo, S.; Orjala, J. Chemodiversity in freshwater and terrestrial cyanobacteria—A source for drug discovery. *Curr. Drug Targets* **2011**, *12*, 1654–1673. [CrossRef]

65. Gerwick, W.H.; Fenner, A.M. Drug discovery from marine microbes. *Microb. Ecol.* **2013**, *65*, 800–806.

..., J.; Menakha, M. Pharmaceutical applications of cyanobacteria-A review. *J. Acute Med.* **2015**, *5*, 15–23. [CrossRef]

...n, S.A.; Akhter, N.; Auckloo, B.N.; Khan, I.; Lu, Y.; Wang, K.; Wu, B.; Guo, Y.W. Structural diversity, biological properties and applications of natural products from cyanobacteria. A review. *Mar. Drugs* **2017**, *15*, 354. [CrossRef]

Demay, J.; Bernard, C.; Reinhardt, A.; Marie, B. Natural products from Cyanobacteria: Focus on beneficial activities. *Mar. Drugs* **2019**, *17*, 320. [CrossRef] [PubMed]

69. Leikoski, N.; Liu, L.; Jokela, J.; Wahlsten, M.; Gugger, M.; Calteau, A.; Permi, P.; Kerfeld, C.A.; Sivonen, K.; Fewer, D.P. Genome mining expands the chemical diversity of the cyanobactin family to include highly modified linear peptides. *Chem. Biol.* **2013**, *20*, 1033–1043. [CrossRef] [PubMed]

70. Melo, N.; Wolff, G.H.; Costa-da-Silva, A.L.; Arribas, R.; Triana, M.F.; Gugger, M.; Riffell, J.A.; DeGennaro, M.; Stensmyr, M.C. Geosmin attracts *Aedes aegypti* mosquitoes to oviposition sites. *Curr. Biol.* **2020**, *30*, 127–134. [CrossRef] [PubMed]

71. Agger, S.A.; Lopez-Gallego, F.; Hoye, T.R.; Schmidt-Dannert, C. Identification of sesquiterpene synthases from *Nostoc punctiforme* PCC 73102 and *Nostoc* sp. strain PCC 7120. *J. Bacteriol.* **2008**, *190*, 6084–6096. [CrossRef] [PubMed]

72. Moosmann, P.; Ueoka, R.; Grauso, L.; Mangoni, A.; Morinaka, B.I.; Gugger, M.; Piel, J. Cyanobacterial ent-sterol-like natural products from a deviated ubiquinone pathway. *Angew. Chem. Int. Ed.* **2017**, *56*, 4987–4990. [CrossRef] [PubMed]

73. Bösch, N.M.; Borsa, M.; Greczmiel, U.; Morinaka, B.I.; Gugger, M.; Oxenius, A.; Vagstad, A.L.; Piel, J. Landornamides: Antiviral ornithine-containing ribosomal peptides discovered through genome mining. *Angew. Chem. Int. Ed.* **2020**, *59*, 11763–11768. [CrossRef]

74. Hubrich, F.; Bösch, N.M.; Chepkirui, C.; Morinaka, B.I.; Rust, M.; Gugger, M.; Robinson, S.L.; Vagstad, A.L.; Piel, J. Ribosomally derived lipopeptides containing distinct fatty acyl moieties. *Proc. Natl. Acad. Sci. USA* **2022**, *119*, e2113120119. [CrossRef]

75. Pancrace, C.; Ishida, K.; Briand, E.; Pichi, D.G.; Weiz, A.R.; Guljamow, A.; Scalvenzi, T.; Sassoon, N.; Hertweck, C.; Dittmann, E.; et al. Unique biosynthetic pathway in bloom-forming cyanobacterial genus *Microcystis* jointly assembles cytotoxic aeruginoguani-dines and microguanidines. *ACS Chem. Biol.* **2019**, *14*, 67–75. [CrossRef]

76. Pancrace, C.; Jokela, J.; Sassoon, N.; Ganneau, C.; Desnos-Ollivier, M.; Wahlsten, M.; Humisto, A.; Calteau, A.; Bay, S.; Fewer, D.P.; et al. Rearranged biosynthetic gene cluster and synthesis of hassallidin E in *Planktothrix serta* PCC 8927. *ACS Chem. Biol.* **2017**, *12*, 1796–1804. [CrossRef]

77. Martin, C.; Oberer, L.; Buschdtt, M.; Weckesser, J. Cyanopeptolins, new depsipeptides from the cyanobacterium *Microcystis* sp. PCC 7806. *J. Antibiot.* **1993**, *46*, 1550–1556. [CrossRef] [PubMed]

78. Matern, U.; Oberer, L.; Erhard, M.; Herdmand, M.; Weckesser, J. Hofmannolin, a cyanopeptolin from *Scytonema hofmanni* PCC 7110. *Phytochemistry* **2003**, *64*, 1061–1067. [CrossRef] [PubMed]

79. Matern, U.; Oberer, L.; Falchetto, R.A.; Erhard, M.; König, W.A.; Herdman, M.; Weckesser, J. Scyptolin A and B, cyclic depsipeptides from axenic cultures of *Scytonema hofmanni* PCC 7110. *Phytochemistry* **2001**, *58*, 1087–1095. [CrossRef] [PubMed]

80. Ueoka, R.; Uria, A.R.; Reiter, S.; Mori, T.; Karbaum, P.; Peters, E.E.; Helfrich, E.J.; Morinaka, B.I.; Gugger, M.; Takeyama, H.; et al. Metabolic and evolutionary origin of actin-binding polyketides from diverse organisms. *Nat. Chem. Biol.* **2015**, *11*, 705–712. [CrossRef]

81. Pancrace, C.; Barny, M.A.; Ueoka, R.; Calteau, A.; Scalvenzi, T.; Pédron, J.; Barbe, V.; Piel, J.; Humbert, J.F.; Gugger, M. Insights into the *Planktothrix* genus: Genomic and metabolic comparison of benthic and planktic strains. *Sci. Rep.* **2017**, *7*, 41181. [CrossRef]

82. Helfrich, E.J.N.; Ueoka, R.; Dolev, A.; Rust, M.; Meoded, R.A.; Bhushan, A.; Califano, G.; Costa, R.; Gugger, M.; Steinbeck, C.; et al. Automated structure prediction of trans-acyltransferase polyketide synthase products. *Nat. Chem. Biol.* **2019**, *15*, 813–821. [CrossRef]

83. Brito, Â.; Vieira, J.; Vieira, C.P.; Zhu, T.; Leao, P.N.; Ramos, V.; Lu, X.; Vasconcelos, V.M.; Gugger, M.; Tamagnini, P. Comparative genomics discloses the uniqueness and the biosynthetic potential of the marine cyanobacterium *Hyella patelloides*. *Front. Microbiol.* **2020**, *11*, 1527. [CrossRef]

84. Coates, R.C.; Podell, S.; Korobeynikov, A.; Lapidus, A.; Pevzner, P.; Sherman, D.H.; Allen, E.E.; Gerwick, L.; Gerwick, W.H. Characterization of cyanobacterial hydrocarbon composition and distribution of biosynthetic pathways. *PLoS ONE* **2014**, *9*, e85140. [CrossRef]

85. Mendez-Perez, D.; Begemann, M.B.; Pfleger, B.F. Modular synthase-encoding gene involved in α-olefin biosynthesis in *Synechococcus* sp. strain PCC 7002. *Appl. Environ. Microbiol.* **2011**, *77*, 4264–4267. [CrossRef]

86. Bauersachs, T.; Gugger, M.; Schwark, L. Heterocyte glycolipid diketones: A novel type of biomarker in the N2-fixing heterocytous cyanobacterium *Microchaete* sp. *Org. Geochem.* **2020**, *141*, 103976. [CrossRef]

87. Bauersachs, T.; Miller, S.R.; Gugger, M.; Mudimu, O.; Friedl, T.; Schwark, L. Heterocyte glycolipids indicate polyphyly of stigonematalean cyanobacteria. *Phytochemistry* **2019**, *166*, 112059. [CrossRef] [PubMed]

88. Fan, Q.; Huang, G.; Lechno-Yossef, S.; Wolk, C.P.; Kaneko, T.; Tabata, S. Clustered genes required for synthesis and deposition of envelope glycolipids in *Anabaena* sp. strain PCC 7120. *Mol. Microbiol.* **2005**, *58*, 227–243. [CrossRef] [PubMed]

89. Liaimer, A.; Jenke-Kodama, H.; Ishida, K.; Hinrichs, K.; Stangeland, J.; Hertweck, C.; Dittmann, E. A polyketide interferes with cellular differentiation in the symbiotic cyanobacterium *Nostoc punctiforme*. *Environ. Microbiol. Rep.* **2011**, *3*, 550–558. [CrossRef]

90. Liaimer, A.; Helfrich, E.J.N.; Hinrichs, K.; Guljamow, A.; Ishida, K.; Hertweck, C.; Dittmann, E. Nostopeptolide plays a governing role during cellular differentitaion of the symbiotic cyanobacterium *Nostoc punctiforme*. *Proc. Natl. Acad. Sci. USA* **2015**, *112*, 1862–1867. [CrossRef]

91. Moosmann, P.; Ueoka, R.; Gugger, M.; Piel, J. Aranazoles: Extensively chlorinated nonribosomal peptide-polyketide hybrids from the cyanobacterium *Fischerella* sp. PCC 9339. *Org. Lett.* **2018**, *20*, 5238–5241. [CrossRef] [PubMed]

92. Schmidt, E.W.; Nelson, J.T.; Rasko, D.A.; Sudek, S.; Eisen, J.A.; Haygood, M.G.; Ravel, J. Patellamide A and C biosynthesis by a microcin-like pathway in *Prochloron didemni*, the cyanobacterial symbiont of *Lissoclinum patella*. *Proc. Natl. Acad. Sci. USA* **2005**, *102*, 7315–7320. [CrossRef]

93. Rounge, T.B.; Rohrlack, T.; Tooming-Klunderud, A.; Kristensen, T.; Jakobsen, K.S. Comparison of cyanopeptolin genes in Planktothrix, Microcystis, and Anabaena strains: Evidence for independent evolution within each genus. *Appl. Environ. Microbiol.* **2007**, *73*, 7322–7330. [CrossRef]

94. Patterson, G.M.L.; Smith, C.D.; Kimura, L.H.; Britton, B.A.; Carmeli, S. Action of tolytoxin on cell morphology, cytoskeletal organization, and actin polymerization. *Cell Motil. Cytoskelet.* **1993**, *24*, 39–48. [CrossRef]

95. Zhang, X.; Minale, L.; Zampella, A.; Smith, C.D. Microfilament depletion and circumvention of multiple drug resistance by sphinxolides. *Cancer Res.* **1997**, *57*, 3751–3758.

96. Smith, C.D.; Carmeli, S.; Moore, R.E.; Patterson, G.M. Scytophycins, novel microfilament-depolymerizing agents which circumvent P-glycoprotein-mediated multidrug resistance. *Cancer Res.* **1993**, *53*, 1343–1347.

97. Patterson, G.M.; Carmeli, S. Biological effects of tolytoxin (6-hydroxy-7-O-methyl-scytophycin b), a potent bioactive metabolite from cyanobacteria. *Arch. Microbiol.* **1992**, *157*, 406–410. [CrossRef] [PubMed]

98. Dilsizoglu Senol, A.; Pepe, A.; Grudina, C.; Sassoon, N.; Ueoka, R.; Bousset, L.; Melki, R.; Piel, J.; Gugger, M.; Zurzolo, C. Effect of tolytoxin on tunneling nanotube formation and function. *Sci. Rep.* **2019**, *9*, 5741. [CrossRef] [PubMed]

99. Thiery, I.; Nicolas, L.; Rippka, R.; Tandeau de Marsac, N. Selection of cyanobacteria isolated from mosquito breeding sites as a potential food source for mosquito larvae. *Appl. Environ. Microbiol.* **1991**, *57*, 1354–1359. [CrossRef] [PubMed]

100. Briand, E.; Reubrecht, S.; Mondeguer, F.; Sibat, M.; Hess, P.; Amzil, Z.; Bormans, M. Chemically mediated interactions between Microcystis and Planktothrix: Impact on their growth, morphology and metabolic profiles. *Environ. Microbiol.* **2019**, *21*, 1552–1566. [CrossRef] [PubMed]

101. Briand, E.; Bormans, M.; Gugger, M.; Dorrestein, P.C.; Gerwick, W.H. Changes in secondary metabolic profiles of *Microcystis aeruginosa* strains in response to intraspecific interactions. *Environ. Microbiol.* **2016**, *18*, 384–400. [CrossRef]

Review

Anaerobes and Toxins, a Tradition of the Institut Pasteur

Michel R. Popoff [1,*] and Sandra Legout [2]

[1] Bacterial Toxins, Institut Pasteur, 75724 Paris, France
[2] Centre de Ressources en Information Scientifique, Institut Pasteur, 75015 Paris, France
* Correspondence: popoff2m@gmail.com

Abstract: Louis Pasteur, one of the eminent pioneers of microbiology, discovered life without oxygen and identified the first anaerobic pathogenic bacterium. Certain bacteria were found to be responsible for specific diseases. Pasteur was mainly interested in the prevention and treatment of infectious diseases with attenuated pathogens. The collaborators of Pasteur investigated the mechanisms of pathogenicity and showed that some bacterial soluble substances, called toxins, induce symptoms and lesions in experimental animals. Anaerobic bacteriology, which requires specific equipment, has emerged as a distinct part of microbiology. The first objectives were the identification and taxonomy of anaerobes. Several anaerobes producing potent toxins were associated with severe diseases. The investigation of toxins including sequencing, mode of action, and enzymatic activity led to a better understanding of toxin-mediated pathogenicity and allowed the development of safe and efficient prevention and treatment (vaccination with anatoxins, specific neutralizing antisera). Moreover, toxins turned out to be powerful tools in exploring cellular mechanisms supporting the concept of cellular microbiology. Pasteurians have made a wide contribution to anaerobic bacteriology and toxinology. The historical steps are summarized in this review.

Keywords: Pasteur; anaerobe; toxin; anatoxin; vaccine

Key Contribution: For the bicentenary of Louis Pasteur's birth, this review retraces the historical aspects of the contribution of Louis Pasteur and Pasteurians in the development of anaerobic bacteriology and toxinology.

Citation: Popoff, M.R.; Legout, S. Anaerobes and Toxins, a Tradition of the Institut Pasteur. *Toxins* 2023, 15, 43. https://doi.org/10.3390/toxins15010043

Received: 8 December 2022
Revised: 29 December 2022
Accepted: 3 January 2023
Published: 5 January 2023

1. Introduction

Louis Pasteur was one of the pioneers of microbiology. He showed the role of bacteria in the fermentation processes and discovered that some bacteria can live without oxygen. Bacteria were found also to be responsible for specific diseases. L. Pasteur described the first anaerobic pathogenic bacterium. His main interest was the prevention and treatment of infectious diseases based on attenuated pathogens. Then, the collaborators of L. Pasteur, and the community of scientists at Institut Pasteur and outside developed anaerobic bacteriology and investigated the mechanisms of bacterial pathogenicity. Soluble factors secreted by certain bacteria were found to be involved in the onset of symptoms and lesions and were termed toxins. This short review summarizes the historical aspects of anaerobic bacteriology and toxinology developed by Pasteurians and collaborators. The data are presented in chronological order.

2. "Poisons" before the Pasteur's Era

Poisons have been known from the most ancient times. Very early on, humans had to deal with dangerous substances produced by certain plants or secreted by some animals. The knowledge of these dangers based on cumulative observations of the fortuitous ingestion of certain plants/mineral compounds and accidental encounters with venomous animals was likely transmitted to individuals of the clan/tribe and then from generation to generation. An initial use of poisons was probably for hunting and fishing, and poisons have also been widely employed for war and criminal purposes. It was speculated that hunters of the Mesolithic era (about 9000–1000 BC) used poisonous weapons. Early archeological evidence is from Ancient Egypt (about 2000 BC). Arrows were found to contain poisonous substances inducing strophanthin- or curare-like symptoms in mice, probably from plant origin [1]. In the Antiquity period, poisons are mentioned in several texts such as those from Homer (The Odyssey and The Iliad, about 800 BC), Hippocrates, Democritus, who investigated the effects of several poisons, and Aristotle, who described arrow poisons [1,2]. The term "toxicon" derives from the Greek word τοξικον formed from the word τοξον meaning bow. Thus, toxicon designated arrow poisons. All drugs including harmful substances or substances used for treating diseases were designated as "pharmakon" or "pharmaka". The Romans used the word "toxicum" for any poison [1,2]. The terms poison and potion derive from the Latin word "potionem" meaning drink, as they designate substances generally absorbed by the oral route. Then, poison was used for the designation of harmful substances and potions for medicines.

The most ancient documents related to poisons and medicines are from the Sumerian (as early as 4000 B.C.E.) and Egyptian (1500 B.C.E.) civilizations [2]. The distinction between poison and medicine is sometimes subtle. Paracelsus, in the early Renaissance period (1493–1541), argued that "the dose makes the poison" [2].

In the 19th century, putrefaction products (pus material, putrid meat, putrid vegetables) were found to be lethal when injected into animals (reviewed in [3]). The origin and nature of the "putrid poisons" (Peter Panum 1856) were a matter of debate. They were supposed to derive from the decomposition of organic material or from a bacterial process. The term "ptomaïnes" (Francesco Selmi 1878) was used to designate cadaveric alkaloids. Ludwig Brieger (1885–1886) further characterized putrescine and cadaverine and showed that bacteria are able to produce specific toxic compounds. He introduced the specific name "toxins" for bacterial toxic substances [4].

3. Anaerobes and Toxins in Pasteur's Era

The major contribution of Louis Pasteur and the German physician Robert Koch is the demonstration that specific bacteria are responsible for specific diseases. First, Casimir J. Davaine isolated and R. Koch identified the "bactéridie charbonneuse", then called *Bacillus anthracis*, as the causative agent of anthrax [5]. At this period, the mechanism of bacterial pathogenicity was obscure. It was questionable whether the whole live bacteria or only a fraction of the bacteria were required to promote the symptoms and lesions of the disease. R. Koch (1883) who isolated the agent of cholera (*Vibrio cholerae*) suggested that the disease was due to a "poison" produced by the bacteria. The preliminary observations of the Italian physician Arnaldo Cantani (1886) showed that the cultures of *V. cholera* injected in dogs were toxic. Independently, Richard Pfeiffer (1892–1894) and Nicolas Gamaleïa (1892) confirmed the production of the cholera toxin by the agent of cholera [3,6]. N. Gamaleïa was a Russian microbiologist who worked in the laboratory of L. Pasteur (1886) and with Elie Metchnikoff. Friedrich Löffler (1884), a collaborator of R. Koch, isolated the causative agent of diphtheria that was identified one year before by the German medical doctor Edwin Klebs. Löffler argued that the tissue damages result from a toxic compound synthesized by the diphtheria bacillus that diffuses from the localized infection site throughout the host [3,7].

In 1888, Emile Roux (collaborator of L. Pasteur from 1878 to 1888 and then-Director of the Institut Pasteur (1904–1933)) and Alexandre Yersin (collaborator of L. Pasteur and E. Roux from 1885–1890) at the Institut Pasteur, Paris, demonstrated the presence of the diphtheria toxin in culture filtrates of the diphtheria bacillus that was able to induce similar lesions to those observed in natural disease and death in experimental animals [8]. The cholera toxin and diphtheria toxin were the two first bacterial toxins that were identified. A. Cantini and R. Pfeiffer distinguished between endotoxins that are linked to the bacterial cells and exotoxins that are released in the culture medium [6].

Although L. Pasteur recognized the existence of bacterial poisons, he was doubtful about the role of toxins in diseases [9]. He was focused on the prevention of infectious diseases by attenuated bacteria that he called vaccination in honor of Edward Jenner who prevented smallpox through the inoculation of cowpox. L. Pasteur discovered that an old culture of the causative agent of fowl cholera (*Pasteurella multocida*) was avirulent and can protect against the inoculation of virulent challenges. This was the first bacterial vaccine. Then, he was interested in the prevention of anthrax which was a severe disease widely spread in cattle and sheep. He tried several protocols to inactivate *B. anthracis*. In 1881, in the famous experiment at Pouilly Le-Fort, L. Pasteur succeeded in demonstrating the efficacy of vaccination against anthrax with an attenuated *B. anthracis* strain and became a great celebrity. However, L. Pasteur was not the first discoverer of the anthrax vaccine. A few months earlier, Henry Toussaint, a French scientist, and William Smith Greenfield, a British veterinarian, showed that attenuated *B. anthracis* can protect against anthrax [5,9,10]. It is noteworthy that the two first vaccines against bacterial infectious diseases were performed with toxigenic bacteria that were attenuated in the production of toxins. However, the mechanism of vaccination remained mysterious. L Pasteur admitted that a bacterial soluble substance might induce the protection, but he was more convinced that attenuated bacteria act by competition with pathogens by depleting the host of essential nutrients required for bacterial viability and growth (discussed in this issue [9]).

During his studies on fermentation (1857–1877), L. Pasteur discovered life without oxygen. First, he identified that sugar fermentation results from an aerobic yeast alcohol fermentation followed by an anaerobic yeast lactic fermentation. Then, in 1861, he found that butyric fermentation was associated with a motile, spore-forming microorganism able to grow in the absence of free oxygen. He called this "infusoire", "butyric vibrio", probably from the *Clostridium butyricum*/*Clostridium beijerinckii*/*Clostridium acetobutylicum* group [9,11]. In 1863, L. Pasteur distinguished aerobic and anaerobic microorganisms, anaerobes being microorganisms growing in the absence of air and for which air is lethal [11]. In 1865, L. Pasteur and his collaborator Jules Joubert identified the first pathogenic anaerobe from a sheep that died of septicemia supposed to be anthrax. In 1877, they succeeded in isolating and cultivating the microorganism in anaerobic conditions and termed it "septic vibrio" [12].

L. Pasteur developed a basic method for culture in anaerobic conditions. A glass balloon containing liquid culture medium was boiled to release all the air, and the glass tapered end was closed with a flame [13,14] (Figure 1). In 1887, E. Roux described several techniques for culturing microorganisms in anaerobiosis [15]. From around 1895, a novel technique without vacuum or air replacement was introduced by Adrien Veillon (1864–1931) when he was a young scientist in the laboratory of Jacques-Joseph Grancher, Director of the "Hôpital des Enfants Malades" (Paris) and collaborator of L. Pasteur. This method consisted of deep agar in long glass tubes (Veillon's tubes). A. Veillon investigated the microorganisms responsible for suppuration. He showed that many anaerobic bacteria are saprophytes in the oral cavity of humans and distinguished between the endogenous and exogenous microflora [16]. In 1900, A. Veillon joined the laboratory of the "Microbie technique" of E. Roux, and one year later, he was a physician at the hospital of Institut Pasteur. During the First World War, he was involved in the treatment and prevention of war wounds, notably gas gangrene due to *C. perfringens*. He amended the fight against gangrene by creating laboratories of bacteriology in military hospitals of surgery [17].

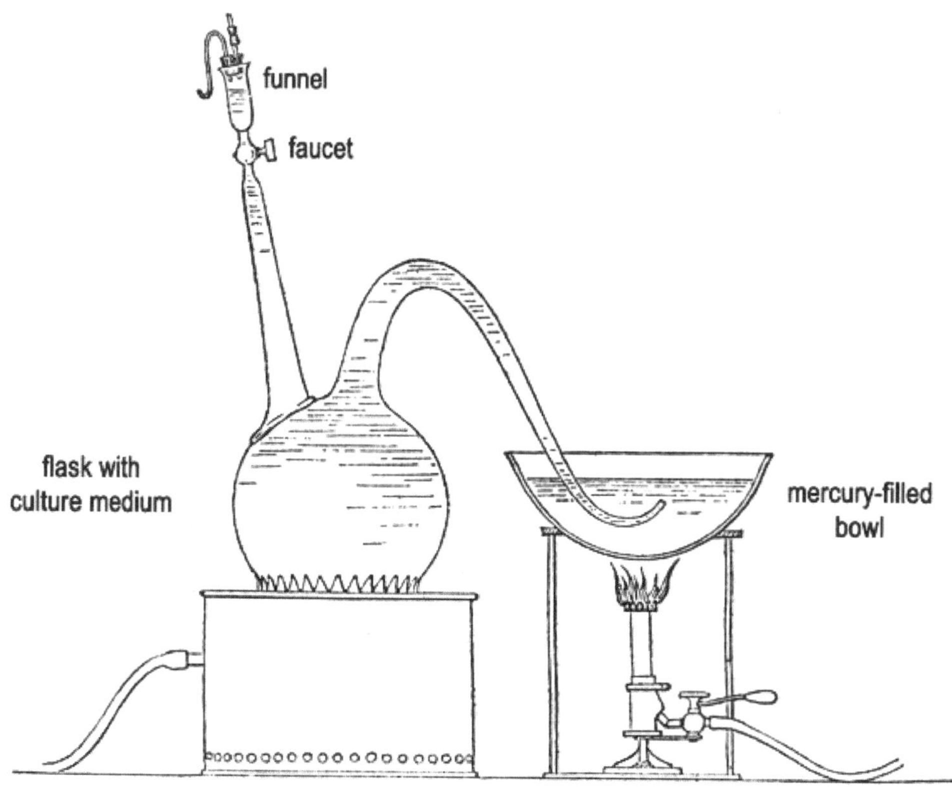

Figure 1. Device developed by L. Pasteur for anaerobic culture [14] (Institut Pasteur/Musée Pasteur). A 6 L flask containing the culture medium is boiled for at least 30 min to eliminate the dissolved oxygen. After cooling at 25–30 °C the curved tube is connected to a mercury-filled bowl. The funnel is flushed with carbon dioxide and simultaneously filled with 10 mL culture inoculum. The flask is inoculated by opening the faucet with precaution to keep a small volume of inoculum in the funnel, avoiding the introduction of air.

4. Period 1900–1940: The "Microbie Technique"

During the period of about 1904–1914, several scientists worked on anaerobes and bacterial toxins in the service of the "Microbie technique" of E. Roux: Jean Binot, Edouard Dujardin-Baumetz, Constatin Levaditi, René Legroux, Auguste Charles Marie, Victor Morax, Maurice Nicolle, Aleaxandre Salimbeni, Adrien Veillon, and Michel Weinberg.

The agent of botulism was identified by Emile van Ermengem in Belgium (*Bacillus botulinus*, then *Clostridium botulinum*) [18]. In France, the first reported botulism outbreak in humans was provided by the medical doctor Octave Du Mesnil in 1875 [19]. Human botulism was rare in France until the Second World War [20]. During the period of 1940–1944, the incidence of botulism was high as reported by R. Legroux and collaborators at the Institut Pasteur, mainly due to poor hygiene in the preparation of homemade preserved food [21–23] (Figure 2). In 1926, René Legroux and André-Pierre Marie prepared concentrated botulinum toxin. Unfortunately, A.-P. Marie died from botulism, probably by eye contamination with a tiny particle of concentrated and dried toxin [24,25]. Another collaborator, Colette Jeramec, was also contaminated, but less severely. In 1934,

R. Legroux, with Lucien Second, succeeded in producing liquid botulinum anatoxin and anti-botulinum serum in horses, which was used for the treatment of patients with botulism during the Second World War [24]. Serotherapy against botulism was further analyzed in guinea pigs [26]. Legroux and collaborators investigated experimental botulism in rabbits and horses. Notably, they explored the influence of the route of toxin inoculation on the induction of the disease [27,28].

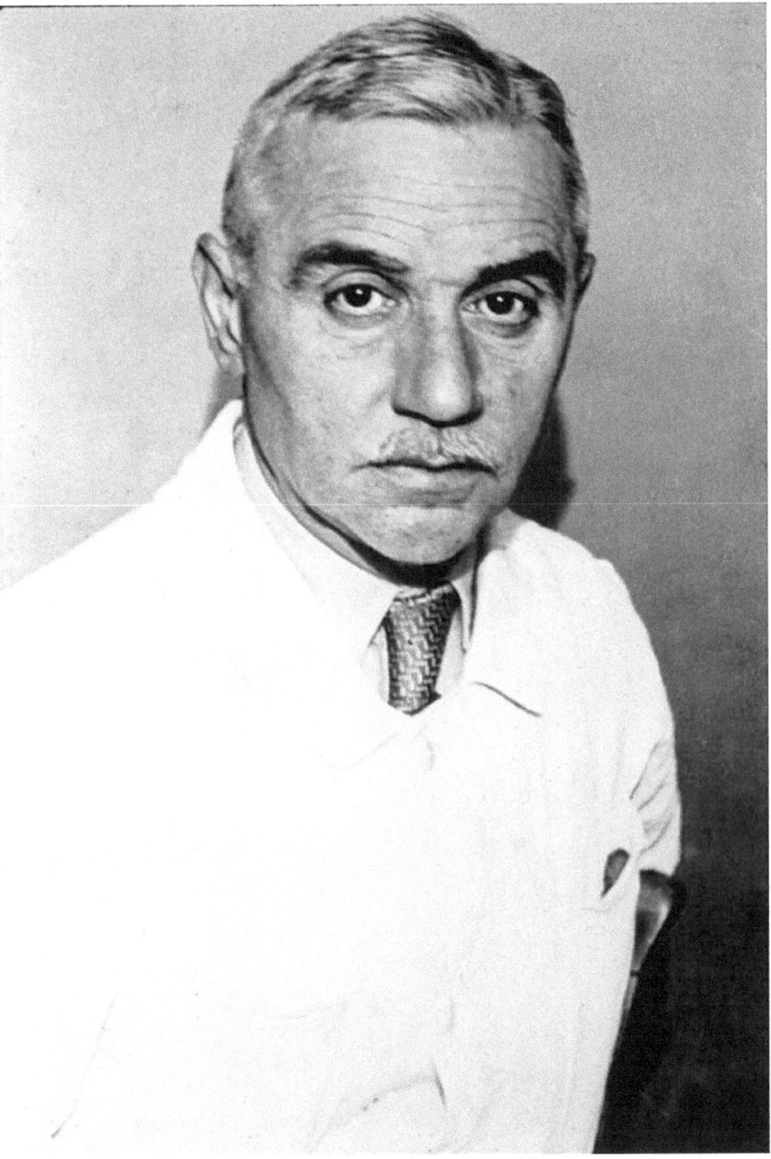

Figure 2. René Legroux (1877–1951) portrait of about 1930 (Institut Pasteur/Musée Pasteur). R. Legroux investigated human botulism during the Second World War and prepared botulinum anatoxins and anti-sera.

Michel Weinberg (1866–1940), a Russian physician born in Odessa, performed his medical studies in Paris (1892–1898), and he joined the Metchnikoff's service of "Microbie Morphologique" at the Institut Pasteur in 1900 (Figure 3). He studied the role of helminths in the translocation of bacterial pathogens through the intestinal mucosa, notably in the onset of appendicitis. During the First World War, he worked on the agents of gangrenes that affected numerous wounded soldiers. He found that the microbes responsible for gangrenes are diverse, and he identified novel species such as *Bacillus oedematiens* (*Clostridium novyi*), *B. fallax* (*C. fallax*), and *B. aerofetidus*. In 1918, he published a book with Pierre Seguin on gangrenes, in 1927 a book with B. Ginsbourg on the taxonomy of anaerobes and their role in pathology, and an updated one on anaerobes with R. Nativelle and A. R. Prévot in 1937 [29–31]. Then, he developed sera against the agents of gangrenes that were used by numerous surgeons. His work has been the subject of abundant publications in scientific journals [32].

Figure 3. Michel Weinberg's (1868–1940) portrait of about 1930. (Institut Pasteur/Musée Pasteur). M. Weinberg studied the agents of gangrene and wrote with B. Ginsbourg a book on the taxonomy of anaerobes.

The agent of tetanus was first identified by Arthur Nicolaier in 1884 and successfully isolated and cultivated by Shibasaburo Kitasato in the Emil von Behring laboratory (Berlin) in 1889 [9]. This agent was found to be an anaerobic bacterium with typical round terminal spores giving the appearance of drumsticks (*Bacillus tetani*, then *Clostridium tetani*), and it was demonstrated that the symptoms of tetanus are induced by a toxin (Tetanus toxin, TeNT) released in culture filtrates [9]. Von Behring and Kitasato showed that the sera from immunized animals against tetanus toxin were protective [33]. E. Roux and Louis Vaillard (1893), a French military physician, further investigated tetanus in experimental animals and developed iodinated TeNT as an immunogen. They confirmed that sera against TeNT can prevent the disease and are curative when administrated soon after the onset of symptoms [34]. Auguste-Charles Marie (1864–1935), a French physician and scientist, joined the Institut Pasteur in the service of E. Metchnikoff and then of E. Roux. He analyzed the trafficking of TeNT in rabbits, mice, and guinea pigs. He showed with Victor Morax, a Swiss physician and biologist who joined the laboratory of Microbie Technique, that the toxin, when administrated by the intravenous route, remains only a short time in the blood circulation and is trapped by the nervous system. In 1902–1903, A-C. Marie and V. Morax suggested that TeNT is transported by the peripheral neurons to the central nervous system [35,36]. The retrograde axonal transport of TeNT by motorneurons to the central nervous system was experimentally confirmed by several scientific teams in the period 1955–1979 (reviewed in [37,38]). More recently, Schiavo et al. further analyzed the movement of TeNT in cultured neurons [39].

A great step in the preparation of antigens was accomplished by the Pasteurian Gaston Ramon (1886–1963), a French veterinarian and biologist, who was in charge of the production of antisera in horses at the annex of Garches. In 1923, G. Ramon found that diphtheria toxin treated with low doses of formalin was a safe and potent immunogen and coined the term "anatoxin" [40]. The same procedure was successfully developed to prepare the tetanus anatoxin in 1925 [9,41]. Ramon observed a flocculation phenomenon when diphtheria toxin was mixed with a corresponding antiserum and developed an in vitro assay for the titration of antigens by flocculation [42,43]. The flocculation method for the titration of tetanus anatoxin is still a reference method in the Pharmacopeia.

The scientific activity at the Institut Pasteur was spread in several laboratories without specific denominations and not organized in scientific departments that were set up in 1967. In the scientific report of 1934 (ARCH. DUJ.C.1), E. Roux mentioned that the activity of M. Weinberg and his students on anaerobic microbes was of great importance and that they succeeded in obtaining anatoxins of botulism and gas gangrene agents based on the method developed by G. Ramon. It was recognized that the works of M. Weinberg were determinant for obtaining antigangrene sera that were used during the First World War. In the re-organization of the Institut Pasteur in 1934, a laboratory dedicated to anaerobes was created and called the "Service des microbes anaérobies". It was directed by M. Weinberg until his death in 1940. In 1936, an extension of the laboratory spaces was attributed to M. Weinberg for his activities on anaerobes.

It was noted in the report on the Institut Pasteur reorganization of 1942 by René Dujarric de la Riviere, Secretary general of the Institut Pasteur (ARCH. DUJ.C.1), that "since the nice works of Pasteur, followed by those of Veillon and Weinberg, the study of anaerobes has taken rightly an important place at Institut Pasteur. The species identification by cultural aspects, lesions induced in animals, immunological properties are now completed with biochemical investigations".

5. Period 1941–1968: The "Service des Anaérobies"

André-Romain Prévot (1894–1982) obtained a certificate in mineralogy at the Faculty of Sciences of Lille and then studied medicine in Paris (Figure 4). During the First World War, as an auxiliary military officer, he was confronted with many cases of gangrene and tetanus. In 1924, he wrote his medical thesis on anaerobic streptococci and his PhD on anaerobic cocci in 1933. A.-R. Prévot joined the laboratory of M. Weinberg dedicated to anti-gangrene serotherapy at the Institut Pasteur in 1922. He was named the deputy head of the laboratory of tetanus in 1939 and the head of the Anaerobe Laboratory (Service des Anaérobies) in 1941. A.-R. Prévot developed a specific laboratory on anaerobes until his retirement in 1966 [44]. A.-R. Prévot brought a great contribution to the taxonomy and physiology of anaerobes. He analyzed the respiratory types of bacteria and distinguished seven groups from strict anaerobic to strict aerobic bacteria. The bacterial classification was based on cultural, morphological, and biochemical properties. He isolated and characterized numerous novel anaerobe species. In addition to his numerous scientific publications, he wrote two voluminous books on anaerobes, one already mentioned by M. Weinberg and R. Nativelle in 1937 and another one with André Turpin and Paul Kaiser in 1967 which were reference books in the classification of these bacteria, as well as 13 other books on the biology of anaerobes, their metabolism, and their role in pathology and in the environment [31,45]. A.-R. Prévot was a member of the international committee of bacterial nomenclature, and he contributed actively to the definition of bacterial species and their classification. In the 1940s–1960s, the Institut Pasteur was recognized as an international reference center for the identification of anaerobes. Walter Edward Cladek Moore (1927–1996), a US microbiologist, was interested in anaerobic bacteria. W. E. Moore visited twice the service of A.-R. Prévot at the Institut Pasteur in the 1960s, and A.-R. Prévot gave him anaerobe samples from his entire collection. W. E. Moore took away about 2000 strains and created the anaerobe laboratory at the Virginia Polytechnic Institute, Blacksburg, which was the reference center for anaerobes in the US until his death in 1996 [46]. W. E. Moore with Lillian V. Holdeman and Elizabeth P. Cato edited the Anaerobe Laboratory Manual which was a recognized reference in the field of anaerobe identification [47].

(A)   (B)

Figure 4. André-Romain Prévot (1894–1982) (**A**) portrait of about 1930–1935. A.-R. (**B**) Prévot in his laboratory monitoring a continuous feed centrifuge. A.-R. Prévot developed anaerobic bacteriology and taxonomy, and he investigated botulism in France (Institut Pasteur/Musée Pasteur).

A.-R. Prévot and collaborators identified the first cases of botulism in cattle and horses due to *C. botulinum* C and D in France [48,49]. They investigated experimental botulism in horses and developed an anatoxin against botulinum toxins C and D for animal vaccination [50]. Moreover, they described the first identification of human botulism type E in France [51] and an exceptional outbreak of human botulism type D in Chad which is the unique human botulism of this type reported in the literature [52,53]. Indeed, *C. botulinum* D is essentially involved in animal botulism. It is noteworthy that the strain isolated from this atypical outbreak and characterized by A.-R. Prévot is recognized as the reference strain of *C. botulinum* D [52,54].

In 1960, the Service des Anaérobies managed by A.-R. Prévot and M. Raynaud encompassed 20 laboratories with animal facilities, industrial production, and research laboratories in the Institut Pasteur, Paris, and the annexes of Garches.

Among the Pasteurian scientists working on anaerobes and toxins during this period, we have to mention Marcel Raynaud, André Turpin, Edgar-Hans Relyveld, and Marcel Rouyer.

Marcel Raynaud (1911–1974), a medical doctor (1940) and PhD (1946) on "*Clostridium sporogenes* soluble toxic substances", joined the Institut Pasteur in 1942 in the biochemical laboratory of M. Macheboeuf and then the "Service des anaérobies" of A.-R. Prévot. In 1947, he became the head of the bacteriology and immunology laboratory. M. Raynaud developed multiple activities in bacteriology and immunology including protein toxins, endotoxins, toxin–antitoxin reactions, and horse immunoglobulins. Notably, he contributed to the production/purification of TeNT and the characterization of the TeNT anatoxin [55–57].

André Turpin (1920–1977), a French Doctor of Pharmacy, joined the laboratory of tetanus in Garches in 1947. He was the head of the laboratory of tetanus in 1964, which was renamed the laboratory of anaerobic toxins in 1970. A. Turpin performed the production of TeNT as well as of other toxins from anaerobes and collaborated with B. Bizzini and M. Raynaud on the deciphering of the TeNT structure [55,58–60] and on the study of anaerobes with A.-R. Prévot [45].

E.-H. Relyveld, a French scientist, spent his entire career (1952–1988) at the Institut Pasteur Garches where he was involved in the preparation of toxins and anatoxins and the production of vaccines. In addition to his activity on diphtheria toxin and staphylococcal toxins, he participated with M. Raynaud and A. Turpin in the preparation of TeNT and anti-tetanus serum [56,61].

Marcel Rouyer (1898–1981), a French medical doctor, entered the Institut Pasteur in 1940 and was successively the head of laboratory (1946) and the head of service (1956). M. Rouyer managed the laboratory of tetanus, then the laboratory of diphtheria, and finally the "Service des Anaérobies" from 1967 to 1968 following the retirement of A.-R. Prévot.

6. Period 1969–2000s: Towards the Genetics of Anaerobes and Molecular Mode of Action of Toxins

Madeleine Sebald (1930–) supported her medical thesis in 1957 and PhD in 1962 on the taxonomy of Gram-negative anaerobes. She joined the Service des Anaérobies of A.-R. Prévot in 1957 as a trainee and then a research assistant. From 1969 to 1996, M. Sebald was the Director of the Laboratory of Anaerobes, which was renamed the Unit of Anaerobes in 1977. M. Sebald markedly contributed to the taxonomy of anaerobes, more especially of Gram-negative anaerobes by introducing DNA-based methods such as GC% content that were novel approaches at this period. Then, M. Sebald was interested in the physiology and genetics of anaerobes including sporulation and spore germination in some *Clostridium* species. Thereby, she investigated the solventogenesis in *Clostridium acetobutylicum*, which regained industrial interest for the production of solvents from biomasses in 1981 after the first oil shock.

The genetic resistance to antibiotics has been analyzed in clostridia and in *Bacteroides fragilis*. Resistance to tetracycline, chloramphenicol, and macrolides has been determined to be supported by plasmids or transposons in *C. perfringens* and in *B. fragilis*. For the first time, the transferability of plasmids carrying antibiotic resistance genes between anaerobes such as between *C. perfringens* strains, *C. perfringens*/*Clostrioides difficile* (formerly *Clostridium difficile*), and *B. fragilis* strains has been demonstrated [62–64]. This opened the way to build shuttle vectors as tools for genetic investigation in anaerobes [65]. The main difficulty was to transfer recombinant plasmids into anaerobe recipients. The transformation of *C. acetobutylicum* protoplasts with plasmid DNA was an efficient method, but the regeneration of protoplasts into viable bacteria was problematic. Finally, electroporation was a more suitable method of transformation in anaerobes. An additional insight into antibiotic resistance concerned the 5-nitroimidazole, which is widely used in the treatment of infections caused by anaerobes. However, some *B. fragilis* strains are resistant to this antibiotic. The mechanism of resistance to 5-nitroimidazole has been elucidated, and the genes (*nim*) responsible for the resistance have been characterized [66,67].

M. Sebald contributed actively to the investigation of human botulism in France. In total, 660 cases were confirmed in the period from 1970–1996. In 1974, the Laboratory of Anaerobes was recognized as the National Reference Center (NRC) of Anaerobes including the survey of botulism and was renamed the NRC of Anaerobes and Botulism in 2002. Since 1973, the laboratory's diagnosis of botulism has been improved by the detection of the toxin in the serum of patients [68]. Until this period, the confirmation of botulism was only based on an investigation of suspected foodstuffs but only when they were available. In the 1990s, the first identifications of botulism outbreaks in waterbirds and farmed birds due to *C. botulinum* type C were performed in France [69]. No incidence of the increasing cases of avian botulism on human health was reported.

In the 1980s, the emergence of *C. difficile* antibiotic-associated diarrheas and pseudomembranous colitis prompted the NRC of Anaerobes to initiate the diagnosis of these infections in France by toxin detection based on an assay for cytotoxicity in patient's feces and to develop bacteriological investigations of this pathogen. In 2008, a laboratory on *C. difficile* managed by Bruno Dupuy was created and was mainly involved in the regulation of toxinogenesis and the genetic aspects of this pathogen (see the article of B. Dupuy in this issue).

Bernard Bizzini joined the service of Immunochemistry managed by M. Raynaud in the annex of Garches in 1952 and supported his PhD on TeNT in 1970. In collaboration with A. Turpin and M. Raynaud, B. Bizzini developed the production and purification of TeNT [58]. Then, B. Bizzini moved to the Institut Pasteur, Paris, to the department of Protein Chemistry in 1974 and then to the unit of Protein Immunochemistry (1980–1992) where his main activity was in TeNT. B. Bizzini contributed to deciphering the TeNT structure by a biochemical approach based on dissection by proteases and in determining the biological and immunological properties of TeNT fragments [38,70]. Thus, in collaboration with Klaus Stoeckel and Manfred Schwab, he showed that a fragment of the TeNT heavy chain mediates the binding to ganglioside and retrograde axonal transport [71]. Through multiple collaborations, B. Bizzini participated in the elucidation of the TeNT structure function.

Joseph Alouf (1929–2014), was born in Lebanon and has performed his pharmaceutical and scientific studies in Paris. In 1956, J. Alouf joined the service of Bacterial Chemistry at Garches that was managed by M. Raynaud in the Service des Anaérobies of A.-R. Prévot. His interest was in the investigation of hemolysins. He first analyzed the physicochemical and biological properties of streptolysin O from *Streptococcus pyogenes*, and then he investigated related hemolysins such as perfringolysin from *C. perfringens* and listeriolysins from *Listeria* sp. which are now known as the family of cholesterol-dependent cytolysins. In 1972, he moved to the Institut Pasteur Paris as the head of the Laboratory of Bacterial Toxins within the Laboratory of Anaerobes directed by M. Sebald. In 1977, a novel laboratory space was attributed to J. Alouf, and he was promoted to head of the Bacterial Antigens unit, which became the unit of Microbial Toxins in 1992. J. Alouf combined toxinology and immunology, notably regarding bacterial toxins as superantigens. Besides his research activity, he was involved in teaching in particular as the director of the General Immunology course at the Institut Pasteur for 20 years [72]. In addition to his scientific articles, J. Alouf was an editor of several books on bacterial toxins such as the Comprehensive Sourcebook of Bacterial Protein Toxins (four editions between 1991 and 2015) [73].

Numerous scientists joined or visited the Bacterial Antigens/Microbial Toxins unit. This period, from about 1980 to 2000, which witnessed the genetic characterization and discovery of the molecular mode of action of many bacterial toxins was very exciting.

Michele Mock and her group investigated the genetics of *Bacillus anthracis* and anthrax toxins, notably the edema toxin that is an adenylate cyclase [74,75] (see the article of Pierre Goossens in this issue). Patrice Boquet performed his postdoctoral fellowship in the laboratory of Alwin Max Pappenheimer, Harvard University, Boston, which played a central contribution in the discovery of the enzymatic activity of the diphtheria toxin ADP-ribosylation of the elongation factor 2 leading to an inhibition of protein synthesis. P. Boquet was interested in the mode of entry of the diphtheria toxin into cells. In 1978, he joined J. Alouf's group at the Institut Pasteur. In addition to his work on diphtheria toxin entry into cells and the genetic characterization of the diphtheria toxin gene [76,77], he showed that the hydrophobic N-terminal domain of the TeNT heavy chain forms channels into lipid membranes at acidic pH [78,79]. Thus, his research supported the notion that the diphtheria toxin and TeNT use a similar entry pathway into cells based on pore formation through the endosomal membrane at acidic pH. In 1985–1986, Michael Gill a fellow of the Papenheimer's group who discovered the ADP-ribosylating activity of the cholera toxin [80], came to the Institut Pasteur for a sabbatical year. His project was to analyze the ADP-ribosylating activity of *C. botulinum* C and D. Indeed, Klauss Aktories and colleagues demonstrated that C2 toxin from *C. botulinum* C ADP-ribosylates cellular actin [81]. As the already identified ADP-ribosylating toxins at this time recognized a G-protein (GTP-binding protein) as a substrate, M. Gill suspected that a functional substrate of the C2 toxin was a G-protein, probably of small size, but not actin that is an ATP-binding protein. Experiments performed at the Institut Pasteur showed that supernatants of *C. botulinum* C and D exhibited ADP-ribosylating activity towards a small molecule of about 21 kDa in addition to actin. However, the p21 ADP-ribosylation was not due to the C2 toxin but to a distinct toxin or enzyme called C3. K. Aktories, who was aware of the scientific activity at the Institut Pasteur, requested a *C. botulinum* D strain from M. Sebald and promptly published the ADP-ribosylation of the C3 enzyme [82]. Then, the group of Narumiya identified the substrate of the C3 enzyme as a small G-protein homologous to Ras protein called Rho (Ras-homologous) [83]. Pierre Chardin, P. Boquet, M. Gill, et al. showed that C3 through the ADP-ribosylation of Rho induces a disorganization of the actin cytoskeleton [84]. Later, Alan Hall's group demonstrated that Rho and related G-proteins (Rac, Cdc42) are key molecules involved in the regulation of actin cytoskeleton structures [85,86]. Thereby, the C3 enzyme, which is specific to the Rho protein, is a powerful tool in cell biology and was intensively used to unravel the regulation of actin cytoskeleton assembly.

The discovery of the C3 enzyme prompted research to identify other possible ADP-ribosylating toxins in clostridia. Thus, many clostridial strains have been tested, and other toxins from the C2 family were identified such as *Clostridium spiroforme* toxin and *C. difficile* transferase (CDT) [87,88]. Interestingly, the *C. difficile* strains which produce CDT in addition to the two large molecular size toxins, toxin A (TcdA) and toxin B (TcdB), and which acquire fluoroquinolone resistance, were recognized as human epidemic strains such as ribotypes 027, 078 [89]. The cloning of the Iota toxin genes from *C. perfringens*, a toxin related to the C2 toxin, revealed a high level of similarity between the binding component of the Iota toxin (Ib) and the protective antigen (PA) of the anthrax toxins [90]. Later, both components were found to share a similar structure and mode of action in pore formation through endosome membranes facilitating the translocation of the enzymatic components into the cytosol [91].

In the late 1980s, the TeNT and BoNT/A genes were cloned and sequenced by the German group of Heiner Niemann [92,93]. In the Bacterial Toxin unit and in collaboration with Niemann's group, the sequencing of the neurotoxin gene of *C. botulinum* C, D, E, and *C. butyricum* E was been performed (reviewed in [94]). The neurotoxin sequencing revealed that all the TeNT and BoNT types contained a signature of zinc-dependent proteases. The enzymatic activity of clostridial neurotoxins was subsequently characterized by Cesare Montecucco, Giampietro Schiavo, et al. [95]. Then, the first complete botulinum locus including the genes of BoNT and non-toxic proteins which associate with the neurotoxin to form botulinum complexes was characterized in *C. botulinum* C [96]. The role in the neurotoxin synthesis of the regulatory genes (*botR*) from the botulinum loci and *tetR* upstream of the *tent* gene has been investigated. These genes encode for alternative sigma factors that positively control neurotoxin production (review in [94]).

From 1997 to 2019, the Anaerobic Bacteria and Toxins unit at the Institut Pasteur investigated several clostridial toxins (*C. perfringens* epsilon, beta2, delta toxins, *C. sordellii* lethal toxin, *C. difficile* toxins, *C. septicum* alpha toxin) and further characterized clostridial neurotoxins in collaboration, for some aspects, with external scientists such as Klaus Aktories, Holger Barth, Roland Benz, Juan Blasi, Holger Brüggemann, Christian Lévèque, Jordi Molgo, Bernard Poulain, Bradley Stiles, and Richard Titball [97].

In 2019, the unit was renamed Bacterial Toxins and was directed by Emmanuel Lemichez. This unit is dedicated to clostridial neurotoxins and other bacterial toxins such as *Escherichia coli* CNF (see the article of Lemichez on this issue).

7. Concluding Remarks

Since the founding of the Institut Pasteur in 1887 for developing a rabies vaccination, microbiology, including research and training, has been a continuous and central purpose of this institute. The microbiology research activity was performed in a common laboratory called the "Microbie technique". Then, anaerobic bacteriology and toxinology emerged as distinct activities which required specific equipment and were developed in more specialized laboratories. First, taxonomy and the identification of pathogens were the main tasks. Subsequent investigations on the mechanisms of pathogenicity showed that some bacteria secrete soluble toxic factors that are responsible for symptoms and lesions. In-depth analysis of toxins over time according to the development of technology (biochemistry, enzymology, molecular biology, cellular biology, crystallography) brought a better understanding of certain diseases and allowed the development of specific preventions and treatments (vaccines based on anatoxins, serotherapy, toxin inhibitors). Moreover, toxins were found to be efficient therapeutic drugs (for example, therapeutic applications of BoNTs) and also powerful tools in dissecting specific cellular processes. Thus, toxins contribute to the emergence of the concept of cellular microbiology [98]. Throughout its history, the Institut Pasteur has made relevant contributions in anaerobic bacteriology and toxinology.

Author Contributions: M.R.P. and S.L. wrote the manuscript. All authors have read and agreed to the published version of the manuscript.

Funding: This review received no external funding.

Institutional Review Board Statement: Not applicable.

Informed Consent Statement: Not applicable.

Data Availability Statement: Not applicable.

Acknowledgments: We thank Michaël Davy, Photothèque Institut Pasteur, for providing the figures and photos.

Conflicts of Interest: The authors declare no conflict of interest.

References

1. Bisset, N.G. Arrow and dart poisons. *J. Ethnopharmacol.* **1989**, *25*, 1–41. [CrossRef] [PubMed]
2. Hayes, A.N.; Gilbert, S.G. Historical milestones and discoveries that shaped the toxicology sciences. *EXS* **2009**, *99*, 1–35. [CrossRef] [PubMed]
3. Cavaillon, J.M. Historical links between toxinology and immunology. *Pathog. Dis.* **2018**, *76*, 4923027. [CrossRef] [PubMed]
4. Brieger, L. *Weitere Untersuchungen über Ptomaine*; August Hirschwald: Berlin, Germany, 1885; Volume 1.
5. Schwartz, M. Dr. Jekyll and Mr. Hyde: A short history of anthrax. *Mol. Asp. Med.* **2009**, *30*, 347–355. [CrossRef]
6. Pollitzer, R.; Swaroop, S.; Burrows, W. Cholera. *Monogr. Ser. World Health Organ.* **1959**, *58*, 1001–1019.
7. Collier, R.J. Diphtheria toxin: Mode of action and structure. *Bacteriol. Rev.* **1975**, *39*, 54–85. [CrossRef]
8. Roux, E.; Yersin, A. Contribution à l'étude de la diphtérie. *Ann. Inst. Pasteur.* **1888**, *2*, 629–661.

9.   Cavaillon, J.M. From Bacterial Poisons to Toxins: The Early Works of Pasteurians. *Toxins* **2022**, *14*, 759. [CrossRef]
10.   Berche, P. Louis Pasteur, from crystals of life to vaccination. *Clin. Microbiol. Infect.* **2012**, *18*, 1–6. [CrossRef]
11.   Pasteur, L. Recherches sur la putréfaction. *CR Acad. Sci.* **1861**, *LVI*, 1189–1194.
12.   Pasteur, L.; Joubert, J.; Chamberland, C. La théorie des germes et ses applications à la médecine et à la chirurgie. *Bull. Acad. Med.* **1878**, *VII*, 432–453.
13.   Pasteur, L. Expériences et vues nouvelles sur la nature des fermentations. *CR Acad. Sci.* **1861**, *LII*, 1260–1264.
14.   Pasteur, L. *Oeuvres de Pasteur Collected by Pasteur Vallery-Radot*; Masson, Ed.; Masson et Cie: Paris, France, 1992; Volume Tome V Etudes sur la Bière, p. 412.
15.   Roux, E. Sur la culture des microbes anaérobies. *Ann. Inst. Pasteur.* **1887**, *2*, 49–62.
16.   Veillon, A.; Zuber, M. Recherches sur quelques microbes strictement anaérobies et leur rôle en pathologie. *Arch. Med. Exp.* **1898**, *10*, 517–545.
17.   Roux, E. André Veillon. *Ann. Inst. Pasteur.* **1931**, *XLVII*, 1–3.
18.   van Ermengem, E. Classics in infectious diseases. A new anaerobic bacillus and its relation to botulism. E. van Ermengem. Originally published as "Ueber einen neuen anaeroben Bacillus und seine Beziehungen zum Botulismus" in Zeitschrift fur Hygiene und Infektionskrankheiten 26: 1–56, 1897. *Rev. Infect. Dis.* **1979**, *1*, 701–719.
19.   Du Mesnil, O. Empoisonnement par de la viande de conserve. *Ann. Santé. Publique.* **1875**, *43*, 472–478.
20.   Rasetti-Escargueil, C.; Lemichez, E.; Popoff, M.R. Human Botulism in France, 1875–2016. *Toxins* **2020**, *12*, 338. [CrossRef]
21.   Legroux, R.; Jeramec, C.; Levaditi, J.C. Statistique du botulisme de l'occupation 1940–1944. *Bull. Acad. Med.* **1945**, *129*, 643–645.
22.   Legroux, R.; Levaditi, J.C.; Jéramec, C. Le botulisme en France pendant l'occupation. *Presse. Med.* **1947**, *57*, 109–110.
23.   Legroux, R.; Jeramec, C. Le botulisme et les jambons salés. *Bull. Acad. Natl. Med.* **1944**, *128*, 174–178.
24.   Anonymous. René Legroux. *Ann. Inst. Pasteur.* **1951**, *80*, 332–336.
25.   Legroux, R. André-Pierre Marie. *Ann. Inst. Pasteur.* **1929**, *43*, 959.
26.   Legroux, R.; Levaditi, J.C. On the serotherapy of experimental botulinum intoxication of the guinea pig. *Ann. Inst. Pasteur.* **1946**, *72*, 216–221.
27.   Legroux, R.; Levaditi, J.C.; Lamy, R. Le botulisme experimental du cheval et la question du botulisme naturel. *Ann. Inst. Pasteur.* **1946**, *72*, 545–552.
28.   Legroux, R.; Lavaditi, J.C.; Jeramec, C. Influence des voies d'introduction de la toxine sur le botulisme expérimental du lapin. *Ann. Inst. Pasteur.* **1945**, *71*, 490–493.
29.   Weinberg, M.; Ginsbourg, B. *Données Récentes sur les Microbes Anaérobies et leur Role en Pathologie*; Masson et Cie: Paris, France, 1927.
30.   Weinberg, M.; Seguin, P. *La Gangrene Gazeuse*; Masson et Cie: Paris, France, 1918.
31.   Weinberg, M.; Nativelle, R.; Prévot, A.R. *Les Microbes Anaérobies*; Masson et Cie: Paris, France, 1937; p. 1186.
32.   Anonymous. Michel Weinberg. *Ann. Inst. Pasteur.* **1940**, *64*, 461–465.
33.   Behring, E.; Kitasato, S. Über das Zustandekommen der Diphtherie-Immunität und der Tetanus-Immunität bei Thieren. *Dtsch. Med. Wochenschrift.* **1890**, *49*, 1113–1114.
34.   Roux, E.; Vaillard, L. Contribution à l'étude du tétanos, Prévention et Traitement par le Serum Antitoxique. *Ann. Inst. Pasteur.* **1893**, *7*, 65–140.
35.   Marie, A.C.; Morax, V. Sur l'absorption de la toxine tétanique. *Ann. Inst. Pasteur.* **1902**, *16*, 818–832.
36.   Morax, V.; Marie, A.C. Recherches sur l'absorption de la toxine tétanique. *Ann. Inst. Pasteur.* **1903**, *17*, 335–342.
37.   Wellhöner, H.H. Tetanus and Botulinum Neurotoxins. In *Selective Neurotoxicity*; Herken, H., Hucho, F., Eds.; Springer: Berlin/Heidelberg, Germany, 1992; Volume 102, pp. 357–417.
38.   Bizzini, B. Axoplasmic transport and transynaptic movement of tetanus toxin. In *Botulinum Neurotoxin and Tetanus Toxin*; Simpson, L.L., Ed.; Academic Pres: San Diego, CA, USA, 1989; pp. 203–229.
39.   Surana, S.; Tosolini, A.P.; Meyer, I.F.G.; Fellows, A.D.; Novoselov, S.S.; Schiavo, G. The travel diaries of tetanus and botulinum neurotoxins. *Toxicon* **2018**, *147*, 58–67. [CrossRef]
40.   Ramon, G. Sur le pouvoir floculant et sur les propriétés immunisantes d'une toxine diphtérique rendue anatoxique (anatoxine). *CR Acad. Sci.* **1923**, *177*, 1338–1340.
41.   Ramon, G.; Descombey, M.P. Sur l'immunisation antitétanique et sur la production de l'antitoxine tétanique. *CR Soc. Biol.* **1925**, *93*, 508–509.
42.   Ramon, G. Floculation dans un mélange neutre toxine-antitoxine diphtériques. *CR Soc. Biol.* **1922**, *86*, 661–663.
43.   Ramon, G.; Descombey, M.P. Sur l'appéciation de la valeur antigène de la toxine et de l'anatoxine tétaniques par la méthode de floculation. *CR Soc. Biol.* **1926**, *95*, 434–436.
44.   Anonymous. André-Romain Prévot. *Bull. Institut. Pasteur.* **1983**, *81*, 1–3.
45.   Prévot, A.R.; Turpin, A.; Kaiser, P. *Les Bactéries Anaérobies*; Dunod: Paris, France, 1967; p. 2188.
46.   Strotter, W.H.; Wallenstein, P. *From VPI to State University—President T. Marshall Hahn Jr. and the Transformation of Virginia Tech, 1962–1974*; Mercer University Press: Macon, GA, USA, 2004.
47.   Holdeman, L.V.; Cato, E.P.; Moore, W.E.C. *Anaerobe Laboratory Manual*, 4th ed.; Virginia Polytechnic Institute and State University: Blacksbourg, VA, USA, 1977.
48.   Prévot, A.R.; Brygoo, E.R. Etude de la première souche française de *Cl. botulinum* D. *Ann. Inst. Pasteur.* **1950**, *78*, 274–276.

49. Prévot, A.R.; Huet, M.; Tardiieux, P. Etude de vingt-cinq foyers récents de botulisme animal. *Bull. Acad. Vet. Fr.* **1950**, *23*, 481–487. [CrossRef]

50. Jacquet, J.; Prevot, A.R. Recherches sur le botulisme equin experimental. *Ann. Inst. Pasteur.* **1951**, *81*, 334–337.

51. Prévot, A.R.; Huet, M. Existence in France of human botulism due to fish and to Clostridium botulinum E. *Bull. Acad. Natl. Med.* **1951**, *135*, 432–435. [PubMed]

52. Demarchi, J.; Mourgues, C.; Orio, J.; Prévot, A.R. Existence of type D botulism in man. *Bull. Acad. Natl. Med.* **1958**, *142*, 580–582.

53. Rasetti-Escargueil, C.; Lemichez, E.; Popoff, M.R. Public Health Risk Associated with Botulism as Foodborne Zoonoses. *Toxins* **2020**, *12*, 17. [CrossRef]

54. Prévot, A.R.; Sillioc, R. Activation of botulinum D toxinogenesis by Bacillus licheniformis. *Ann. Inst. Pasteur.* **1959**, *96*, 630–632.

55. Raynaud, M.; Bizzini, B.; Turpin, A. Heterogeneity of tetanus anatoxin. *Ann. Inst. Pasteur.* **1971**, *120*, 801–815.

56. Raynaud, M.; Relyveld, E.H.; Turpin, A.; Mangalo, R. Preparation of highly purified diphtheric, tetanic & staphylococcic anatoxins absorbed on calcium phosphate (brushite). *Ann. Inst. Pasteur.* **1959**, *96*, 60–71.

57. Raynaud, M.; Turpin, A.; Bizzini, B. Existence of tetanus toxin under several aggregation conditions. *Ann. Inst. Pasteur.* **1960**, *99*, 167–172.

58. Bizzini, B.; Turpin, A.; Raynaud, M. Production et purification de la toxine tétanique. *Ann. Inst. Pasteur.* **1969**, *116*, 686–712.

59. Turpin, A.; Raynaud, M. Tetanus toxin. *Ann. Inst. Pasteur.* **1959**, *97*, 718–732.

60. Turpin, A.; Raynaud, M.; Rouyer, M. The purification of tetanus toxin and antitoxin. *Ann. Inst. Pasteur.* **1952**, *82*, 299–309.

61. Raynaud, M.; Turpin, A.; Relyveld, E.H.; Corvazier, R.; Girard, O. Preparation of tetanus antitoxin by immunization of horses with tetanus anatoxins of high purity. *Ann. Inst. Pasteur.* **1959**, *96*, 649–658.

62. Brefort, G.; Magot, M.; Ionesco, H.; Sebald, M. Characterization and transferability of *Clostridium perfringens* plasmids. *Plasmid* **1977**, *1*, 52–66. [CrossRef] [PubMed]

63. Privitera, G.; Dublanchet, A.; Sebald, M. Transfer of multiple antibiotic resistance between subspecies of Bacteroides fragilis. *J. Infect Dis.* **1979**, *139*, 97–101. [CrossRef] [PubMed]

64. Privitera, G.; Sebald, M.; Fayolle, F. Common regulatory mechanism of expression and conjugative ability of a tetracycline resistance plasmid in Bacteroides fragilis. *Nature* **1979**, *278*, 657–659. [CrossRef] [PubMed]

65. Truffaut, N.; Hubert, J.; Reysset, G. Construction of shuttle vectors useful for transforming *Clostridium acetobutylicum*. *FEMS Microbiol. Lett.* **1989**, *49*, 15–20. [CrossRef] [PubMed]

66. Reysset, G.; Haggoud, A.; Sebald, M. Genetics of resistance of î to 5-nitroimidazole. *Clin Infect Dis.* **1993**, *16*, S401–S403. [CrossRef]

67. Trinh, S.; Haggoud, A.; Reysset, G.; Sebald, M. Plasmids pIP419 and pIP421 from *Bacteroides*: 5-nitroimidazole resistance genes and their upstream insertion sequence elements. *Microbiology* **1995**, *141*, 927–935. [CrossRef]

68. Sebald, M.; Saimot, G. Toxémie botulique. Intérêt de sa mise en évidence dans le diagnostic du botulisme humain de type B. *Ann. Microbiol.* **1973**, *124*, 61–69.

69. Sebald, M.; Duchemin, J.; Desenclos, J.C. Le risque botulique C chez l'homme. *Bull. Epidémiol. Hebd* **1997**, *2*, 8–9.

70. Bizzini, B. Tetanus toxin. *Microbiol. Rev.* **1979**, *43*, 224–240. [CrossRef]

71. Bizzini, B.; Stoeckel, K.; Schwab, M. An antigenic polypeptide fragment isolated from tetanus toxin: Chemical characterization, binding to gangliosides and retrograde axonal transport in various neuron systems. *J. Neurochem.* **1977**, *28*, 529–542. [CrossRef] [PubMed]

72. Cavaillon, J.M. Joseph Alouf (1929–2014). *FEMS Microbiol. Lett.* **2014**, *355*, 90–91. [CrossRef] [PubMed]

73. *The Comprehensive Sourcebook of Bacterial Protein Toxins*, 4th ed.; Alouf, J.; Ladant, D.; Popoff, M.R. (Eds.) Elsevier: Amsterdam, The Netherlands, 2015.

74. Mock, M.; Fouet, A. Anthrax. *Annu. Rev. Microbiol.* **2001**, *55*, 647–671. [CrossRef] [PubMed]

75. Mock, M.; Ullmann, A. Calmodulin-activated bacterial adenylate cyclases as virulence factors. *TIMS* **1993**, *1*, 187–192. [CrossRef]

76. Kaczorek, M.; Delpeyroux, F.; Chenciner, N.; Streeck, R.E.; Murphy, J.R.; Boquet, P.; Tiollais, P. Nucleotide sequence and expression of the diphtheria tox228 gene in Escherichia coli. *Science* **1983**, *221*, 855–858. [CrossRef]

77. Lemichez, E.; Bomsel, M.; Devilliers, G.; vanderSpek, J.; Murphy, J.R.; Lukianov, E.V.; Olsnes, S.; Boquet, P. Membrane translocation of diphtheria toxin fragment A exploits early to late endosome trafficking machinery. *Mol. Microbiol.* **1997**, *23*, 445–457. [CrossRef]

78. Boquet, P.; Duflot, E. Tetanus toxin fragment forms channels in lipid vesicles at low pH. *Proc. Natl. Acad. Sci. USA* **1982**, *79*, 7614–7618. [CrossRef]

79. Roa, M.; Boquet, P. Interaction of tetanus toxin with lipid vesicles at low pH. *J. Biol. Chem.* **1985**, *260*, 6827–6835. [CrossRef]

80. Gill, D.M.; Merens, R. ADP-ribosylation of membrane proteins catalyzed by cholera toxin: Basis of the activation of adenylate cyclase. *Proc. Natl. Acad. Sci. USA* **1978**, *75*, 3050–3054. [CrossRef]

81. Aktories, K.; Bärmann, M.; Ohishi, I.; Tsuyama, S.; Jakobs, K.H.; Habermann, E. Botulinum C2 toxin ADP-ribosylates actin. *Nature* **1986**, *322*, 390–392. [CrossRef]

82. Aktories, K.; Weller, U.; Chhatwal, G.S. *Clostridium botulinum* type C produces a novel ADP-ribosyltransferase distinct from botulinum C2 toxin. *FEBS Lett.* **1987**, *212*, 109–113. [CrossRef]

83. Narumiya, S.; Sekine, A.; Fujiwara, M. Substrate for botulinum ADP-ribosyltransferase, Gb, has an amino acid sequence homologous to a putative rho gene product. *J. Biol. Chem.* **1988**, *263*, 17255–17257. [CrossRef] [PubMed]

84. Chardin, P.; Boquet, P.; Madaule, P.; Popoff, M.R.; Rubin, E.J.; Gill, D.M. The mammalian G protein *rho*C is ADP-ribosylated by *Clostridium botulinum* exoenzyme C3 and affects actin microfilaments in Vero Cells. *EMBO J.* **1989**, *8*, 1087–1092. [CrossRef] [PubMed]
85. Nobes, C.D.; Hall, A. Rho, Rac, and Cdc42 GTPases regulate the assembly of multimolecular focal complexes associated with actin stress fibers, lamellipodia, and filopodia. *Cell* **1995**, *81*, 53–62. [CrossRef] [PubMed]
86. Peterson, M.F.; Self, A.J.; Garrett, M.D.; Just, I.; Korus, A.; Aktories, K.; Hall, A. Microinjection of recombinant p^{21RHO} induces rapid changes in cell morphology. *J. Cell Biol.* **1990**, *111*, 1001–1007. [CrossRef]
87. Popoff, M.R.; Boquet, P. *Clostridium spiroforme* toxin is a binary toxin which ADP-ribosylates cellular actin. *Biochem. Biophys. Res. Commun.* **1988**, *152*, 1361–1368. [CrossRef]
88. Popoff, M.R.; Rubin, E.J.; Gill, D.M.; Boquet, P. Actin-specific ADP-ribosyltransferase produced by a *Clostridium difficile* strain. *Infect. Immun.* **1988**, *56*, 2299–2306. [CrossRef]
89. Gerding, D.N.; Johnson, S.; Rupnik, M.; Aktories, K. *Clostridium difficile* binary toxin CDT: Mechanism, epidemiology, and potential clinical importance. *Gut Microbes.* **2014**, *5*, 15–27. [CrossRef]
90. Perelle, S.; Gibert, M.; Boquet, P.; Popoff, M.R. Characterization of *Clostridium perfringens* iota-toxin genes and expression in *Escherichia coli. Infect. Immun.* **1993**, *61*, 5147–5156, Correction in *Escherichia coli. Infect. Immun.* **1995**, *5163*, 4967. [CrossRef]
91. Stiles, B.G.; Wigelsworth, D.J.; Popoff, M.R.; Barth, H. Clostridial binary toxins: Iota and C2 family portraits. *Front. Cell. Infect. Microbiol.* **2011**, *1*, 2–13. [CrossRef]
92. Binz, T.; Kurazono, H.; Wille, M.; Frevert, J.; Wernars, K.; Niemann, H. The complete sequence of botulinum neurotoxin type A and comparison with other clostridial neurotoxins. *J. Biol. Chem.* **1990**, *265*, 9153–9158. [CrossRef] [PubMed]
93. Eisel, U.; Jarausch, W.; Goretzki, K.; Henschen, A.; Engels, J.; Weller, U.; Hudel, M.; Habermann, E.; Niemann, H. Tetanus toxin: Primary structure, expression in *E. coli*, and homology with botulinum toxins. *EMBO J.* **1986**, *5*, 2495–2502. [CrossRef] [PubMed]
94. Popoff, M.R.; Marvaud, J.C. Structural and genomic features of clostridial neurotoxins. In *The Comprehensive Sourcebook of Bacterial Protein Toxins*, 2nd ed.; Alouf, J.E., Freer, J.H., Eds.; Academic Press: London, UK, 1999; Volume 2, pp. 174–201.
95. Rossetto, O.; Deloye, F.; Poulain, B.; Pellizzari, R.; Schiavo, G.; Montecucco, C. The metallo-proteinase activity of tetanus and botulism neurotoxins. *J. Physiol. Paris* **1995**, *89*, 43–50. [CrossRef] [PubMed]
96. Hauser, D.; Eklund, M.W.; Boquet, P.; Popoff, M.R. Organization of the botulinum neurotoxin C1 gene and its associated non-toxic protein genes in *Clostridium botulinum* C468. *Mol. Gen. Genet.* **1994**, *243*, 631–640. [CrossRef] [PubMed]
97. Popoff, M.R.; Bouvet, P. Clostridial toxins. *Future Microbiol.* **2009**, *4*, 1021–1064. [CrossRef]
98. Cossart, P.; Boquet, P.; Normark, S.; Rappuoli, R. *Cellular Microbiology*; ASM Press: Washington, DC, USA, 2000.

Review

Regulation of Clostridial Toxin Gene Expression: A Pasteurian Tradition

Bruno Dupuy

Institut Pasteur, Université Paris-Cité, UMR-CNRS 6047, Laboratoire Pathogenèse des Bactéries Anaérobies, F-75015 Paris, France; bdupuy@pasteur.fr

Abstract: The alarming symptoms attributed to several potent clostridial toxins enabled the early identification of the causative agent of tetanus, botulism, and gas gangrene diseases, which belongs to the most famous species of pathogenic clostridia. Although *Clostridioides difficile* was identified early in the 20th century as producing important toxins, it was identified only 40 years later as the causative agent of important nosocomial diseases upon the advent of antibiotic therapies in hospital settings. Today, *C. difficile* is a leading public health issue, as it is the major cause of antibiotic-associated diarrhea in adults. In particular, severe symptoms within the spectrum of *C. difficile* infections are directly related to the levels of toxins produced in the host. This highlights the importance of understanding the regulation of toxin synthesis in the pathogenicity process of *C. difficile*, whose regulatory factors in response to the gut environment were first identified at the Institut Pasteur. Subsequently, the work of other groups in the field contributed to further deciphering the complex mechanisms controlling toxin production triggered by the intestinal dysbiosis states during infection. This review summarizes the Pasteurian contribution to clostridial toxin regulation studies.

Keywords: *Clostridioides difficile*; toxins; regulation; metabolism

Key Contribution: As part of the bicentenary of Louis Pasteur's birth, this review summarizes the contribution of Pasteurians to the identification and understanding of the mechanisms involved in the regulation of clostridial toxins.

Citation: Dupuy, B. Regulation of Clostridial Toxin Gene Expression: A Pasteurian Tradition. *Toxins* **2023**, *15*, 413. https://doi.org/10.3390/toxins15070413

Received: 9 May 2023
Revised: 13 June 2023
Accepted: 19 June 2023
Published: 26 June 2023

1. Introduction

As for the other pathogenic clostridia, the disease associated with *Clostridioides* (formerly *Clostridium*) *difficile*, a Gram-positive spore-forming anaerobic bacterium, is strictly related to the production of potent exotoxins. *C. difficile* is the major pathogen responsible for nosocomial diarrhea in adults with disturbed gut microbiota due to broad-spectrum antibiotics. The clinical manifestations of *C. difficile* infections (CDI) may extend from mild diarrhea to severe life-threatening pseudomembranous colitis, a sometimes fatal gastrointestinal disease [1]. These symptoms are generally caused by the production of two toxins (TcdA and TcdB) that glucosylate members of the Rho family GTPases in host cells, thus inducing the disorganization of the actin cytoskeleton, cell death, and an acute inflammatory response [2,3]. Due to their glucosyltransferase activity, both toxins belong to the "large clostridial glucosylating toxins" family (LCGTs), encompassing lethal and hemorrhagic toxins (TcsL and TcsH, respectively) from *Paeniclostridium (formerly Clostridium) sordellii*, alpha-toxin (TcnA) from *Clostridium novyi* and the TpeL toxin from *Clostridium perfringens* [4]. In 1987, Wren's group observed a relationship between the symptoms of antibiotic-associated diarrhea, the *C. difficile* strain, and its ability to produce toxins [5]. Subsequently, a correlation was observed between toxin levels and the severity of CDI [6], which was reinforced in the early 2000s with the emergence in North America and Europe of epidemic and hypervirulent *C. difficile* strains NAPI/027 [7,8]. These strains were responsible for a significant increase in CDI incidence and associated death, and synthesize higher levels of toxins A and B than non-epidemic strains. Therefore, diseases caused by *C. difficile* depend not only on the toxins produced, but also on the control of their synthesis and secretion, which is crucial in the pathogenicity process of *C. difficile*. Thus, deciphering the regulatory mechanisms of toxin production is important for understanding the complex responses triggered by *C. difficile* to the particular nutritional states encountered in the dysbiotic gut during infection.

Numerous studies have been conducted over the past 30 years to better understand the biochemical mode of action of the *C. difficile* toxins [3]. However, little was known about the regulation of *C. difficile* toxins when I joined Linc Sonenshein's laboratory in 1995 to work in this area, mainly due to the difficulty of genetically manipulating this bacterium. Together with von Eichel-Streiber's laboratory, we showed that the expression of toxin genes was dependent on the growth phases (i.e., inhibited during exponential growth and activated when cells enter the stationary phase) [9,10]. Moreover, we and others found that many environmental changes and growth conditions influence toxin levels, in which the nutritional signals with modified concentrations following gut dysbiosis are the most important environmental cues. Thus, it has been shown that limited concentrations of biotin, trehalose, or high amounts of short-chain fatty acids such as butyric acid in the culture medium stimulate toxin production [11,12], while rapidly metabolizable sugars like glucose or amino acids such as cysteine, proline, and branched-chain amino acids (BCAAs) significantly reduce toxin yields [9,13–17]. To date, several environmental stresses and nutritional signals have been reported to also control toxin gene expression [18]. This suggests that regulation of toxin production must be an essential strategy for the adaptation of *C. difficile* to the environmental conditions encountered during gut colonization and infection.

Looking for the molecular mechanisms that control *C. difficile* toxin gene expression depending on environmental signals was the major goal of my research when I joined Stewart Cole's group at the Institut Pasteur in the early 2000s and later when I managed my own group from 2008.

1.1. In the Beginning, There Was the Pathogenicity Locus (PaLoc)

One major advance in the understanding of the mechanism of toxin gene regulation came from the molecular investigation of a 19.6 kb chromosomal region known as the pathogenicity locus (Paloc) that is only found in toxigenic strains of *C. difficile* [19]. The PaLoc contains the genes encoding TcdA (*tcdA*) and TcdB (*tcdB*), and three additional accessory genes, called *tcdR*, *tcdE*, and *tcdC* (Figure 1A). In most *C. difficile* strains, the PaLoc locus is located at the same genomic position and is replaced in the non-toxigenic strains by a non-coding highly conserved 115/75 bp region [19,20]. However, we recently isolated strains with PaLoc loci integrated in different ectopic genomic sites, distant from the usual, unique Paloc integration site considered to date, suggesting that the PaLoc locus have been probably acquired by horizontal transfer [21]. Such atypical organization of the Paloc integration was reinforced in the same year by the work of Janezic et al. [22]. Except for *tcdC*, the PaLoc genes are all coordinately expressed at the entry into stationary phase [10] and we showed that the levels of *tcdA* mRNA were approximately twofold higher compared to those of *tcdB* [9]. This was in agreement with larger amounts of TcdA analyzed after toxin purification [23]. Both *tcdA* and *tcdB* are transcribed mainly from their identified promoters [9,10], while they can also be transcribed by a polycistronic transcript from an upstream promoter [5,9]. We observed that the promoter regions of these toxin genes were not similar to the canonical σ^{70} consensus promoters of prokaryotes, but rather showed strong similarities to each other, as well as to some promoters of other toxin and bacteriocin genes from several *Clostridium* species [24–26], the regulators involved in the transcriptional initiation of which have similarities (see below). PaLoc-like regions are conserved in *P. sordellii* [24], *C. novyi* and *C. perfringens* [27,28] containing the LCGT-encoding genes together with *tcdR*- and *tcdE*-like genes, which supports that the LCGT genes are located within PaLoc-like loci in multiple clostridia species.

1.2. Toxin Genes Are Specifically Transcribed by TcdR, an Alternative Sigma Factor Negatively Controlled by the Anti-Sigma Factor TcdC

Regulation of toxin synthesis is a multifactorial and complex process that allows adaptation of *C. difficile* virulence to external conditions. This currently involves several regulators and sigma factors including first those present in the PaLoc, (i.e., TcdC and TcdR), with opposite roles in toxin expression. While TcdR is a positive regulator of toxin synthesis [29–32], TcdC represses their expression [33]. The *tcdR* gene, located upstream of *tcdB* within the PaLoc (Figure 1A), encodes a small basic protein of 22 kDa, which contains a typical C-terminal helix–turn–helix (HTH) DNA-binding motif [19]. Moncrief et al. presented the first evidence that TcdR was a positive regulator of the *C. difficile* toxin genes [29] and with Linc Sonenshein's laboratory, we showed using genetic and biochemical approaches that TcdR is required for specific transcriptional initiation of the *tcdA* and *tcdB* genes as an alternative sigma factor for RNA polymerase (RNAP) [30]. Interestingly, Ranson et al. [31] showed that a bimodal expression of toxin expression in cell is controlled by the bistability of the TcdR promoter that governs the decision between toxin-On and toxin-OFF status in a subset of cells in the population.

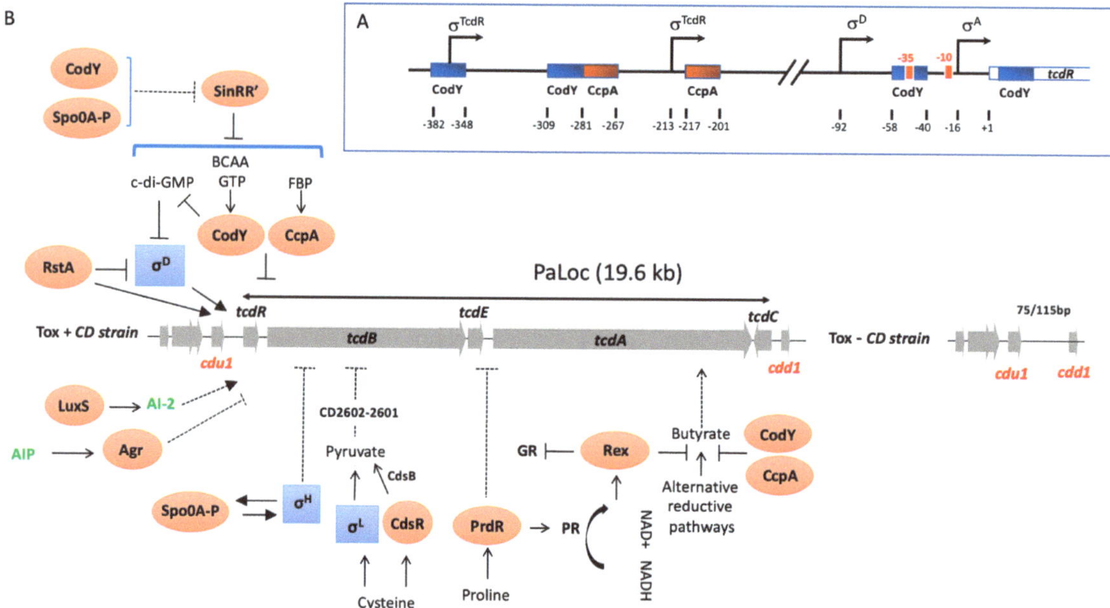

Figure 1. (**A**) Schematic of the promoter regions of *tcdR* denoting the relative locations of the transcriptional start sites experimentally demonstrated [9,32,34]. Blue and red boxes approximate CodY- and CcpA-binding sites within the toxin gene promoters, respectively [15,35,36]. (**B**) Direct and indirect PaLoc regulators and metabolic inputs. Activating metabolites include FBP, Fructose-1,6-bisphosphate; BCAAs, branch chain amino acids; NAD, Nicotinamide adenine dinucleotide; AI-2, auto-inducer 2, AIP, autoinducer peptide, c-di-GMP, cyclic di-guanosyl-5′monophosphate and CdsB, a cysteine desulfidase. Alternative reductive pathways include the Stickland glycine reductase (GR) pathway, succinate utilization pathway and butyrate production and square boxes correspond to alternative σ factors while oval boxes are transcriptional regulators. Arrowed lines indicate positive controls while lines ending with a bar across correspond to negative controls. Dashed arrows indicate mechanisms that are not fully understood.

We demonstrated in addition that TcdR not only activates the initiation of *tcdA* and *tcdB* transcription, but also positively regulates its expression in an autoregulatory manner [32].

This was in agreement with the presence in the region upstream of the *tcdR* gene (Figure 1A) of two potential promoters with the −35 consensus sequence similar to those of the toxin gene promoters [32]. Transcription of the *tcdR* gene is not only positively controlled by TcdR, but also by SigD, a sigma factor that regulates flagellar gene expression, which is consistent with the presence of a SigD-dependent promoter in the promoter region of *tcdR* (Figure 1B) [34].

TcdR belongs to a new sub-group of the σ^{70} family that also encompasses other alternative σ factors of pathogenic clostridia required for the transcription of genes encoding the bacteriocin and cytotoxin of *C. perfringens* (UviA and TpeR, respectively) [25,27], the botulinum and tetanus neurotoxins (BotR and TetR, respectively) [26], the lethal and hemorrhagic toxin genes of *P. sordellii* (TcsR) [24] and the alpha-toxin (TcnA) from *C. novyi* (*TcnR*) [24]. While these sigma factors show similarity to the extracytoplasmic function (ECF) sigma factor family (group IV of the σ^{70}-family), they differ slightly in structure and function, thus classifying them in a distinct phylogenetic sub-family of the σ^{70} family of σ factors. Moreover, we showed that TcdR-related σ factors can substitute for one another, but not for the ECF sigma factor SigW [37], supporting the idea that the TcdR-like proteins can be assigned to an unique group of σ factors (Group V) distinct from the ECF group [24,27,37].

The *tcdC* gene, which is located downstream of the Paloc genes on the opposite strand, is highly expressed during the exponential growth phase. However, its expression is strongly repressed at the onset of stationary growth phase, concomitantly with the transcription start of the other *tcd* genes from their own promoters in a TcdR-dependent manner [10]. In addition, we showed in vitro that expression of TcdC specifically prevents *tcdA* transcription, suggesting that TcdC is likely a negative regulator of toxin gene expression [33]. TcdC is an acidic protein with a predicted molecular weight of 26 kDa [19]. It is a membrane-associated protein [38] that is able to form dimers [33], which is consistent with a coiled-coil domain found in the central region of the protein. Such structural features support the notion that TcdC controls toxin gene transcription through modulation of TcdR activity in an anti-σ factor manner. We showed using genetic and biochemical approaches that TcdC negatively regulates *C. difficile* toxin gene expression by interfering with the ability of the TcdR-containing RNAP holoenzyme to interact with *tcdA* and *tcdB* promoters [33]. However, TcdC can also interact directly with the core RNAP, suggesting that TcdC acts by competing with TcdR to bind to the RNAP core and thereby impairs the formation of TcdR-core complexes [33]. Although these in vitro experiments clearly demonstrated that TcdC interferes with the TcdR-dependent transcription of toxin genes, other in vitro and in vivo studies have shown contradictory results on the involvement of TcdC on toxin gene expression. For instance, chromosomal complementation of the strain R20291 lacking a functional *tcdC* gene, as observed in all NAPI/027 strains, with a functional *tcdC* gene, did not change the toxin titers in vitro [39]. In addition, while *tcdC* genes are widespread among clinical isolates, the presence of *tcdC* cannot predict the hyperproduction of toxins in these strains [39–41]. These conflicting data may be related to the experimental variations between studies including the strains and the growth conditions used that may in part impact TcdC expression or activity. In vivo investigation of isogenic *C. difficile* strains was a prerequisite to clarifying the role of TcdC. This was performed with Dena Lyras's group, who generated an isogenic strain of the *C. difficile* NAPI/027 strain expressing TcdC. We showed that expression of TcdC within the native host downregulates toxin production and attenuates the virulence in the hamster model of infection [42]. Further studies are still required to elucidate the role of TcdC in toxin regulation.

1.3. Toxin Synthesis Is under the Control of Global Metabolic Regulators

One of the most important types of environmental signal controlling toxin production is nutritional compounds such as carbon sources or certain amino acids [9,13,43]. Overall, bacteria have developed mechanisms to uptake carbon and energy sources in the most beneficial and economical way for the cell. This regulation passes through a hierarchy of carbohydrate use. Thus, the presence of a rapidly metabolizable carbon source, such as glucose, inhibits the production of enzymes required for the transport and metabolism of other sugars. This phenomenon is called carbon catabolite repression (CCR). We showed that glucose, as well as other rapidly metabolizable carbon sources like fructose and mannitol, repress PaLoc gene expression [9]. These sugars are usually taken up via the phosphoenolpyruvate (PEP)–phosphotransferase transport system (PTS), a complex carbohydrate transport mechanism found in many Gram-positive and Gram-negative bacteria [44]. Generally, the regulation of gene transcription by such carbon sources involves the CCR system [9]. The CCR mechanism in Gram-positive bacteria, particularly well described in *Bacillus subtilis*, involve three main components. The first is called the catabolite responsive element (*cre*), a *cis*-acting DNA sequence located upstream or in the $5'$ part of catabolic-regulated genes, whose modifications lead to an absence of CCR. The second component is the catabolite control protein A (CcpA), a member of the LacI/GalR family of transcriptional regulators, which in the presence of glucose, binds to the *cre* site of the catabolic-regulated genes or operons modulating their expression [45]. The third component of CCR is the phosphocarrier HPr protein phosphorylated at the regulatory residue Ser-46 (HPr-Ser46-P) by a HPr-kinase/phosphorylase, which interacts with CcpA increasing the affinity of this regulator to *cre* sites [45]. All genes encoding components of the CCR system are present in the *C. difficile* genome. Moreover, potential *cre* sites were found inside promoter regions of PaLoc genes (Figure 1B). Based on *C. difficile* mutant strains defective in the *pstI* gene of the PTS or in *ccpA*, we showed that both uptake of glucose and the global regulator CcpA are required for glucose-dependent repression of toxin genes [15]. However, we observed that the level of toxin production in the *ccpA* mutant grown without glucose was lower than in the parental strain, indicating that CcpA regulated other regulators involved in toxin gene transcription, such as Rex and CodY, as we showed by the transcriptomic analysis of the *ccpA* mutant [35] and below. Furthermore, we demonstrated that CcpA mediates glucose-dependent repression of toxin production by interfering directly with the promoter region or the $5'$ ends of several PaLoc genes, with the strongest affinity for the promoter region of *tcdR* [35]. This is in agreement with the presence of two potential *cre* sites upstream of the transcriptional start of *tcdR* (Figure 1A; [44]). In addition, neither HPr nor HPr-Ser-64-P stimulated CcpA binding to its targets, while FBP alone did, which is somehow different from the standard mode of action of CCR in *B. subtilis*. Glucose also represses the synthesis of LCGT produced by *P. sordellii* and *C. perfringens* [24,27]. Both *P. sordellii* and *C. perfringens* encode CcpA homologs, but their role in the glucose-dependent regulation of toxin production still requires further experimental validation. A recent study showed that the ability of certain hypervirulent *C. difficile* strains, such as 027, to metabolize low levels of trehalose, a glucose disaccharide, increases disease severity through a significant increase in TcdB levels [46], although the mechanism involved is not yet known.

The PaLoc genes are transcribed in a coordinated manner according to the growth phase [10]. In *B. subtilis*, the regulator CodY monitors the nutrient sufficiency of the environment. CodY represses genes that are superfluous in nutrient-rich conditions and releases their repression when nutrients become limited in stationary phase. GTP and BCAAs, such as isoleucine and valine, act as co-repressors of CodY by increasing CodY affinity to its DNA targets [47]. Both isoleucine and valine significantly reduced *C. difficile* toxin synthesis [13]. CodY is conserved in several low-G+C Gram-positive bacteria, where it regulates not only stationary-phase genes, but also virulence factors [47]. In *C. difficile*, CodY acts as a repressor of *tcdR* gene transcription by interacting directly with its promoter region (Figure 1A), leading downstream effects on *tcdB*, *tcdE* and *tcdA* gene expression [36]. As with CcpA, in addition to its direct control of toxin gene expression, CodY also regulates master regulators such as Spo0A and metabolic pathways such as butyrate synthesis involved in toxin production [48]. Thus, regulation of toxin synthesis by both CcpA and CodY provides a molecular link between the metabolic status of the cell and *C. difficile* pathogenicity.

1.4. Toxin Synthesis Is also under the Control of Specific Metabolic Regulators

Among the amino acid pools modified during dysbiosis, cysteine and proline have the strongest effects on toxin production [13]. We showed that cysteine-dependent repression of toxin production is not mediated by a global nutritional regulator involved in toxin repression like Fur, CcpA or CodY, or that it acts as a reducing agent [13]. However, it requires SigL, a sigma factor belonging to the σ^{54} family primarily involved in nitrogen metabolic genes and known to play an important role in the metabolism and virulence of Gram-positive bacteria [49,50]. We have demonstrated that cysteine-dependent repression of toxin production occurs indirectly through accumulation of pyruvate, a direct by-product of cysteine catabolism controlled by SigL [17]. In contrast, addition of the pyruvate by-products, such as formate and acetate did not affect PaLoc gene transcription [17], indicating that pyruvate and sulphide, rather than cysteine, are likely the main signals modulating toxin production. This has been confirmed by the impact of the cysteine desulfidase CdsB, the main enzyme involved in the cysteine degradation whose inactivation prevent the cysteine-repression of toxin production [51]. We recently showed that the regulation of toxins by pyruvate is controlled by a two-component system (CD2602-2601) similar to the *E. coli* YpdA/YpdB TCS system [17]. The presence of proline in the medium not only represses toxin expression but also controls the major pathways of the Stickland reactions (co-fermentation of pairs of amino acids) used by *C. difficile* to produce ATP and regenerate NAD^{+} [16]. In fact, proline induces the regulator PrdR which stimulates synthesis of the proline reductase, one of the key Stickland enzymes. We have shown by global transcriptomic analysis that proline also represses through the proline reductase activity, transcription of alternative NAD^{+}-generating pathways, such as the succinate utilization and butyrate production subsequently to intracellular levels of NADH/NAD+. This suggested that the regulation of toxin expression by proline was probably more related to the redox status than the direct action of PrdR. In several Gram-positive bacteria, the global redox-sensing regulator Rex directly senses changes in the redox status. Rex is only active as a DNA-binding protein when the intracellular $NADH/NAD^{+}$ ratio is low. In a *rex* null mutant of *C. difficile*, the addition of proline did not repress fermentation pathways producing butyryl-CoA from acetyl-CoA or succinate [52]. We demonstrated that, in addition to proline-responsive expression of these alternative reductive pathways, Rex also facilitates the proline-dependent repression of toxin gene expression, probably through the regulation of butyrate known to activate toxin production [52].

1.5. Overall, Toxin Production Is Controlled by a Complex Regulatory Network

As of today, a large panel of regulators have been identified that control toxin production in response to the physiological lifestyle and host environmental stresses, and several reviews summarizing our current knowledge of the regulation of toxin expression have been recently published [53,54]. Among them, it is worth noting that toxin production is monitored by regulatory mechanisms established at the onset of stationary phase that also control the initiation of sporulation. Indeed, SigH, involved in the transcription of major genes of the transition phase and Spo0A, the master regulator of sporulation initiation, indirectly controls toxin gene expression at the onset of the stationary phase. While its impact is strain specific, Spo0A negatively regulates toxin production probably through regulation of *sinRR'* transcription [55,56]. In *C. difficile*, the *sin* locus encodes two regulators (SinR and SinR') which work antagonistically to control motility, sporulation and toxin production in response to growth phase and environmental signals with a clear impact of SinR' that negatively regulates toxin production and motility by interacting with SinR, inhibiting its influence on toxin production by a mechanism not yet known [55,56]. For SigH, we showed that it can repress toxin expression [57], presumably by coordinating the transcription of a gene encoding a repressor of toxin gene transcription such as Spo0A, since SigH is required for its expression [57]. Recently, RstA, a transcriptional regulator of the RNPP generally involved in the quorum sensing [58], was identified as a regulator positively controlling sporulation initiation and negatively impacting mobility and toxin production. It regulates toxin expression by directly binding the promoters of the toxin and *tcdR* genes, as well as the promoter for the *sigD* gene repressing expression of SigD, known to directly control *tcdR* transcription [59]. For many bacterial pathogens, virulence factors are synthesized at high cell density through quorum signaling systems. This cell–cell communication system involves the *agr* quorum-sensing locus, *agrACDB*, which is either complete (*agr2* locus) or incomplete (*agr1 locus*) according to the *C. difficile* strains. In *C. difficile* strains containing the *agr2* locus, inactivation of response regulator encoding genes *agrA* results in the decrease in toxin production presumably through the control of flagellar synthesis and the signaling molecule cyclic di-guanosyl-5′monophosphate (c-di-GMP) metabolism [60]. Indeed, an artificial increase in the intracellular levels of c-di-GMP in *C. difficile* led to switch ON/OFF expression of the *flgB* flagellar operon, including the flagellar alternative σ factor SigD, resulting in repression of *tcdR* transcription and toxin gene expression [61]. In addition to the *agr* system, the quorum sensing molecule AI-2 produced by the *luxS* gene upregulated the expression of PaLoc genes by a mechanism not yet defined [62].

Finally, the global repressor LexA and the recombinase RecA, known to control the SOS regulatory network, plays a key role in the bacterial response to DNA damage [63]. Upon DNA damage, the activated form of RecA facilitates LexA inactivation, resulting in expression of SOS genes [64]. In *C. difficile*, antibiotics known to trigger SOS responses enhanced toxin production when added at subinhibitory doses [65]. In agreement, *C. difficile* LexA not only controls DNA damage but also monitors other biological functions including regulation of toxin A production. Indeed, the production of TcdA but not TcdB increases in a *lexA* mutant compared to the wild-type strain, which is consistent with the ability of LexA to bind to the *tcdA* promoter region, containing a LexA binding motif [66].

2. Conclusions

Since the identification of the Paloc genomic region in 1986 [19], much work has been carried out, in addition to our contribution, on the regulation of *C. difficile* toxin gene expression, which appears to be highly complex, influenced by multiple environmental factors and involving a wide panel of regulators. In the murine model, we showed that toxin synthesis is expressed late during infection [67]. Thus, in vivo toxin production by clostridia must result from a complex regulatory network established along transitional phase and in response to nutrient limitation and stress during gut dysbiosis. Therefore, the mechanisms and niches associated with their toxin upregulation must also be considered as virulence factors in their own right.

Funding: This review received no external funding.

Institutional Review Board Statement: Not applicable.

Informed Consent Statement: Not applicable.

Data Availability Statement: Not applicable.

Acknowledgments: I thank Allyson Holmes for proofreading the manuscript.

Conflicts of Interest: The author declares no conflict of interest.

References

1. Rupnik, M.; Wilcox, M.H.; Gerding, D.N. *Clostridium difficile* infection: New developments in epidemiology and pathogenesis. *Nat. Rev. Microbiol.* **2009**, *7*, 526–536. [CrossRef] [PubMed]
2. Just, I.; Selzer, J.; Wilm, M.; von Eichel-Streiber, C.; Mann, M.; Aktories, K. Glucosylation of Rho proteins by *Clostridium difficile* toxin B. *Nature* **1995**, *375*, 500–503. [CrossRef] [PubMed]
3. Di Bella, S.; Ascenzi, P.; Siarakas, S.; Petrosillo, N.; di Masi, A. *Clostridium difficile* Toxins A and B: Insights into Pathogenic Properties and Extraintestinal Effects. *Toxins* **2016**, *8*, 134. [CrossRef] [PubMed]
4. Jank, T.; Belyi, Y.; Aktories, K. Bacterial glycosyltransferase toxins. *Cell. Microbiol.* **2015**, *17*, 1752–1765. [CrossRef] [PubMed]
5. Wren, B.; Heard, S.R.; Tabaqchali, S. Association between production of toxins A and B and types of *Clostridium difficile*. *J. Clin. Pathol.* **1987**, *40*, 1397–1401. [CrossRef] [PubMed]

6.	Akerlund, T.; Svenungsson, B.; Lagergren, A.; Burman, L.G. Correlation of disease severity with fecal toxin levels in patients with *Clostridium difficile*-associated diarrhea and distribution of PCR ribotypes and toxin yields in vitro of corresponding isolates. *J. Clin. Microbiol.* **2006**, *44*, 353–358. [CrossRef]

7.	Warny, M.; Pepin, J.; Fang, A.; Killgore, G.; Thompson, A.; Brazier, J.; Frost, E.; McDonald, L.C. Toxin production by an emerging strain of *Clostridium difficile* associated with outbreaks of severe disease in North America and Europe. *Lancet* **2005**, *366*, 1079–1084. [CrossRef] [PubMed]

8.	Merrigan, M.; Venugopal, A.; Mallozzi, M.; Roxas, B.; Viswanathan, V.K.; Johnson, S.; Gerding, D.N.; Vedantam, G. Human hypervirulent *Clostridium difficile* strains exhibit increased sporulation as well as robust toxin production. *J. Bacteriol.* **2010**, *192*, 4904–4911. [CrossRef] [PubMed]

9.	Dupuy, B.; Sonenshein, A.L. Regulated transcription of *Clostridium difficile* toxin genes. *Mol. Microbiol.* **1998**, *27*, 107–120. [CrossRef]

10.	Hundsberger, T.; Braun, V.; Weidmann, M.; Leukel, P.; Sauerborn, M.; von Eichel-Streiber, C. Transcription analysis of the genes tcdA-E of the pathogenicity locus of *Clostridium difficile*. *Eur. J. Biochem.* **1997**, *244*, 735–742. [CrossRef]

11.	Yamakawa, K.; Karasawa, T.; Ikoma, S.; Nakamura, S. Enhancement of *Clostridium difficile* toxin production in biotin-limited conditions. *J. Med. Microbiol.* **1996**, *44*, 111–114. [CrossRef] [PubMed]

12.	Karlsson, S.; Lindberg, A.; Norin, E.; Burman, L.G.; Akerlund, T. JToxins, butyric acid and other short-chain fatty acids are coordinately expressed and down regulated by cysteine in *Clostridium difficile*. *Infect. Immun.* **2000**, *68*, 5881–5888. [CrossRef]

13.	Karlsson, S.; Burman, L.G.; Akerlund, T. Suppression of toxin production in *Clostridium difficile* VPI 10463 by amino acids. *Microbiology* **1999**, *145 Pt 7*, 1683–1693. [CrossRef] [PubMed]

14.	Karlsson, S.; Burman, L.G.; Akerlund, T. Induction of toxins in *Clostridium difficile* is associated with dramatic changes of its metabolism. *Microbiology* **2008**, *154*, 3430–3436. [CrossRef] [PubMed]

15.	Antunes, A.; Martin-Verstraete, I.; Dupuy, B. CcpA-mediated repression of *Clostridium difficile* toxin gene expression. *Mol. Microbiol.* **2011**, *79*, 882–899. [CrossRef] [PubMed]

16.	Bouillaut, L.; Self, W.T.; Sonenshein, A.L. Proline-dependent regulation of *Clostridium difficile* Stickland metabolism. *J. Bacteriol.* **2013**, *195*, 844–854. [CrossRef]

17.	Dubois, D.; Dancer-Thibonnier, T.; Monot, M.; Hamiot, M.; Bouillaut, L.; Soutourina, O.; Martin-Verstraete, I.; Dupuy, B. Control of *Clostridium difficile* physiopathology in response to cysteine availability. *Infect. Immun.* **2016**, *84*, 2389–2405. [CrossRef] [PubMed]

18.	Bouillaut, L.; Dubois, T.; Sonenshein, A.L.; Dupuy, B. Integration of metabolism and virulence in *Clostridium difficile*. *Res. Microbiol. Rev* **2014**, *164*, 375–383. [CrossRef] [PubMed]

19.	Braun, V.; Hundsberger, T.; Leukel, P.; Sauerborn, M.; von Eichel-Streiber, C. Definition of the single integration site of the pathogenicity locus in *Clostridium difficile*. *Gene* **1996**, *181*, 29–38. [CrossRef]

20.	Dingle, K.E.; Elliott, B.; Robinson, E.; Griffiths, D.; Eyre, D.W.; Stoesser, N.; Vaughan, A.; Golubchik, T.; Fawley, W.N.; Wilcox, M.H.; et al. Evolutionary history of the *Clostridium difficile* pathogenicity locus. *Genome Biol. Evol.* **2014**, *6*, 36–52. [CrossRef]

21.	Monot, M.; Eckert, C.; Lemire, A.; Hamiot, A.; Dubois, T.; Tessier, C.; Dumoulard, B.; Hamel, B.; Petit, A.; Lalande, V.; et al. *Clostridium difficile*: New Insights into the Evolution of the Pathogenicity Locus. *Sci. Rep.* **2015**, *5*, 15023. [CrossRef] [PubMed]

22.	Janezic, S.; Marin, M.; Martin, A.; Rupnik, M. A new type of toxin A-negative, toxin B-positive *Clostridium difficile* strain lacking a complete tcdA gene. *J. Clin. Microbiol.* **2015**, *53*, 692–695. [CrossRef]

23.	Von Eichel-Streiber, C.; Harperath, U.; Bosse, D.; Hadding, U. Purification of two high molecular weight toxins of *Clostridium difficile* which are antigenically related. *Microb. Pathog.* **1987**, *2*, 307–318. [CrossRef]

24.	Sirigi Reddy, A.R.; Girinathan, B.P.; Zapotocny, R.; Govind, R. Identification and characterization of *Clostridium sordellii* toxin gene regulator. *J. Bacteriol.* **2013**, *195*, 4246–4254. [CrossRef] [PubMed]

25.	Dupuy, B.; Mani, N.; Katayama, S.; Sonenshein, A.L. Transcription activation of a UV-inducible Clostridium perfringens bacteriocin gene by a novel sigma factor. *Mol. Microbiol.* **2005**, *55*, 1196–1206. [CrossRef] [PubMed]

26.	Raffestin, S.; Dupuy, B.; Marvaud, J.C.; Popoff, M.R. BotR/A and TetR are alternative RNA polymerase sigma factors controlling the expression of the neurotoxin and associated protein genes in *Clostridium botulinum* type A and *Clostridium tetani*. *Mol. Microbiol.* **2005**, *55*, 235–249. [CrossRef] [PubMed]

27.	Carter, G.P.; Larcombe, S.; Li, L.; Jayawardena, D.; Awad, M.M.; Songer, J.G.; Lyras, D. Expression of the large clostridial toxins is controlled by conserved regulatory mechanisms. *Int. J. Med. Microbiol.* **2014**, *304*, 1147–1159. [CrossRef]

28.	Saadat, A.; Melville, S. Holin-Dependent Secretion of the Large Clostridial Toxin TpeL by *Clostridium perfringens*. *J. Bacteriol.* **2021**, *203*, e00580-20. [CrossRef]

29.	Moncrief, J.S.; Barroso, L.A.; Wilkins, T.D. Positive regulation of *Clostridium difficile* toxins. *Infect. Immun.* **1997**, *65*, 1105–1108. [CrossRef]

30.	Mani, N.; Dupuy, B. Regulation of toxin synthesis in *Clostridium difficile* by an alternative RNA polymerase sigma factor. *Proc. Natl. Acad. Sci. USA* **2001**, *98*, 5844–5849. [CrossRef]

31.	Ransom, E.M.; Kaus, G.M.; Tran, P.M.; Ellermeier, C.D.; Weiss, D.S. Multiple factors contribute to bimodal toxin gene expression in *Clostridioides (Clostridium) difficile*. *Mol. Microbiol.* **2018**, *110*, 533–549. [CrossRef] [PubMed]

32.	Mani, N.; Lyras, D.; Barroso, L.; Howarth, P.; Wilkins, T.; Rood, J.I.; Sonenshein, A.L.; Dupuy, B. Environmental response and autoregulation of *Clostridium difficile* TxeR, a sigma factor for toxin gene expression. *J. Bacteriol.* **2002**, *184*, 5971–5978. [CrossRef]

33. Matamouros, S.; England, P.; Dupuy, B. *Clostridium difficile* toxin expression is inhibited by the novel regulator TcdC. *Mol. Microbiol.* **2007**, *64*, 1274–1288. [CrossRef]

34. El Meouche, I.; Peltier, J.; Monot, M.; Soutourina, O.; Pestel-Caron, M.; Dupuy, B.; Pons, J.L. Characterization of the SigD regulon of *Clostridium difficile* and its positive control of toxin production through the regulation of *tcdR*. *PLoS ONE* **2013**, *8*, e83748. [CrossRef] [PubMed]

35. Antunes, A.; Camiade, E.; Monot, M.; Courtois, E.; Barbut, F.; Sernova, N.V.; Rodionov, D.A.; Martin-Verstraete, I.; Dupuy, B. Global transcriptional control by glucose and carbon regulator CcpA in *Clostridium difficile*. *Nucleic Acids Res.* **2012**, *40*, 10701–10718. [CrossRef] [PubMed]

36. Dineen, S.S.; Villapakkam, A.C.; Nordman, J.T.; Sonenshein, A.L. Repression of *Clostridium difficile* toxin gene expression by CodY. *Mol. Microbiol.* **2007**, *66*, 206–219. [CrossRef] [PubMed]

37. Dupuy, B.; Raffestin, S.; Matamouros, S.; Mani, N.; Popoff, M.R.; Sonenshein, A.L. Regulation of toxin and bacteriocin gene expression in *Clostridium* by interchangeable RNA polymerase sigma factors. *Mol. Microbiol.* **2006**, *60*, 1044–1057. [CrossRef]

38. Govind, R.; Vediyappan, G.; Rolfe, R.D.; Fralick, J.A. Evidence that *Clostridium difficile* TcdC is a membrane-associated protein. *J. Bacteriol.* **2006**, *188*, 3716–3720. [CrossRef]

39. Cartman, S.T.; Kelly, M.L.; Heeg, D.; Heap, J.T.; Minton, N.P. Precise manipulation of the *Clostridium difficile* chromosome reveals a lack of association between the *tcdC* genotype and toxin production. *Appl. Environ. Microbiol.* **2012**, *78*, 4683–4690. [CrossRef]

40. Curry, S.R.; Marsh, J.W.; Muto, C.A.; O'Leary, M.M.; Pasculle, A.W.; Harrison, L.H. *tcdC* genotypes associated with severe TcdC truncation in an epidemic clone and other strains of *Clostridium difficile*. *J. Clin. Microbiol.* **2007**, *45*, 215–221. [CrossRef]

41. Murray, R.; Boyd, D.; Levett, P.N.; Mulvey, M.R.; Alfa, M.J. Truncation in the *tcdC* region of the *Clostridium difficile* PathLoc of clinical isolates does not predict increased biological activity of Toxin B or Toxin A. *BMC Infect. Dis.* **2009**, *9*, 103. [CrossRef] [PubMed]

42. Carter, G.P.; Douce, G.R.; Govind, R.; Howarth, P.M.; Mackin, K.E.; Spencer, J.; Buckley, A.M.; Antunes, A.; Kotsanas, D.; Jenkin, G.A.; et al. The anti-sigma factor TcdC modulates hypervirulence in an epidemic BI/NAP1/027 clinical isolate of *Clostridium difficile*. *PLoS Pathog.* **2011**, *7*, e1002317. [CrossRef] [PubMed]

43. Kazamias, M.T.; Sperry, J.F. Enhanced fermentation of mannitol and release of cytotoxin by *Clostridium difficile* in alkaline culture media. *Appl. Environ. Microbiol.* **1995**, *61*, 2425–2427. [CrossRef]

44. Deutscher, J.; Francke, C.; Postma, P.W. How phosphotransferase system-related protein phosphorylation regulates carbohydrate metabolism in bacteria. *Microbiol. Mol. Biol. Rev.* **2006**, *70*, 939–1031. [CrossRef]

45. Stulke, J.; Hillen, W. Regulation of carbon catabolism in *Bacillus* species. *Annu. Rev. Microbiol.* **2000**, *54*, 849–880. [CrossRef] [PubMed]

46. Collins, J.; Robinson, C.; Danhof, H.; Knetsch, C.W.; van Leeuwen, H.C.; Lawley, T.D.; Auchtung, J.M.; Britton, R.A. Dietary trehalose enhances virulence of epidemic *Clostridium difficile*. *Nature* **2018**, *553*, 291–294. [CrossRef]

47. Sonenshein, A.L. CodY, a global regulator of stationary phase and virulence in Gram-positive bacteria. *Curr. Opin. Microbiol.* **2005**, *8*, 203–207. [CrossRef]

48. Dineen, S.S.; McBride, S.M.; Sonenshein, A.L. Integration of metabolism and virulence by *Clostridium difficile* CodY. *J. Bacteriol.* **2010**, *192*, 5350–5362. [CrossRef]

49. Dalet, K.; Briand, C.; Cenatiempo, Y.; Hechard, Y. The *rpoN* gene of *Enterococcus faecalis* directs sensitivity to subclass IIa bacteriocins. *Curr. Microbiol.* **2000**, *41*, 441–443. [CrossRef]

50. Okada, Y.; Okada, N.; Makino, S.; Asakura, H.; Yamamoto, S.; Igimi, S. The sigma factor RpoN (sigma54) is involved in osmotolerance in *Listeria monocytogenes*. *FEMS Microbiol. Lett.* **2006**, *263*, 54–60. [CrossRef]

51. Gu, H.; Yang, Y.; Wang, M.; Chen, S.; Wang, H.; Li, S.; Ma, Y.; Wang, J. Novel Cysteine Desulfidase CdsB Involved in Releasing Cysteine Repression of Toxin Synthesis in *Clostridium difficile*. *Front. Cell. Infect. Microbiol.* **2018**, *7*, 531. [CrossRef] [PubMed]

52. Bouillaut, L.; Dubois, T.; Francis, M.B.; Daou, N.; Monot, M.; Sorg, J.A.; Dupuy, B.; Sonenshein, A.L. Role of the global regulator Rex in control of NAD+ -regeneration in *Clostridioides (Clostridium) difficile*. *Mol Microbiol.* **2019**, *111*, 1671–1688. [CrossRef] [PubMed]

53. Martin-Verstraete, I.; Peltier, J.; Dupuy, B. The Regulatory Networks That Control *Clostridium difficile* Toxin Synthesis. *Toxins* **2016**, *8*, 153. [CrossRef]

54. Chandra, H.; Sorg, J.A.; Hassett, D.J.; Sun, X. Regulatory transcription factors of *Clostridioides difficile* pathogenesis with a focus on toxin regulation. *Crit. Rev. Microbiol.* **2022**, *7*, 1–16. [CrossRef] [PubMed]

55. Girinathan, P.B.; Ou, J.; Dupuy, B.; Govind, R. Pleiotropic roles of *Clostridium difficile* sin locus. *PLoS Pathog.* **2018**, *14*, e1006940. [CrossRef]

56. Dhungel, B.A.; Govind, R. SpoOA Suppresses *sin* Locus Expression in *Clostridioides difficile*. *mSphere* **2020**, *5*, e00963-20. [CrossRef]

57. Saujet, L.; Monot, M.; Dupuy, B.; Soutourina, O.; Martin-Verstraete, I. The key sigma factor of transition phase, SigH, controls sporulation, metabolism, and virulence factor expression in *Clostridium difficile*. *J. Bacteriol.* **2011**, *193*, 3186–3196. [CrossRef]

58. Edwards, A.N.; Tamayo, R.; McBride, S.M. A novel regulator controls *Clostridium difficile* sporulation, motility and toxin production. *Mol. Microbiol.* **2016**, *100*, 954–971. [CrossRef] [PubMed]

59. Edwards, N.A.; Anjuwon-Foster, B.R.; McBride, S.M. RstA Is a major regulator of *Clostridioides difficile* toxin production and motility. *mBio* **2019**, *10*, e01991-18. [CrossRef]

60. Martin, M.J.; Clare, S.; Goulding, D.; Faulds-Pain, A.; Barquist, L.; Browne, H.P.; Pettit, L.; Dougan, G.; Lawley, T.D.; Wren, B.W. The *agr* locus regulates virulence and colonization genes in *Clostridium difficile* 027. *J. Bacteriol.* **2013**, *195*, 3672–3681. [CrossRef]
61. McKee, R.W.; Mangalea, M.R.; Purcell, E.B.; Borchardt, E.K.; Tamayo, R. The second messenger cyclic Di-GMP regulates *Clostridium difficile* toxin production by controlling expression of *sigD*. *J. Bacteriol.* **2013**, *195*, 5174–5185. [CrossRef] [PubMed]
62. Lee, A.S.Y.; Song, K.P. LuxS/autoinducer-2 quorum sensing molecule regulates transcriptional virulence gene expression in *Clostridium difficile*. *Biochem. Biophys. Res. Commun.* **2005**, *335*, 659–666. [CrossRef] [PubMed]
63. Butala, M.; Zgur-Bertok, D.; Busby, S.J. The bacterial LexA transcriptional repressor. *Cell. Mol. Life Sci.* **2009**, *66*, 82–93. [CrossRef]
64. Erill, I.; Campoy, S.; Barbe, J. Aeons of distress: An evolutionary perspective on the bacterial SOS response. *FEMS Microbiol. Rev.* **2007**, *31*, 637–656. [CrossRef]
65. Walter, B.M.; Cartman, S.T.; Minton, N.P.; Butala, M.; Rupnik, M. The SOS response master regulator LexA is associated with sporulation, motility and biofilm formation in *Clostridium difficile*. *PLoS ONE* **2015**, *10*, e0144763. [CrossRef] [PubMed]
66. Walter, B.M.; Rupnik, M.; Hodnik, V.; Anderluh, G.; Dupuy, B.; Paulic, N.; Zgur-Bertok, D.; Butala, M. The LexA regulated genes of the *Clostridium difficile*. *BMC Microbiol.* **2014**, *14*, 88. [CrossRef]
67. Janoir, C.; Denève, C.; Bouttier, S.; Barbut, F.; Hoys, S.; Caleechum, L.; Chapetón-Montes, D.; Pereira, F.C.; Henriques, A.O.; Collignon, A.; et al. Adaptive strategies and pathogenesis of *Clostridium difficile* from *in vivo* transcriptomics. *Infect. Immun.* **2013**, *81*, 3757–3769. [CrossRef] [PubMed]

 *toxins*

Article

Does Clostridium Perfringens Epsilon Toxin Mimic an Auto-Antigen Involved in Multiple Sclerosis?

Marie-Lise Gougeon [1,*], Valérie Seffer [1], Cezarela Hoxha [2], Elisabeth Maillart [3] and Michel R. Popoff [2]

[1] Unité Immunité Innée et Virus, Institut Pasteur, 25-28 rue du Dr. Roux, 75724 Paris, Cedex 15, France; valerie.seffer@pasteur.fr

[2] Unité des Toxines Bactériennes, Institut Pasteur, Université Paris Cité, CNRS UMR 2001 INSERM U1306, 75015 Paris, France; cezarela.hoxha@pasteur.fr (C.H.); michel-robert.popoff@pasteur.fr (M.R.P.)

[3] Département de Neurologie, AP-HP, Hôpital Pitié-Salpêtrière, Multiple Sclerosis Center, 75015 Paris, France; elisabeth.maillart@aphp.fr

* Correspondence: marie-lise.gougeon@pasteur.fr

Abstract: Multiple sclerosis (MS) is a chronic immune-mediated neurological disorder, characterized by progressive demyelination and neuronal cell loss in the central nervous system. Many possible causes of MS have been proposed, including genetic factors, environmental triggers, and infectious agents. Recently, *Clostridium perfringens* epsilon toxin (ETX) has been incriminated in MS, based initially on the isolation of the bacteria from a MS patient, combined with an immunoreactivity to ETX. To investigate a putative causative role of ETX in MS, we analyzed the pattern of antibodies reacting to the toxin using a sensitive qualitative assay. This prospective observational study included one hundred patients with relapsing remitting multiple sclerosis (RRMS), all untreated, and ninety matched healthy controls. By assessing the isotypic pattern and serum concentration of ETX-reacting antibodies, our study shows a predominant IgM response over IgG and IgA antibody responses both in MS patients and controls, and significantly higher levels of IgM reacting to ETX in MS patients compared to the control group. A longitudinal follow-up of ETX-specific antibody response in a subgroup of MS patients did not show any correlation with disease evolution. Overall, these unexpected findings are not compatible with a specific recognition of ETX by serum antibodies from MS patients. They rather argue for a cross immunological reactivity with an antigen, possibly an autoantigen, mimicking ETX. Thus, our data argue against the hypothesis of a causal relationship between *C. perfringens* ETX and MS.

Keywords: multiple sclerosis; *Clostridium perfringens*; epsilon toxin ETX; IgM; IgG; IgA antibodies; EDSS

Key Contribution: *Clostridium perfringens* epsilon toxin (ETX) is a demyelinating toxin that has been proposed as a causative agent of multiple sclerosis in humans. Our investigation shows a high prevalence of antibodies against ETX, predominantly IgM over IgG and IgA, in a cohort of one hundred patients as well as in ninety matched healthy controls, arguing for a cross immunological reactivity likely with an autoantigen mimicking ETX.

Received: 15 November 2024
Revised: 29 December 2024
Accepted: 3 January 2025
Published: 7 January 2025

Citation: Gougeon, M.-L.; Seffer, V.; Hoxha, C.; Maillart, E.; Popoff, M.R. Does Clostridium Perfringens Epsilon Toxin Mimic an Auto-Antigen Involved in Multiple Sclerosis? *Toxins* **2025**, *17*, 27. https://doi.org/10.3390/toxins17010027

1. Introduction

Multiple sclerosis (MS) is considered as a chronic immune-mediated disease of the central nervous system (CNS). MS commonly affects adults between the ages of 20 and 50 years, and more often women (female to male sex ratio about 3:1) [1,2]. MS is characterized by progressive inflammation, demyelination, and neuronal cell loss in CNS [1,3,4]. Pathogenesis typically involves perivenular inflammation, blood brain barrier (BBB) opening, demyelination, and a late phase of tissue repair lasting for weeks to months [4,5]. T cells, notably CD4$^+$ T cells, are considered as the main players in the genesis of MS lesions. The immunological processing in MS is further supported by increased levels of cytokines in the serum and cerebrospinal fluid (CSF) of patients. A balance between pro-inflammatory and anti-inflammatory cytokines seems crucial for the progression of the disease [6–9]. Several factors can trigger and/or modulate the immune system and subsequent inflammatory responses. Thus, changes in gut microbiota have been observed in MS patients. Gut dysbiosis can alter the intestinal permeability and facilitate the passage of microbial antigens, leading to immunomodulation and neuroinflammation in CNS [10–13]. However, the precise pathways from gut to neuroinflammation are still poorly understood and the initial cause of MS remains mysterious. A comprehensive description of the aetio-pathogenesis remains to be presented, but environmental factors such as smoking, vitamin D, and certain viral infections such as Epstein–Barr virus (EBV), now firmly established as a risk factor for MS [14], and rubella virus, also associated with increased risk of MS [15], are involved in disease development. Multiple candidate antigens include myelin autoantigens, notably the myelin basic protein and proteolipid protein, neuron- and astrocyte-derived antigens, viral antigens, and various gut microbial antigens or metabolites [8,13,14].

More recently, *Clostridium perfringens* epsilon toxin (ETX) has been incriminated in MS. An ETX-producing *C. perfringens* type B strain has been isolated from a MS patient, and immunoreactivity to ETX was found 10 times more prevalent in people with MS than in healthy controls (HCs) [16]. Another study showed that anti-ETX antibodies were about twice as frequent in MS patients than in the control group [17]. ETX-producing *C. perfringens* was detected more frequently in the gut microbiota of MS patients than in HCs, but at very low levels [18]. ETX is a pore-forming toxin produced by *C. perfringens* type B and D [19–21]. ETX is responsible for an acute and rapidly fatal disease in sheep, goats, and more rarely cattle, termed enterotoxemia [19,22,23]. Overgrowth of ETX-producing *C. perfringens*, mainly *C. perfringens* type D, in the intestine of these animal species results in secretion of high levels of ETX, which crosses the intestinal barrier without enterocyte alteration, disseminates through blood circulation, passes across the BBB, and targets specific cells in CNS such as granule cells, oligodendrocytes, and brain endothelial vascular cells [24–26]. This leads to excessive release of glutamate associated to neurological symptoms of convulsions and opisthotonos, and to brain lesions of perivascular edema and necrosis, which are observed in sheep enterotoxemia [27]. In experimental animals, mouse and rat, sublethal doses of ETX induce a demyelination [18,26,28] and ETX-induced experimental autoimmune encephalomyelitis has been proposed as a suitable model of MS [18]. ETX recognizes the myelin and lymphocyte (MAL) protein receptor, which is expressed in several tissues in humans, but most of them are not affected in MS [29,30]. Thus, a possible role of ETX in MS is questionable [31]. To further investigate if ETX is associated with MS, we analyzed anti-ETX antibodies in the sera of MS patients and healthy controls by qualitative and quantitative tests.

2. Results

2.1. Study Population

Demographics and clinical features are summarized in Table 1. The MS cohort included 100 patients with relapsing remitting MS (RRMS), all untreated at time of sampling. The median age was 36 years, IQR = 30–42, and 77% of MS patients were females. The mean EDSS score was 4.0 ± 1.4 (range 0–7.5), and the mean duration of disease was 9.3 ± 6.6 years (range 1 month–32.5 years). The dynamics of the different parameters and their possible relationship with the expansion or reduction of ETX-specific antibody response were studied in a 3- to 7-year period in a group of 10 patients, whose clinical characteristics are summarized in Table 1. Samples from 90 HCs were used to characterize serum reactivity to ETX in healthy subjects. For univariate analysis of indicated variables between MS and HC, sex- and age-matched controls were used from this cohort.

Table 1. Demographic data and clinical characteristics of the study population.

Study Subjects	n	Female n (%)	MS Subtype	Age at First Sampling (Y) Median (Range)	Disease Duration (Y) (Median, Range)
All MS	100	77 (77.0)	RRMS	36 (18–58)	8.3 (0–32.5)
P1	1	F	RRMS	47	16.7
P2	1	F	RRMS	41	11.3
P3	1	F	RRMS	33	8.3
P4	1	F	RRMS	35	14.3
P5	1	F	RRMS	25	4.3
P6	1	F	RRMS	30	7.5
P7	1	F	RRMS	21	5
P8	1	F	RRMS	21	0.5
P9	1	F	RRMS	32	3.3
P10	1	F	RRMS	41	5.4
Healthy controls	90	47 (52.0)	NA	44 (20–78)	NA

2.2. Immunological Response Against ETX as Tested by Western Blotting

Sera from MS patients were tested by Western blotting with native ETX and 1000-fold serum dilution, as used by Wagley et al. [17]. Typical patterns of immunoreactivity of the sera are shown in Figure 1.

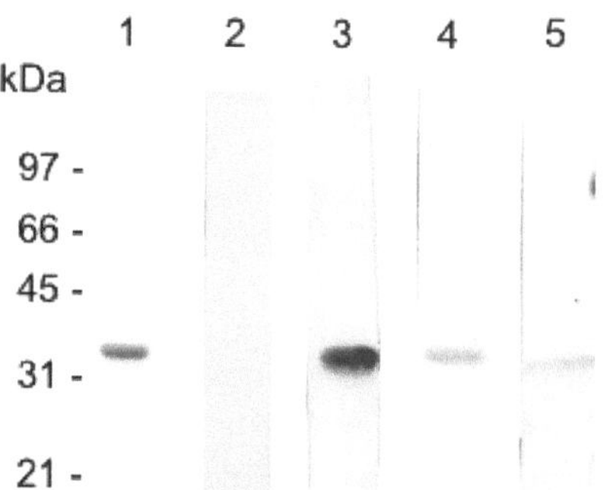

Figure 1. Western blots showing different immunoreactivities of sera to ETX. Lane 1, activated ETX (2.5 µg) stained with Coomassie blue. Lane 2, no reactivity with ETX. Lane 3, strong, Lane 4, medium, and Lane 5, weak immunoreactivity to ETX.

We found that 65.1% of 89 MS sera reacted with ETX as well as 77.7% of healthy controls (Table 2). The strong immune-reactive responses to ETX were slightly similar in both populations. Most MS patients showed a weak immuno-reactivity against ETX, whereas a medium response was found in the sera of healthy controls. However, the distinction between weak and medium reactivity in Western blotting is subjective and not highly precise.

Table 2. Immunoreactivity of the sera with ETX by Western blotting.

Immunoreactivity	Sera from MS Patients n (%)	Sera from Healthy Controls n (%)
strong	10 (14.3%)	10 (11.1%)
medium	14 (15.7%)	36 (40.0%)
weak	34 (38.2%)	24 (26.6%)
negative	31 (34.8%)	20 (22.2%)
Total positive/total number	58/89 (65.1%)	70/90 (77.7%)

2.3. Absence of Immunological Neutralization Response Against ETX

Any of the sera from MS patients (n = 100) and HCs (n = 90) showed a neutralization activity against ETX, as tested at 1:10 dilution in a MDCK cytotoxicity assay monitored by the entry of propidium iodide (PI) (cf. Section 4).

2.4. ETX-Specific Antibody Response in Healthy Controls and MS Patients

To obtain a more precise picture of ETX immunoreactivity in human sera, we used a quantitative ELISA to determine the concentrations of ETX-specific antibodies, as well as their IgM, IgG, and IgA isotypes. We tested sera from 90 HCs and 100 MS patients. Figure 2A shows that the great majority of sera from HCs exhibited antibodies that reacted with ETX, and all three isotypes were detected. The breadth of ETX antibody response was high, varying from 0.1 to 10 μg/mL for IgM, and 0.05 to 5 μg/mL for IgG and IgA. The IgM antibody response was predominant and significantly higher than the IgG and IgA responses. A very similar pattern of antibodies reacting with ETX was detected in samples from MS patients. The median concentration of IgM reacting with ETX (1.42; IQR: 0.92–2.15 μg/mL) was significantly higher than the median concentration of IgG (0.25; IQR: 0.17–0.32) and IgA (0.19; IQR: 0.14–0.32), and it reached very high levels, ranging from 0.38 μg/mL to 43.0 μg/mL. Comparison of ETX-reacting antibodies in MS samples vs. sex- and age-matched healthy controls shows that only the IgM levels were significantly higher in MS patients compared to HCs (MS: median 1.41 μg/mL; IQR: 0.73–2.20 vs. HCs: median 0.84 μg/mL; IQR: 0.40–1.57, $p < 0.001$) (Figure 2A).

Considering that women are more at risk for MS than men and that they account for 77% of our cohort, we addressed the question of the impact of gender on ETX immunoreactivity. Analysis of the data, stratified according to gender within each group, showed that in both MS patients and HCs, the magnitude of ETX-specific IgM and IgA responses was similar in women compared to men, while IgG response was significantly increased in men (Figure 2B). When ETX immunoreactivity was compared between MS patients and HCs after gender stratification, a strong and highly significant increase in ETX-specific IgM response was detected in women with MS as compared to control women, whereas no difference was detected in men (Figure 2C). IgG and IgA responses were similar in women with MS vs. control women.

2.5. ETX-Specific Antibody Response According to Age

The onset of MS most commonly occurs between 20 and 40 years. It was therefore worthwhile analyzing a possible association between ETX-specific responses and age. Age stratification shows that IgM response in MS patients was at the highest level at the age of 30 to 39, then slowly declined while maintaining a consistent level (Figure 3A). This was not observed for ETX IgG or IgA responses. A different pattern was observed in HCs with a significantly higher IgM level at the age of 20 to 29 (Figure 3A).

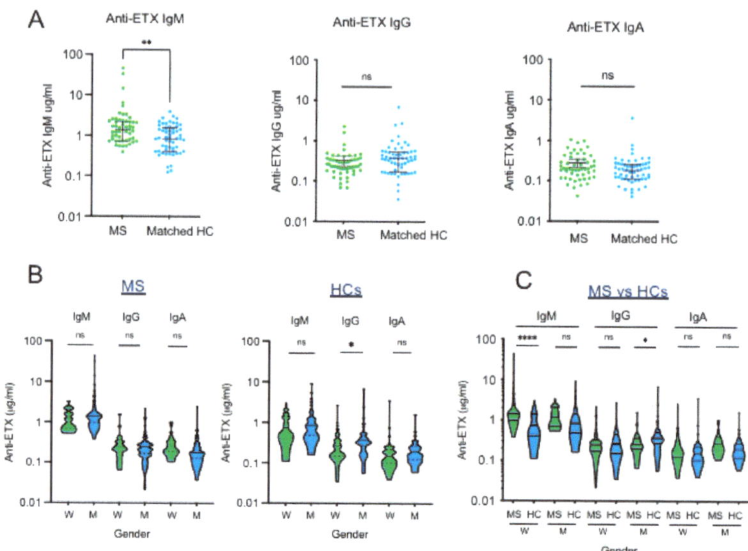

Figure 2. ETX-specific antibody response in MS patients vs. healthy controls. (**A**) Comparison of ETX-reacting IgM, IgG and IgA antibody responses in MS vs. matched HCs. (**B**) Comparison of ETX-reacting IgM, IgG, and IgA antibodies in MS and HCs according to gender. (**C**) Comparison of ETX-reacting IgM, IgG, and IgA antibody responses in MS vs. matched HCs according to gender. * $p < 0.05$; ** $p < 0.01$; **** $p < 0.0001$; ns, not significant.

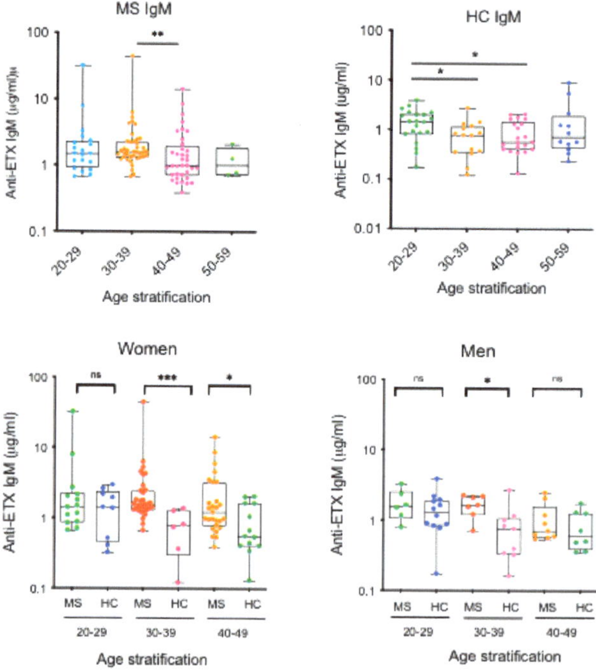

Figure 3. ETX-specific antibody response according to age. (**A**) Serum concentration of ETX-reacting IgM in MS patients and HCs according to age. (**B**) Comparison of serum concentration of ETX-reacting IgM in MS vs. HCs according to age in women and men. * $p < 0.05$; ** $p < 0.01$ *** $p < 0.001$; ns, not significant.

We then focused our analyses on females and asked whether the levels of ETX-IgM would be increased in MS vs. HCs for specific age groups. Figure 3B shows that for both 30–39 and 40–49 age groups, IgM reacting with epsilon toxin were significantly higher in MS compared to HCs ($p < 0.001$ and $p = 0.03$, respectively). The same analysis performed on males shows a significant increase in IgM reacting with epsilon toxin for the 30–39 age group ($p = 0.03$) only, compared to HCs (Figure 3B). Altogether, these data highlight that MS is associated with significantly increased levels of IgM reacting with *C. perfringens* ETX, particularly detected in young women.

2.6. Reactivity of MS Sera with a Linear Peptide Spanning the Amino Acid Sequence of Epsilon Toxin

In their study investigating for serum antibodies against ETX in UK patients with MS, Wagley et al. tested the sera for reactivity with linear overlapping peptides spanning the amino acid sequence of ETX. They found antibodies directed against the membrane insertion loop of domain 2 and especially against the TGVSLTTSYSFANTN peptide in sera from MS patients but not in sera from controls [17]. We tested the reactivity of sera from MS patients and HCs to this peptide. Figure 4A shows that the majority of sera, whether from patients or controls, reacted to this peptide, with a predominant IgM response, significantly higher than IgG and IgA responses. As expected, peptide-specific antibody response was significantly lower than ETX-specific response for the three isotypes (Figure 4B). However, a strong correlation was found between peptide-specific and ETX-specific antibody responses in sera from both MS patients and HCs, but this was observed only for IgM (Figure 4C). Such a correlation was not observed for IgG and IgA responses. Comparison of MS vs. HCs responses to the peptide shows that the level of IgM reacting with epsilon toxin peptide was significantly higher in MS compared to HCs ($p < 0.001$) (Figure 4D), as observed for ETX-specific antibody responses (Figure 2A). In contrast, peptide-specific IgG and IgA responses were similar in MS and HCs (Figure 4D).

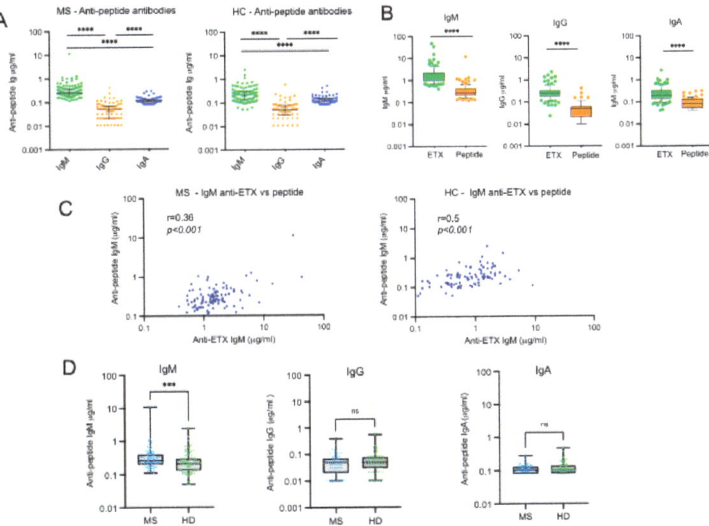

Figure 4. Reactivity of MS sera with ETX linear peptide TGVSLTTSYSFANTN. (**A**) Serum concentration of peptide-reacting IgM, IgG, and IgA antibodies in MS patients (left panel) and HCs (right panel). (**B**) Serum concentration of IgM, IgG, and IgA antibodies reacting to peptide vs. ETX in MS. (**C**) Linear regression of correlation between ETX vs. peptide IgM response in MS patients and HCs. (**D**) Comparison of peptide-reacting IgM, IgG, and IgA responses in MS vs. HCs. *** $p < 0.001$; **** $p < 0.0001$; ns, not significant.

2.7. Correlation Between ETX-Specific Antibody Response and MS Disease Evolution

To understand the meaning of increased levels of ETX-specific IgM in sera from MS patients, we took into consideration several parameters characterizing disease evolution, such as duration of disease, EDSS, and occurrence of relapses. While duration of disease was positively correlated with age ($p < 0.0001$), no correlation was found between disease evolution and ETX-specific IgM or IgG serum concentrations (Figure 5A). Similarly, no correlation was found between peptide-specific IgM and disease evolution. The expanded disability status scale (EDSS) is an ordinal clinical rating scale ranging from 0 (normal neurologic examination) to 10 (death due to MS). In the study MS patients, EDSS ranged from 0 to 7.5. Figure 5B shows that anti-ETX IgM or IgG levels did not vary with EDSS increase, arguing against a correlation between antibody response to epsilon toxin and EDSS. Among 100 MS patients diagnosed with RRMS, 33 had a relapse at time of sampling. Figure 5C shows that the levels of ETX-specific IgM or IgG did not differ between patients in remission and patients experiencing a relapse within this subgroup population.

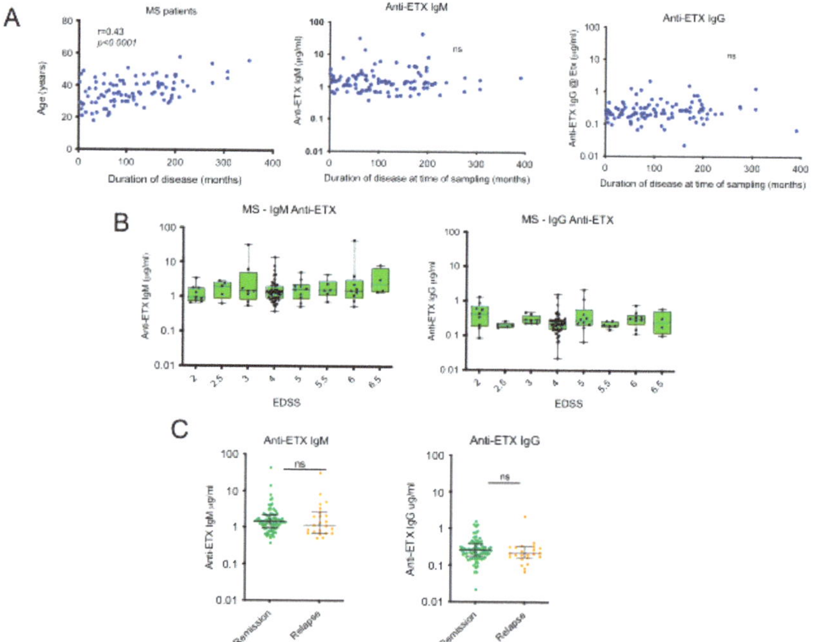

Figure 5. ETX-specific antibody responses and MS disease evolution. (**A**) Linear regression of correlation between age, ETX IgM, ETX IgG, and duration of disease evolution in all samples from MS patients. (**B**) Evolution of ETX-specific IgM (left panel) and IgG (right panel) responses according to EDSS stratification. (**C**) Comparison of anti-ETX IgM and IgG responses in MS patients in remission vs. relapse. ns, not significant.

2.8. Longitudinal Evolution of ETX-Specific Antibody Response in Relation with Disease Evolution

A prospective longitudinal study could be performed over a 2 to 5-year follow-up for 10 patients. Longitudinal serum samples were available at baseline and at three follow-up visits. Clinical characteristics of these patients (P1 to P10) are detailed in Table 1. They were all female, non-treated RRMS, ranging from 21 to 47 years old, whose duration of disease varied from 6 months to 16.7 years, and EDSS ranged from 2.5 to 6.5 at sampling. Except for P10, baseline sampling was performed during a relapse. ETX-specific IgM and IgG antibody response kinetics, combined with EDSS, are shown in Figure 6. At baseline, the levels of IgM reacting with ETX were variable, ranging from 0.53 to 31.5 μg/mL, including levels exceeding 4 μg/mL for four patients (P1, P4, P6, P7), and even reaching > 30 μg/mL for one patient (P7). Apart from one exception (P2), IgG levels at baseline were much lower than IgM levels. Two profiles emerged regarding the kinetics of ETX-specific IgM. A rather flat response curve was observed for P2, P3, P4, P5, P8, and P10, while a steady decline was observed for P6, P7, and P9. We analyzed the relationship between changes in IgM levels and EDSS. In P6 and P7, the decline over time of IgM was associated with a slight decrease in EDSS. A transient decrease of IgM could be associated with a decrease in EDSS (P2, P4), or not (P1). Conversely, a transient increase of IgM could be associated with an increase in EDSS (P4, P5), or not (P1, P8). In P8, it can be noted that the important drop of EDSS (from 4 to 0) was associated with an increase in IgM levels. Overall, these data suggest that there is no evident correlation between the evolution over time of ETX-specific IgM response and the variations of EDSS.

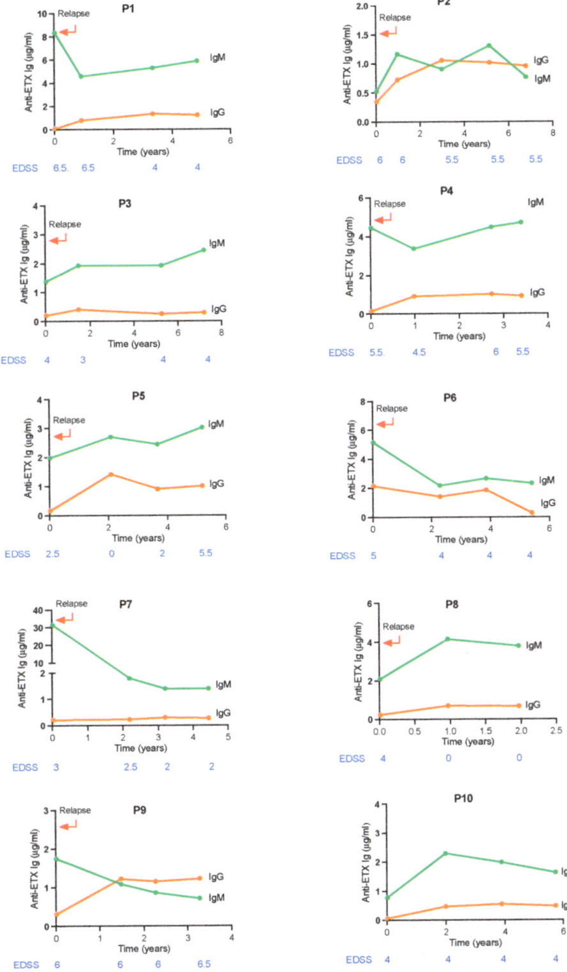

Figure 6. Longitudinal evolution of ETX-specific antibody response in relation to disease evolution. Prospective longitudinal study over a 2 to 5-year follow-up for 10 patients (P1 to P10). Baseline sampling was performed during a relapse, except for P10. The evolution of anti-ETX IgM and IgG responses, combined with EDSS, is shown over time from baseline to 2 to 5-year of follow-up.

3. Discussion

The hypothesis that *C. perfringens* ETX might be an environmental trigger for MS was raised after *C. perfringens* type B was isolated for the first time from a patient at clinical presentation of MS [16]. This finding, together with the known CNS-tropism of *C. perfringens* ETX and binding to oligodendrocytes/myelin, prompted the initiation of serological surveys to assess the immunoreactivity against ETX in MS patients. The survey performed by Rumah et al. in a US population [16] reported immunoreactivity to ETX in about 10% of people with MS and 1% of healthy controls. Another investigation in a UK population of clinically definite multiple sclerosis, performed by Wagley et al. [17], detected seroreactivity to ETX in 24% of the patients and 10% of matched healthy controls. In that study, seroreactivity was also tested against linear overlapping peptides spanning the amino acid sequence of ETX, where 33% of patients' sera reacted to at least one peptide, as compared to 16% in the control group [17]. These two serological surveys used conventional Western blot techniques with methodological limitations for the detection of low-abundance proteins and limited specificity. Against this background, we used a sensitive quantitative ELISA assay to determine the isotype pattern and concentrations of anti-ETX antibodies in sera from untreated RRMS patients and matched controls.

First, using Western blotting, our data confirmed the reactivity to ETX of sera from RRMS patients. However, the overall incidence was higher (65%) than the incidence reported in previous studies [16,17] and the incidence was just as high in the control group. The discrepant results of ETX antibodies in patient's sera by Western blotting are likely related to the different methods used in the distinct studies, including different amounts of ETX antigen and different serum dilutions. In addition, ETX-specific antibodies had no neutralizing activity, in agreement with previous findings [17]. Thus, the detection of antibodies that react to ETX in healthy controls suggests that this reactivity does not characterize an abnormal response linked to MS. The failure of the antibodies to neutralize ETX argues against a toxin-specific antibody response. Indeed, it was reported that *C. perfringens* ETX induces high levels of neutralizing antibodies, either during natural human infection with *C. perfringens* [32] or after immunization of rabbits, sheep, or cattle with *C. perfringens* epsilon-toxin vaccine candidate [33]. Moreover, it is noteworthy that low concentrations of native or recombinant ETX antigen are sufficient to induce high levels of neutralizing antibodies [34,35].

By assessing the isotypic pattern and serum concentration of ETX-reacting antibodies, our study brings new findings that reveal a predominant IgM response over IgG and IgA antibody responses both in MS patients and controls and, moreover, significantly higher levels of IgM reacting to ETX in MS patients compared to controls. As a whole, these findings were unexpected and not consistent with a presumed ETX-specific response. Assuming that ETX production is episodic, due to brief cycles of *C. perfringens* growth followed by long periods of quiescence, as suggested by Ma et al. [18], a dominant IgG antibody response should be detected in MS patients. On the other hand, Huss et al. investigated, by direct detection, the occurrence of ETX and anti-ETX antibodies in MS patients, but neither ETX nor antibodies against it were detected in serum samples, arguing against the hypothesis of a causal relationship between *C. perfringens* ETX and MS [36].

Unspecific immunoreactivity against various antigens is known to occur in MS patients and does not necessarily indicate a primary immune response against a disease related agent [37]. Identified target antigens include neurotropic viruses, such as measles or varicella zoster virus [38], microorganisms such as Chlamydia pneumoniae [39], as well as various self-antigens partly directed against intracellular autoantigens released during tissue destruction [40]. Lipids that are considerably present in the CNS have also been identified as target antigens. In particular, intrathecal IgM recognizing anti-myelin lipids have been detected in MS patients, able to activate complement dependent demyelination and representing a predictor of aggressive evolution in MS [41]. In this line, serum IgM to phosphatidylcholine has recently been identified as a new diagnosis biomarker in RRMS patients and a predictive marker of response to treatment [42,43]. Against this background, we believe that the predominant and long-lasting IgM response that we report in RRMS patients, together with the low affinity and lack of neutralizing activity of the antibodies, is more in favor of a cross-reactivity with an auto-antigen than an ETX-specific immunoreactivity. Autoantibodies are also present in healthy individuals [44,45]. They accumulate throughout life, increase with age [46], and some of them, particularly IgM, are characterized by poly-affinity for various antigens. Natural poly-specific self-reactive IgM antibodies may exert a spectrum of effects from injurious to protective depending upon cellular and molecular context, a role in the regulation of some anti-inflammatory responses [47,48].

Finally, our findings of high prevalence of anti-ETX antibodies in MS patients as well as in HCs contrast with the very low carriage of ETX-producing *C. perfringens* in humans. ETX-producing *C. perfringens* have been detected by a sensitive PCR method in a greater abundance in fecal samples of MS patients compared to healthy people, but at infinitesimal levels. It was speculated that this pathogen could be quiescent for long periods and undergo brief growth phases with ETX production [18]. Despite this, ETX-producing *C. perfringens* was rarely isolated from humans [16,32,49], although it is frequently found in animal species such as sheep [19]. It is noteworthy that MS is more common in urban environments than in rural areas, where the source of human contamination with *C. perfringens* is more accessible [50].

ETX recognizes myelin and lymphocyte protein (MAL) as specific receptor [29,30], which is distributed on endothelial cells in the central nervous system (CNS), oligodendrocytes, peripheral nerve Schwann cells, and mature human T lymphocytes but not on murine T cells [50]. A hallmark of ETX biological activity is ETX targeting oligodendrocytes leading to demyelination, as shown in rat and mice tissue and cell models [26,28,51]. Experimental autoimmune encephalomyelitis (EAE) has been developed in mice and rats and ETX-induced EAE represents a closely related model of MS [18]. ETX can cross the blood brain barrier likely via caveolae-dependent transcytosis through brain endothelial cells and to gain access to brain tissues as investigated in experimental animals upon intravenous ETX administration [25,52]. However, ETX affinity to cells expressing human MAL is about ten times less than cells overexpressing rat of sheep MAL [29,33], suggesting that humans can only develop a mild or chronic form but not an acute ETX disease [21]. However, despite a high ETX affinity for rat MAL, no naturally acquired ETX-induced disease was reported in rats or mice. An initial and crucial step in ETX-producing *C. perfringens* disease is the production of ETX in the intestinal content and passage through the intestinal barrier. Typically, overgrowth of ETX-producing *C. perfringens* in sheep intestinal content subsequent to sudden feeding of large amount of starch-rich food is accompanied by high production of ETX, which passes through the intestinal barrier, disseminates via blood circulation, and enters the CNS leading to characteristic central nervous signs of excitation [19,27]. In contrast, goats, which share an identical MAL protein sequence with sheep MAL (Supplementary Figure S1), mainly develop enterocolitis. It is speculated that ETX is absorbed more readily from sheep intestine than of goats [23,53]. ETX increases intestinal permeability in rats and mice [54], but ETX absorption, through the intestinal barrier according to diverse animal species and humans, remains to be defined. ETX has not yet been detected in fecal and other biological samples of MS patients, suggesting that the ETX antibodies in MS and HCs might result from a cross-reacting antigen with ETX. Moreover, no neutralizing anti-ETX antibodies have been reported in MS patients

([16,17,36] and our study), though ETX-based vaccines induce strong toxin neutralizing response and ETX targets human lymphocytes [33,55].

Gut dysbiosis in MS patients has been identified in numerous studies. Common findings of fecal microbial DNA/RNA investigations in MS consisted of higher ratios of Firmicutes/Bacteroidetes with higher abundance of Streptococcus genus versus lower prevalence of Prevotella genus. However, no increase in Clostridia in the microbiota of MS patients has been shown in these studies [56–59]. The role of gut dysbiosis in MS is not fully understood. However, bacterial metabolites such as short-chain fatty acids and bile acids influence the host immune responses [56]. Gut dysbiosis has been associated not only with MS but with various autoimmune diseases [60]. In this overall context, we believe that the high prevalence of seroreactivity to ETX recognized in MS patients is more compatible with a cross immunological reactivity with an antigen, possibly an autoantigen mimicking ETX, than a toxin-specific response. MS is a complex and heterogeneous disease, traditionally categorized by distinct clinical features, but a new consideration of the course of MS was proposed as a spectrum defined by the relative contributions of overlapping pathological and reparative/compensatory processes [4]. According to Tanja Kuhlmann et al. [4], moving from clinically-based to biologically-based definition of MS progression would contribute to a better understanding of key mechanisms underlying progression and help the implementation of measures to quantify progressive pathology.

4. Conclusions

MS is a multifactorial disease characterized by chronic-immune neurological disorder with progressive demyelination and neuronal cell loss in the central nervous system. *C. perfringens* ETX is a potent toxin that is able to cross the blood brain barrier, and which targets oligodendrocytes and certain neuronal cells by interacting with the MLA receptor. ETX induces demyelination and is responsible for a severe neurological disease in animals, mainly in lambs. ETX has been proposed to be a causative agent of MS in humans. Our investigation on immunological response to ETX by Western blotting and quantitative ELISA shows a high prevalence of anti-ETX antibodies, predominantly IgM over IgG and IgA, in the sera of one hundred MS patients as well as in ninety healthy controls. The wide distribution of anti-ETX antibodies in MS patients as well as in healthy controls supports a cross reactivity of ETX with another antigen, likely an autoantigen, rather than a specific response to ETX as causative agent of MS.

5. Material and Methods

5.1. Patients and Serum Samples

Human biological samples and associated data were obtained from NeuroBioTec (CRB-HCL Hospices Civils de Lyon Biobank BB-0033-00046) and are part of a collection declared at the French Department of Research (DC 2008-72). Informed consent was obtained from all participants. This prospective observational study included patients (n = 100) with relapsing remitting multiple sclerosis (RRMS). Longitudinal serum samples were available at baseline and 4 times over a 3-to-5-year follow-up for 10 patients. None of these patients were treated at the time of study. Healthy controls (n = 90) were selected based on the inclusion criteria of no diagnosis of MS or other neurological disease. They were collected and provided by the Institut Pasteur's Biological Resources Centre (CRBIP). Demographic data and clinical characteristics of patients with MS and healthy controls are described in Table 1.

5.2. Toxin

C. perfringens epsilon prototoxin was produced from *C. perfringens* strain 15246 and purified as previously described [61]. Epsilon prototoxin was activated with TPCK-treated bovine trypsin (Sigma-Aldrich, Sigma Aldrich Chimie S.a.r.l, 38297 Saint-Quentin-Fallavier Cedex, France) for 1 h at room temperature.

5.3. Western Blotting

One μg of ETX was loaded in each well of an SDS-polyacrylamide (10%) gel and then transferred onto nitrocellulose membranes, which were blocked in phosphate-buffered saline containing 0.1% Tween20 (PBST) and 5% dried milk. The membranes were incubated with the sera diluted 1000-fold in PBST overnight at room temperature and then washed with PBST. Detection was performed with goat anti human IgG-peroxidase (Sigma A0170, Sigma Aldrich Chimie S.a.r.l, 38297 Saint-Quentin-Fallavier Cedex, France) 1:3000 in PBST for 1 h at room temperature and enhanced chemiluminescence (Thermo Fisher Scientific, SCI Duguay-Trouin, 44800 Saint-Herblain, France).

5.4. Quantitative ELISA for Epsilon Toxin-Serum and Peptide Antibody Responses

Epsilon toxin or TGVSLTTSYSFANTN peptide (Pepscan, 8243 RC Lelystad, The Netherlands) were used as antigens. Sera were tested by ELISA using a quantitative assay, as previously described [62]. Briefly, 96-well plates (MaxiSorp, Nunc) were coated overnight at 37 °C with 100 μL of epsilon toxin (4 μg/mL) or the peptide (30 μg/mL) diluted in the coating buffer (carbonate/bicarbonate 0.1% *v/v* sodium deoxycholate, pH 9.6). Plates were then washed with the washing buffer (PBS containing 0.1 to 0.5% (*v/v*) Tween®20 (Sigma) to remove unbound antigen. Unsaturated sites were blocked with 200 μL of the saturating buffer (PBS 3% BSA) 1 h at 37 °C. Sera to be tested were diluted with the Multiprobe II® robot in the dilution buffer (PBS 1% BSA 0.1 to 0.5% (*v/v*) Tween®20), and then 100 μL of diluted serum were added in each well and incubated for 1 h at 37 °C. The plates were washed with the washing buffer. Goat anti-human IgG, IgA, or IgM alkaline phosphatase-conjugated antibodies were added and antigen-specific antibodies detected with para-nitrophenyl phosphate (pNPP) substrate. The reaction was then stopped by 100 μL of sodium hydroxide and the optical density was read at 405 nm. The calibration curves were set up using purified polyclonal human IgG, IgA, and IgM (Sigma). The wells were coated with serial dilutions of the corresponding isotypes, ranging from 1.7 to 46 ng/mL for IgM, 0.1 to 18 ng/mL for IgG, and 2 to 70 ng/mL for IgA. The assay was then performed as described above, and the concentrations of serum antibodies were calculated according to the calibration curves. Serum Ig concentrations are expressed in μg/mL.

5.5. Neutralization Assay

Madin–Darby canine kidney (MDCK) cells were grown in Dulbecco's modified Eagle medium (DMEM) supplemented with 10% fetal bovine serum. Confluent cells grown on 96 well plates were incubated with serial dilutions of ETX and propidium iodide (PI) (5 g/mL). At 3 and 18 h, the plates were read with a spectrofluorometer (Fluoroskan II) (excitation 380 nm and emission 620 nm). The results were expressed as the percentage of fluorescence obtained compared to non-intoxicated cells incubated with 0.5% Triton X100 at 37 °C. The last ETX dilution yielding 50% cytotoxicity was considered to contain 1 cytotoxic unit (CU). Sera of MS patients were diluted 1:10 in DMEM containing 4 CU of ETX and PI in a final volume of 100 L and incubated 1 h at 37 °C. Then the mixtures were added to MDCK cells and the lectures were performed at 3 and 18 h.

5.6. Statistical Analysis

Statistical analysis was performed using GraphPad Prism version 9 (GraphPad software, San Diego, CA, USA). Nonparametric measures of associations were used, including the Mann–Whitney U-test, the Wilcoxon signed rank test, linear regression, and Spearman rank correlation. Values of $p < 0.05$ were considered statistically significant.

5.7. Ethics Statement

The study was conducted in accordance with French law relative to clinical non-interventional research. According to the French law on Bioethics (29 July 1994; 6 August 2004; and 7 July 2011, Public Health Code), the patients' written informed consent was collected. Moreover, data confidentiality was ensured in accordance with the recommendations of the French commission for data protection (Commission Nationale Informatique et Liberté, CNIL decision DR-2014-558).

Supplementary Materials: The following supporting information can be downloaded at: https://www.mdpi.com/article/10.3390/toxins17010027/s1, Figure S1: Alignment of MAL protein sequences from sheep (XP_004006224) and goat (XP_017910309). Extracellular loops 1 (ECL1) and 2 (ECL2) which are likely binding receptor sites for ETX according to Rumah, KR, Y Ma, JR Linden, ML Oo, J Anrather, N Schaeren-Wiemers, MA Alonso, VA Fischetti, MS McClain and T Vartanian. The Myelin and Lymphocyte Protein MAL Is Required for Binding and Activity of *Clostridium perfringens* epsilon-Toxin. PLoS Pathog 2015;11:e1004896. [29].

Author Contributions: M.-L.G. and M.R.P. conceived and planned the study. V.S. and C.H. carried out the experiments. M.-L.G., V.S. and M.R.P. analyzed and interpreted the data. M.-L.G. and M.R.P. supervised the project and drafted the manuscript. M.-L.G., V.S., C.H. and E.M. revised and approved the manuscript. All authors have read and agreed to the published version of the manuscript.

Funding: This work received no external funding.

Institutional Review Board Statement: The study was conducted in accordance with the guidelines of the Declaration of Helsinki. Ethical review and approval were waived for this study as it was a retrospective non-interventional study. Data acquisition and recording were approved by the French National Data Protection Commission (Commission Nationale de l' Informatique et des Libertés; reference: DEC20-058).

Informed Consent Statement: Written informed consent was obtained from all subjects involved in this study.

Data Availability Statement: The original contributions presented in this study are included in the article/Supplementary Materials. Further inquiries can be directed to the corresponding author.

Acknowledgments: We wish to thank Richard W Titball for his valuable insights. We also thank NeuroBioTec (CRB-HCL Hospices Civils de Lyon Biobank BB-0033-00046) who provided the biological samples and associated data from the MS patients.

Conflicts of Interest: The authors declare no conflicts of interest.

References

1. Haki, M.; Al-Biati, H.A.; Al-Tameemi, Z.S.; Ali, I.S.; Al-Hussaniy, H.A. Review of multiple sclerosis: Epidemiology, etiology, pathophysiology, and treatment. *Medicine* **2024**, *103*, e37297. [CrossRef]
2. McGinley, M.P.; Goldschmidt, C.H.; Rae-Grant, A.D. Diagnosis and Treatment of Multiple Sclerosis: A Review. *JAMA* **2021**, *325*, 765–779. [CrossRef] [PubMed]
3. Filippi, M.; Preziosa, P.; Rocca, M.A. MRI in multiple sclerosis: What is changing? *Curr. Opin. Neurol.* **2018**, *31*, 386–395. [CrossRef] [PubMed]

4. Kuhlmann, T.; Moccia, M.; Coetzee, T.; Cohen, J.A.; Correale, J.; Graves, J.; Marrie, R.A.; Montalban, X.; Yong, V.W.; Thompson, A.J.; et al. Multiple sclerosis progression: Time for a new mechanism-driven framework. *Lancet Neurol.* **2023**, *22*, 78–88. [CrossRef] [PubMed]

5. Absinta, M.; Sati, P.; Reich, D.S. Advanced MRI and staging of multiple sclerosis lesions. *Nat. Rev. Neurol.* **2016**, *12*, 358–368. [CrossRef] [PubMed]

6. Grunwald, C.; Kretowska-Grunwald, A.; Adamska-Patruno, E.; Kochanowicz, J.; Kulakowska, A.; Chorazy, M. The Role of Selected Interleukins in the Development and Progression of Multiple Sclerosis-A Systematic Review. *Int. J. Mol. Sci.* **2024**, *25*, 2589. [CrossRef]

7. Khan, Z.; Mehan, S.; Gupta, G.D.; Narula, A.S. Immune System Dysregulation in the Progression of Multiple Sclerosis: Molecular Insights and Therapeutic Implications. *Neuroscience* **2024**, *548*, 9–26. [CrossRef] [PubMed]

8. Riedhammer, C.; Weissert, R. Antigen Presentation, Autoantigens, and Immune Regulation in Multiple Sclerosis and Other Autoimmune Diseases. *Front. Immunol.* **2015**, *6*, 322. [CrossRef]

9. Wagner, C.A.; Roque, P.J.; Goverman, J.M. Pathogenic T cell cytokines in multiple sclerosis. *J. Exp. Med.* **2020**, *217*, e20190460. [CrossRef]

10. Christovich, A.; Luo, X.M. Gut Microbiota, Leaky Gut, and Autoimmune Diseases. *Front. Immunol.* **2022**, *13*, 946248. [CrossRef] [PubMed]

11. Hoffman, K.; Doyle, W.J.; Schumacher, S.M.; Ochoa-Reparaz, J. Gut microbiome-modulated dietary strategies in EAE and multiple sclerosis. *Front. Nutr.* **2023**, *10*, 1146748. [CrossRef] [PubMed]

12. Tian, H.; Huang, D.; Wang, J.; Li, H.; Gao, J.; Zhong, Y.; Xia, L.; Zhang, A.; Lin, Z.; Ke, X. The role of the "gut microbiota-mitochondria" crosstalk in the pathogenesis of multiple sclerosis. *Front. Microbiol.* **2024**, *15*, 1404995. [CrossRef] [PubMed]

13. Diaz-Marugan, L.; Kantsjo, J.B.; Rutsch, A.; Ronchi, F. Microbiota, diet, and the gut-brain axis in multiple sclerosis and stroke. *Eur. J. Immunol.* **2023**, *53*, e2250229. [CrossRef] [PubMed]

14. Bjornevik, K.; Cortese, M.; Healy, B.C.; Kuhle, J.; Mina, M.J.; Leng, Y.; Elledge, S.J.; Niebuhr, D.W.; Scher, A.I.; Munger, K.L.; et al. Longitudinal analysis reveals high prevalence of Epstein-Barr virus associated with multiple sclerosis. *Science* **2022**, *375*, 296–301. [CrossRef] [PubMed]

15. Kiani, L. Rubella virus might increase risk of MS. *Nat. Rev. Neurol.* **2024**, *20*, 505. [CrossRef] [PubMed]

16. Rumah, K.R.; Linden, J.; Fischetti, V.A.; Vartanian, T. Isolation of *Clostridium perfringens* type B in an individual at first clinical presentation of multiple sclerosis provides clues for environmental triggers of the disease. *PLoS ONE* **2013**, *8*, e76359. [CrossRef]

17. Wagley, S.; Bokori-Brown, M.; Morcrette, H.; Malaspina, A.; D'Arcy, C.; Gnanapavan, S.; Lewis, N.; Popoff, M.R.; Raciborska, D.; Nicholas, R.; et al. Evidence of *Clostridium perfringens* epsilon toxin associated with multiple sclerosis. *Mult. Scler. J.* **2019**, *25*, 653–660. [CrossRef]

18. Ma, Y.; Sannino, D.; Linden, J.R.; Haigh, S.; Zhao, B.; Grigg, J.B.; Zumbo, P.; Dundar, F.; Butler, D.; Profaci, C.P.; et al. Epsilon toxin-producing *Clostridium perfringens* colonize the multiple sclerosis gut microbiome overcoming CNS immune privilege. *J. Clin. Investig.* **2023**, *133*, e163239. [CrossRef]

19. Popoff, M.R. Epsilon toxin: A fascinating pore-forming toxin. *FEBS J.* **2011**, *278*, 4602–4615. [CrossRef] [PubMed]

20. Rood, J.I.; Adams, V.; Lacey, J.; Lyras, D.; McClane, B.A.; Melville, S.B.; Moore, R.J.; Popoff, M.R.; Sarker, M.R.; Songer, J.G.; et al. Expansion of the *Clostridium perfringens* toxin-based typing scheme. *Anaerobe* **2018**, *53*, 5–10. [CrossRef] [PubMed]

21. Titball, R.W. The Molecular Architecture and Mode of Action of *Clostridium perfringens* epsilon-Toxin. *Toxins* **2024**, *16*, 180. [CrossRef] [PubMed]

22. Uzal, F.A.; Fisher, D.J.; Saputo, J.; Sayeed, S.; McClane, B.A.; Songer, G.; Trinh, H.T.; Fernandez Miyakawa, M.E.; Gard, S. Ulcerative enterocolitis in two goats associated with enterotoxin- and beta2 toxin-positive *Clostridium perfringens* type D. *J. Vet. Diagn. Investig.* **2008**, *20*, 668–672. [CrossRef] [PubMed]

23. Uzal, F.A.; Navarro, M.A.; Li, J.; Freedman, J.C.; Shrestha, A.; McClane, B.A. Comparative pathogenesis of enteric clostridial infections in humans and animals. *Anaerobe* **2018**, *53*, 11–20. [CrossRef] [PubMed]

24. Lonchamp, E.; Dupont, J.L.; Wioland, L.; Courjaret, R.; Mbebi-Liegeois, C.; Jover, E.; Doussau, F.; Popoff, M.R.; Bossu, J.L.; de Barry, J.; et al. *Clostridium perfringens* epsilon toxin targets granule cells in the mouse cerebellum and stimulates glutamate release. *PLoS ONE* **2010**, *5*, e13046. [CrossRef] [PubMed]

25. Soler-Jover, A.; Dorca, J.; Popoff, M.R.; Gibert, M.; Saura, J.; Tusell, J.M.; Serratosa, J.; Blasi, J.; Martin-Satue, M. Distribution of *Clostridium perfringens* epsilon toxin in the brains of acutely intoxicated mice and its effect upon glial cells. *Toxicon* **2007**, *50*, 530–540. [CrossRef] [PubMed]

26. Wioland, L.; Dupont, J.L.; Doussau, F.; Gaillard, S.; Heid, F.; Isope, P.; Pauillac, S.; Popoff, M.R.; Bossu, J.L.; Poulain, B. Epsilon toxin from *Clostridium perfringens* acts on oligodendrocytes without forming pores, and causes demyelination. *Cell. Microbiol.* **2015**, *17*, 369–388. [CrossRef] [PubMed]

27. Finnie, J.W.; Uzal, F.A. Pathology and Pathogenesis of Brain Lesions Produced by *Clostridium perfringens* Type D Epsilon Toxin. *Int. J. Mol. Sci.* **2022**, *23*, 9050. [CrossRef]

28. Linden, J.R.; Ma, Y.; Zhao, B.; Harris, J.M.; Rumah, K.R.; Schaeren-Wiemers, N.; Vartanian, T. *Clostridium perfringens* Epsilon Toxin Causes Selective Death of Mature Oligodendrocytes and Central Nervous System Demyelination. *mBio* **2015**, *6*, e02513. [CrossRef]

29. Rumah, K.R.; Ma, Y.; Linden, J.R.; Oo, M.L.; Anrather, J.; Schaeren-Wiemers, N.; Alonso, M.A.; Fischetti, V.A.; McClain, M.S.; Vartanian, T. The Myelin and Lymphocyte Protein MAL Is Required for Binding and Activity of *Clostridium perfringens* epsilon-Toxin. *PLoS Pathog.* **2015**, *11*, e1004896. [CrossRef]

30. Blanch, M.; Dorca-Arevalo, J.; Not, A.; Cases, M.; Gomez de Aranda, I.; Martinez-Yelamos, A.; Martinez-Yelamos, S.; Solsona, C.; Blasi, J. The Cytotoxicity of Epsilon Toxin from *Clostridium perfringens* on Lymphocytes Is Mediated by MAL Protein Expression. *Mol. Cell. Biol.* **2018**, *38*, 00086-18. [CrossRef]

31. Titball, R.W.; Lewis, N.; Nicholas, R. Is *Clostridium perfringens* epsilon toxin associated with multiple sclerosis? *Mult. Scler. J.* **2023**, *29*, 1057–1063. [CrossRef]

32. Kohn, J.; Warrack, G.H. Recovery of Clostridium welchii type D from man. *Lancet* **1955**, *268*, 385. [CrossRef] [PubMed]

33. Morcrette, H.; Bokori-Brown, M.; Ong, S.; Bennett, L.; Wren, B.W.; Lewis, N.; Titball, R.W. *Clostridium perfringens* epsilon toxin vaccine candidate lacking toxicity to cells expressing myelin and lymphocyte protein. *NPJ Vaccines* **2019**, *4*, 32. [CrossRef] [PubMed]

34. Lobato, F.C.; Lima, C.G.; Assis, R.A.; Pires, P.S.; Silva, R.O.; Salvarani, F.M.; Carmo, A.O.; Contigli, C.; Kalapothakis, E. Potency against enterotoxemia of a recombinant *Clostridium perfringens* type D epsilon toxoid in ruminants. *Vaccine* **2010**, *28*, 6125–6127. [CrossRef]

35. Moreira, G.M.; Salvarani, F.M.; da Cunha, C.E.; Mendonca, M.; Moreira, A.N.; Goncalves, L.A.; Pires, P.S.; Lobato, F.C.; Conceicao, F.R. Immunogenicity of a Trivalent Recombinant Vaccine Against *Clostridium perfringens* Alpha, Beta, and Epsilon Toxins in Farm Ruminants. *Sci. Rep.* **2016**, *6*, 22816. [CrossRef]

36. Huss, A.; Bachhuber, F.; Feraudet-Tarisse, C.; Hiergeist, A.; Tumani, H. Multiple Sclerosis and *Clostridium perfringens* Epsilon Toxin: Is There a Relationship? *Biomedicines* **2024**, *12*, 1392. [CrossRef]

37. Hoftberger, R.; Lassmann, H.; Berger, T.; Reindl, M. Pathogenic autoantibodies in multiple sclerosis—From a simple idea to a complex concept. *Nat. Rev. Neurol.* **2022**, *18*, 681–688. [CrossRef]

38. Reiber, H.; Ungefehr, S.; Jacobi, C. The intrathecal, polyspecific and oligoclonal immune response in multiple sclerosis. *Mult. Scler. J.* **1998**, *4*, 111–117. [CrossRef] [PubMed]

39. Rostasy, K.; Reiber, H.; Pohl, D.; Lange, P.; Ohlenbusch, A.; Eiffert, H.; Maass, M.; Hanefeld, F. Chlamydia pneumoniae in children with MS: Frequency and quantity of intrathecal antibodies. *Neurology* **2003**, *61*, 125–128. [CrossRef] [PubMed]

40. Brandle, S.M.; Obermeier, B.; Senel, M.; Bruder, J.; Mentele, R.; Khademi, M.; Olsson, T.; Tumani, H.; Kristoferitsch, W.; Lottspeich, F.; et al. Distinct oligoclonal band antibodies in multiple sclerosis recognize ubiquitous self-proteins. *Proc. Natl. Acad. Sci. USA* **2016**, *113*, 7864–7869. [CrossRef]

41. Villar, L.M.; Sadaba, M.C.; Roldan, E.; Masjuan, J.; Gonzalez-Porque, P.; Villarrubia, N.; Espino, M.; Garcia-Trujillo, J.A.; Bootello, A.; Alvarez-Cermeno, J.C. Intrathecal synthesis of oligoclonal IgM against myelin lipids predicts an aggressive disease course in MS. *J. Clin. Investig.* **2005**, *115*, 187–194. [CrossRef] [PubMed]

42. Munoz, U.; Sebal, C.; Escudero, E.; Urcelay, E.; Arroyo, R.; Garcia-Martinez, M.A.; Quintana, F.J.; Alvarez-Lafuente, R.; Sadaba, M.C. Serum levels of IgM to phosphatidylcholine predict the response of multiple sclerosis patients to natalizumab or IFN-beta. *Sci. Rep.* **2022**, *12*, 13357. [CrossRef] [PubMed]

43. Sanchez-Vera, I.; Escudero, E.; Munoz, U.; Sadaba, M.C. IgM to phosphatidylcholine in multiple sclerosis patients: From the diagnosis to the treatment. *Ther. Adv. Neurol. Disord.* **2023**, *16*, 17562864231189919. [CrossRef] [PubMed]

44. Avrameas, S.; Selmi, C. Natural autoantibodies in the physiology and pathophysiology of the immune system. *J. Autoimmun.* **2013**, *41*, 46–49. [CrossRef]

45. Merbl, Y.; Zucker-Toledano, M.; Quintana, F.J.; Cohen, I.R. Newborn humans manifest autoantibodies to defined self molecules detected by antigen microarray informatics. *J. Clin. Investig.* **2007**, *117*, 712–718. [CrossRef]

46. Nagele, E.P.; Han, M.; Acharya, N.K.; DeMarshall, C.; Kosciuk, M.C.; Nagele, R.G. Natural IgG autoantibodies are abundant and ubiquitous in human sera, and their number is influenced by age, gender, and disease. *PLoS ONE* **2013**, *8*, e60726. [CrossRef]

47. Gronwall, C.; Silverman, G.J. Natural IgM: Beneficial autoantibodies for the control of inflammatory and autoimmune disease. *J. Clin. Immunol.* **2014**, *34* (Suppl. S1), S12–S21. [CrossRef]

48. Panda, S.; Ding, J.L. Natural antibodies bridge innate and adaptive immunity. *J. Immunol.* **2015**, *194*, 13–20. [CrossRef] [PubMed]

49. Gleeson-White, M.H.; Bullen, J.J. Clostridium welchii epsilon toxin in the intestinal contents of man. *Lancet* **1955**, *268*, 384–385. [CrossRef] [PubMed]

50. Reder, A.T. *Clostridium* epsilon toxin is excessive in multiple sclerosis and provokes multifocal lesions in mouse models. *J. Clin. Investig.* **2023**, *133*, e169643. [CrossRef] [PubMed]

51. Dorca-Arevalo, J.; Soler-Jover, A.; Gibert, M.; Popoff, M.R.; Martin-Satue, M.; Blasi, J. Binding of epsilon-toxin from *Clostridium perfringens* in the nervous system. *Vet. Microbiol.* **2008**, *131*, 14–25. [CrossRef] [PubMed]

52. Linden, J.R.; Flores, C.; Schmidt, E.F.; Uzal, F.A.; Michel, A.O.; Valenzuela, M.; Dobrow, S.; Vartanian, T. *Clostridium perfringens* epsilon toxin induces blood brain barrier permeability via caveolae-dependent transcytosis and requires expression of MAL. *PLoS Pathog.* **2019**, *15*, e1008014. [CrossRef] [PubMed]

53. Uzal, F.A.; Rolfe, B.E.; Smith, N.J.; Thomas, A.C.; Kelly, W.R. Resistance of ovine, caprine and bovine endothelial cells to *Clostridium perfringens* type D epsilon toxin in vitro. *Vet. Res. Commun.* **1999**, *23*, 275–284. [CrossRef]

54. Goldstein, J.; Morris, W.E.; Loidl, C.F.; Tironi-Farinati, C.; McClane, B.A.; Uzal, F.A.; Fernandez Miyakawa, M.E. *Clostridium perfringens* epsilon toxin increases the small intestinal permeability in mice and rats. *PLoS ONE* **2009**, *4*, e7065. [CrossRef] [PubMed]

55. Shetty, S.V.; Mazzucco, M.R.; Winokur, P.; Haigh, S.V.; Rumah, K.R.; Fischetti, V.A.; Vartanian, T.; Linden, J.R. *Clostridium perfringens* Epsilon Toxin Binds to and Kills Primary Human Lymphocytes. *Toxins* **2023**, *15*, 423. [CrossRef] [PubMed]

56. Ouyang, Q.; Yu, H.; Xu, L.; Yu, M.; Zhang, Y. Relationship between gut microbiota and multiple sclerosis: A scientometric visual analysis from 2010 to 2023. *Front. Immunol.* **2024**, *15*, 1451742. [CrossRef]

57. Torres-Chavez, M.E.; Torres-Carrillo, N.M.; Monreal-Lugo, A.V.; Garnes-Rancurello, S.; Murugesan, S.; Gutierrez-Hurtado, I.A.; Beltran-Ramirez, J.R.; Sandoval-Pinto, E.; Torres-Carrillo, N. Association of intestinal dysbiosis with susceptibility to multiple sclerosis: Evidence from different population studies (Review). *Biomed. Rep.* **2023**, *19*, 93. [CrossRef]

58. Zhang, W.; Wang, Y.; Zhu, M.; Liu, K.; Zhang, H.L. Gut flora in multiple sclerosis: Implications for pathogenesis and treatment. *Neural Regen. Res.* **2024**, *19*, 1480–1488. [CrossRef] [PubMed]

59. Dunalska, A.; Saramak, K.; Szejko, N. The Role of Gut Microbiome in the Pathogenesis of Multiple Sclerosis and Related Disorders. *Cells* **2023**, *12*, 1760. [CrossRef]

60. Chang, S.H.; Choi, Y. Gut dysbiosis in autoimmune diseases: Association with mortality. *Front. Cell Infect. Microbiol.* **2023**, *13*, 1157918. [CrossRef]

61. Cole, A.R.; Gibert, M.; Popoff, M.; Moss, D.S.; Titball, R.W.; Basak, A.K. *Clostridium perfringens* epsilon-toxin shows structural similarity to the pore-forming toxin aerolysin. *Nat. Struct. Mol. Biol.* **2004**, *11*, 797–798. [CrossRef] [PubMed]

62. Launay, O.; Sadorge, C.; Jolly, N.; Poirier, B.; Bechet, S.; van der Vliet, D.; Seffer, V.; Fenner, N.; Dowling, K.; Giemza, R.; et al. Safety and immunogenicity of SC599, an oral live attenuated Shigella dysenteriae type-1 vaccine in healthy volunteers: Results of a Phase 2, randomized, double-blind placebo-controlled trial. *Vaccine* **2009**, *27*, 1184–1191. [CrossRef] [PubMed]

Review

Animal Toxins: A Historical Outlook at the Institut Pasteur of Paris

Michel R. Popoff [1,*], **Grazyna Faure** [2], **Sandra Legout** [3] **and Daniel Ladant** [4]

[1] Unité des Toxines Bactériennes, Institut Pasteur, Université Paris Cité, CNRS UMR 2001 INSERM U1306, F-75015 Paris, France
[2] Unité Récepteurs-Canaux, Institut Pasteur, Université Paris Cité, CNRS UMR 3571, F-75015 Paris, France; fgrazyna@pasteur.fr
[3] Centre de Ressources et Information Scientifique, Institut Pasteur, Université Paris Cité, F-75015 Paris, France; sandra.legout@pasteur.fr
[4] Unité de Biochimie des Interactions Macromoléculaires, Institut Pasteur, Université Paris Cité, CNRS UMR 3528, F-75015 Paris, France; daniel.ladant@pasteur.fr
* Correspondence: popoff2m@gmail.com

Abstract: Humans have faced poisonous animals since the most ancient times. It is recognized that certain animals, like specific plants, produce toxic substances that can be lethal, but that can also have therapeutic or psychoactive effects. The use of the term "venom", which initially designated a poison, remedy, or magic drug, is now confined to animal poisons delivered by biting. Following Louis Pasteur's work on pathogenic microorganisms, it was hypothesized that venoms could be related to bacterial toxins and that the process of pathogenicity attenuation could be applied to venoms for the prevention and treatment of envenomation. Cesaire Phisalix and Gabriel Bertrand from the National Museum of Natural History as well as Albert Calmette from the Institut Pasteur in Paris were pioneers in the development of antivenomous serotherapy. Gaston Ramon refined the process of venom attenuation for the immunization of horses using a formalin treatment method that was successful for diphtheria and tetanus toxins. This paved the way for the production of antivenomous sera at the Institut Pasteur, as well as for research on venom constituents and the characterization of their biological activities. The specific activities of certain venom components, such as those involved in blood coagulation or the regulation of chloride ion channels, raises the possibility of developing novel therapeutic drugs that could serve as anticoagulants or as a treatment for cystic fibrosis, for example. Scientists of the Institut Pasteur of Paris have significantly contributed to the study of snake venoms, a topic that is reported in this review.

Keywords: Pasteur; animal toxin; venom; vaccine; serotherapy; therapeutic peptides; phospholipase A_2

Key Contribution: For the bicentennial of Louis Pasteur's birth, this review retraces the historical aspects of the contribution of scientists of the Institut Pasteur of Paris to the study of animal toxins and the development of antivenom serotherapy and novel therapeutic peptides.

Citation: Popoff, M.R.; Faure, G.; Legout, S.; Ladant, D. Animal Toxins: A Historical Outlook at the Institut Pasteur of Paris. *Toxins* **2023**, *15*, 462. https://doi.org/10.3390/toxins15070462

Received: 19 June 2023
Revised: 10 July 2023
Accepted: 14 July 2023
Published: 19 July 2023

1. Introduction

Toxins and poisonings have been known since the most ancient times. During the Paleolithic era in Europe, hunters used poisons. Indeed, paleontologists have discovered unusual grooves on the tips of certain bones used as arrows, suggesting they contained plant or animal poisons. The use of poisons, predominantly from a plant origin, on arrowheads appears to have been widespread among various civilizations across all continents for hunting purposes [1]. The term "toxin" is derived from the Greek word "toxon", which refers to a bow, implying poisoned arrows, while "poison" is a broader term originating from the Latin "potionem" (meaning a drink). Initially, it denoted harmful liquid substances, and later, it encompassed any dangerous substances. The earliest civilizations were familiar with animal poisons, often using them in combination with plant toxins for hunting and fishing. They also recognized toxins for their potential therapeutic or psychoactive effects, capable of inciting fury, trances, love, and ecstasy. Ancient medical texts such as the Indian Vedic books from 1500 BC demonstrate a considerable knowledge of poisonous substances, including animal poisons, during this era [1]. The term "venom" originates from the Latin name Venus, the goddess of love and beauty. Initially, the word was used to indicate a love potion. Later, it adopted a more ambiguous meaning as a remedy, psychoactive drug, or poison [2]. Nowadays, venom refers specifically to animal poisons delivered through biting. There is a remarkable diversity of venomous animals, including approximately 1450 species of fish, 1200 species of marine organisms, 400 snakes, 200 spiders, 75 scorpions, and 60 ticks [1]. This review is focused on the contribution made by some scientists of the Institut Pasteur of Paris to our understanding of animal toxins.

2. Pasteur's Era: The Discovery of Antivenom Serotherapy

Louis Pasteur was aware of the work on bacterial toxins conducted by Peter Ludvig Panum and von Bergman, who focused on toxins produced by putrefying bacteria [3]. However, L. Pasteur was primarily interested in preventing infectious diseases with attenuated microorganisms. For L. Pasteur, the main mechanism of pathogenicity was the development of microbes in the host, causing a depletion of vital substances [3].

In 1888, Cesaire Phisalix (1852–1906), a French military physician and scientist, joined the National Museum of Natural History (MNHN, Museum National d'Histoire Naturelle) in Paris in the laboratory of comparative pathology directed by Auguste Chauveau. The MNHN was created in 1636 as the royal garden for the preservation and training of medicinal plants. It was formally established as the National Museum of Natural History in 1793 during the French Revolution. Its activity extends to research, collection, and training in various fields such as botany, chemistry, mineralogy, and zoology, including the study of venomous animals. C. Phisalix began investigating salamander venom and later collaborated with Gabriel Bertrand on viper venoms. Between 1889 and 1891, C. Phisalix published five articles in which he described the neurotoxic, respiratory, and thermal effects of salamander poison on experimental animals, as well as the possibility of inducing immunity through successive injections of low doses of the poison. G. Bertrand (1867–1962), a French chemist and biologist, entered the MNHN in 1886. He initially served as an assistant in a laboratory of plant physiology applied to agriculture (1889–1890), and subsequently in a laboratory of chemistry applied to organic substances (1890). A strong friendship developed between C. Phisalix and G. Bertrand (Figure 1). G. Bertrand later joined the Institut Pasteur in 1900, where he developed biological chemistry. Encouraged by A. Chauveau, C. Phisalix began studying venoms with the idea that venoms are possibly similar to bacterial toxins. He hypothesized that the methods of toxigenic bacteria attenuation for vaccination purposes could be applied to venoms. Indeed, following the success of L. Pasteur in preventing fowl cholera (1880), anthrax in sheep (1881), and rabies (1885) with the attenuated bacteria or viruses responsible for these diseases, C. Phisalix discovered a way to inactivate venom by heating it at 80 °C for 5 min and subsequently using it to vaccinate guinea pigs. In 1894, C. Phisalix and G. Bertrand presented their work at the Society of Biology in Paris on the immunization of guinea pigs with heat-attenuated venoms, and the subsequent treatment of venom-intoxicated animals with the serum of immunized guinea pigs. This marked the beginning of antivenomous serotherapy [4,5]. Moreover, C. Phisalix and G. Bertrand discovered that, unlike salamanders, vipers contain substances in their blood that confer resistance to their own venom. They also identified a natural resistance to viper venom in two species of grass snakes. Venom inhibitors in the blood of numerous snakes and mammals were later characterized (see below).

Figure 1. Cesaire Phisalix and Gabriel Bertrand. (**A**) C. Phisalix (1852–1906); photo was taken by Marie Phisalix, Musée d'Histoire Naturelle de Moutiers–Haute–Pierre. (**B**) G. Bertrand (1867–1962) around 1905; photo was provided by the Institut Pasteur/Musée Pasteur.

Albert Calmette (1863–1933) (Figure 2), a French military physician, followed the course of "microbie technique" directed by Emile Roux at the Institut Pasteur, Paris, in 1890. The next year, E. Roux sent A. Calmette in Saigon to establish the first Institut Pasteur network site, initially dedicated to the preparation of anti-rabies and anti-smallpox vaccines. During his stay in Indochina, he had the opportunity to investigate snake venoms. Back to France in 1893, he was sent to Lille to install a second Institut Pasteur (Institut Pasteur Lille), which he directed from 1895 to 1919. He tried to develop an immunization against cobra venom by adapting techniques used for bacterial toxins. However, cobra venom is resistant to heating. He then tried inoculating escalating doses, starting from sublethal doses, as well as alternative ways to inactivate snake venom by using various chemical treatments. A. Calmette succeeded in obtaining complete immunization using cobra venom treated with sodium hypochlorite, and he also demonstrated that sera from immunized animals have preventive and therapeutic effects against venom intoxication. He presented his results on antivenomous therapy in 1894 at the same session of the Society of Biology as C. Phisalix and G. Bertrand. A rivalry between the two scientist groups emerged for several years. The French National Academy of Sciences awarded the Monthyon prize to C. Phisalix and G. Bertrand for the discovery of an antivenom serum in 1894. However, A. Calmette claimed that his protocol for venom attenuation was more efficient than that of C. Phisalix. Indeed, snake venoms are more or less thermostable, and the chemical attenuation was more effective. Unfortunately, C. Phisalix died prematurely at age 54, and A. Calmette emphasized the importance of the antivenom serotherapy and his results at the international level. A. Calmette published numerous articles in French, English, and German, and he presented the interest and efficiency of the antivenom serotherapy at the Royal College of Physicians and Surgeons of London in 1896. It was found later that Calmette's protocol of immunization induced highly neutralizing specific IgG antibodies, whereas those of C. Phisalix and G. Bertrand triggered mainly IgM. A. Calmette thought that the sera obtained after immunization with one snake venom could protect against the venom from different snake species. He produced antisera against cobra venom on a large scale at the Institut Pasteur of Lille for therapeutic use in humans. Then, the production of antivenom sera was undertaken in various countries: Brazil (Butantan Institute, 1901), the USA (Philadelphia, 1902), Australia (Sydney, 1902), India (Hafkine Institute of Mumbai, 1903; Kasauli, 1907), South Africa (Johannesburg, 1903), and the UK (London, 1905) [4–9].

Figure 2. Albert Calmette (1863–1933) in his laboratory around 1920–1925. Photo was taken by Henri Manuel, Institut Pasteur/Musée Pasteur.

3. Period of 1923–1978: Development of Antivenom Sera and Initial Characterization of Venom Constituents

Gaston Ramon (1886–1963) (Figure 3), a French veterinarian and biologist, was recruited by E. Roux to produce horse antisera at the Institut Pasteur Garches Annex. The property of Garches, in the Paris suburb of Marnes la Coquette, was gifted to L. Pasteur by the government in 1884 for his studies on rabies. This property, which was part of an old castle (castle of Villeneuve l'Etang), sufficiently distant from residential areas, was more favorable for maintaining a kennel of rabid dogs than Pasteur's laboratory at the Ecole Normale Supérieure in downtown Paris. The Institut Pasteur in Paris was founded through a public international subscription and was inaugurated in 1888. It was dedicated to the treatment of rabies, basic research into infectious diseases, and training on microorganisms (with the first "microbie technique" course taught by E. Roux in 1888). After Pasteur's death in 1895, the Garches Annex was used for the production of antisera in horses and investigations into animal immunization (Figure 4). Louis Martin (1864–1946), a French physician and biologist, was the assistant of E. Roux (1893–1894) and contributed to the development of anti-diphtheria serotherapy. L. Martin was appointed as the deputy head of the laboratory responsible for the production of an anti-diphtheria toxin and, subsequently, an anti-tetanus toxin (1894–1909).

Figure 3. Gaston Ramon (1886–1933) in his laboratory around 1925. Photo was provided by the Institut Pasteur/Musée Pasteur.

G. Ramon was assigned to the serum production laboratory at the Garches Annex by E. Roux from 1911 to 1920 under the supervision of André-Romain Prévot [10]. In 1923, G. Ramon developed an innovative method of toxin inactivation based on a treatment with formalin. Thus, he demonstrated that the diphtheria toxin treated with a low dose of formalin became inactive, yet was able to induce a potent immunological response. He termed this novel form of toxin an "anatoxin" [3]. The idea to use formalin came from the fact that he applied this compound as an antiseptic for the preservation of therapeutic sera. He also used formalin for preserving standardized toxin samples, and he noticed that toxins treated with formalin were harmless, stable, and immunogenic [11]. This inactivation process was applied to other bacterial toxins, such as the tetanus toxin, as well as to venom. Indeed, in 1924, G. Ramon showed that snake venom could be inactivated by formalin to create a product that he referred as an "anavenom". This method, which proved more reliable than a heat treatment or other chemical inactivation methods, was used for the immunization of horses and the production of antivenomous sera. Furthermore, G. Ramon introduced the concept of immunity adjuvants. He showed that the induction of local inflammation with calcium chloride or alum (aluminum hydroxide) could enhance the immune response to anatoxins. From 1926 to 1944, G. Ramon was the director of the Garches Annex, where he coordinated the production of anatoxins and antisera [12,13].

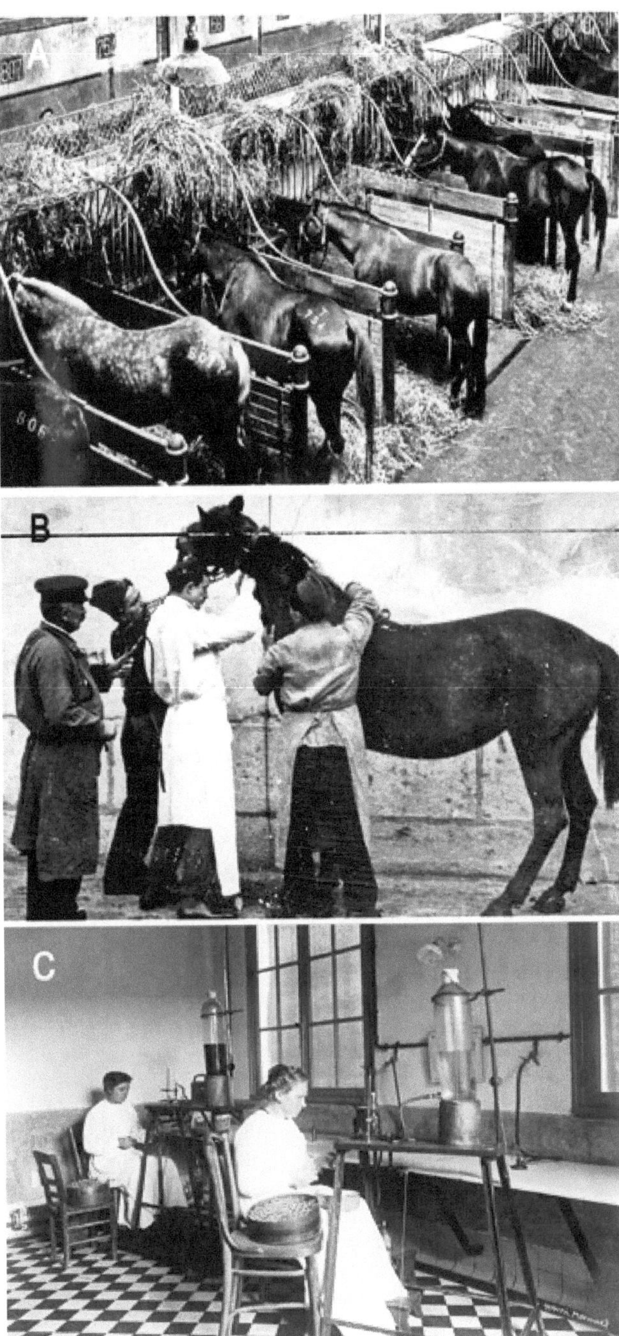

Figure 4. Institut Pasteur of Garches, production of horse antisera. (**A**) Horse stables; photo was provided by the Institut Pasteur/Musée Pasteur. (**B**) Serum sampling in a horse; photo was provided by the Institut Pasteur/Musée Pasteur. (**C**) Serum bottling; photo was taken by Henri Manuel, Institut Pasteur/Musée Pasteur.

Paul Boquet, a French physician and scientist, joined the Institut Pasteur Garches Annex in 1933. He directed a laboratory specializing in antivenomous serotherapy, which was renamed the Laboratory of Venoms and Antivenomous Sera, and he held this position until 1978. P. Boquet conducted extensive research on the composition of snake venoms, notably from various species of Vipers and Naja. He developed novel in vivo and in vitro methods to discern the different toxic properties and enzymatic activities of venoms. The separation of venom constituents was achieved through electrophoresis and liquid chromatography. P. Boquet demonstrated through immunoelectrophoresis that certain toxic constituents are present in the venoms of multiple snake species, facilitating the cross-neutralization of snake venoms with antivenomous sera from different snake species. A key objective was to characterize the unique properties of venom constituents and to select the most effective antigens for creating polyvalent sera through horse immunizations. He proposed the classification of snake venoms into three groups based on their molecular size and serological properties. P. Boquet characterized the major lethal factor of the *Naja* snake's venom, which is a small basic peptide of 61 amino acids called an alpha toxin, and he showed that it is produced by various *Elapidae* snakes. In collaboration with France Tazieff-Depierre, P. Boquet discovered that this factor exhibits activity similar to that of curare. Furthermore, the tritiated alpha toxin from the *Naja* snake (prepared by André Menez, CEA, France) was used to isolate the cholinergic receptor of the electric organ of *Electrophorus* by Jean Pierre Changeux, Institut Pasteur. P. Boquet also investigated the coagulation activity of venoms and phospholipases (PLAs), hypothesizing that PLAs might facilitate the entry of toxins into the nervous system. Moreover, P. Boquet supervised the production and control of antivenomous sera, which the Institut Pasteur distributed to numerous countries, notably Africa and Asia. In 1966, P. Boquet reported that 143 horses were used to produce 4345 L of antivenomous sera, leading to the preparation of 246,176 therapeutic doses of 10 mL each. He also contributed to the development and validation of international methods of antivenomous serum titration as well as the preparation of reference standards of venoms and sera in cooperation with the World Health Organization (WHO) [14–27].

France Tazieff-Depierre (1914–2006), a French pharmacist, entered the Institut Pasteur in 1934 in the Unit of Therapeutic Chemistry and then in the Unit of Protein Chemistry until 1979. Initially, she was interested in curararizing compounds and in the role of Ca^{++} in neurotransmitter release at the neuromuscular junction. She identified several classes of curararizing compounds, including those preventing acetylcholine activity and those with a depolarizing activity. These studies were completed by an investigation into the antagonists of curararizing compounds, such as acetylcholinesterases. F. Tazieff-Depierre showed that Ca^{++} activates acetylcholinesterases in vitro, but prevents their activity when fixed to the acetylcholine receptor. Starting in 1966, she was involved in animal toxins. P. Boquet asked her to analyze the mechanism of the paralytic effects of cobra venom. F. Tazieff-Depierre showed that the alpha toxin from *Naja nigricollis* binds with a high affinity to the acetylcholine receptor and that the paralytic effects are antagonized by anti-acetylcholinesterases, which increase the acetylcholine levels. In contrast, the gamma toxin from *Naja nigricollis* is cardiotoxic and induces muscle paralysis by the excessive Ca^{++} release from muscle fibers. The toxins from scorpion venom share a similar activity on striated muscle. Moreover, toxins from scorpions and sea anemones enhance the release of acetylcholine from neuronal endings through an increased intra-terminal Na^+ concentration and the possible subsequent Ca^{++} release from internal stores, which triggers neurotransmitter release. Interestingly, due to their effect on acetylcholine release, these toxins are able to restore the neuromuscular transmission inhibited by botulinum toxin A. The numerous and pertinent works of F. Tazieff-Depierre shed light on the mode of action of some venom toxins, and also on the molecular mechanism of neurotransmission, notably on the role of Ca^{++} [28–40].

Several other scientists at the Institut Pasteur were also involved in venom research during this period. Camille Delezenne (1868–1932), a French physician, was recruited by Emile Duclaux to manage a physiology laboratory at the Institut Pasteur in 1900. He showed that the hemolytic and coagulase properties of certain snake venoms resulted from specific zinc-dependent enzymatic activities [41]. Marcel Rouvier (1898–1981) joined the Institut Pasteur in 1940 and successively directed the laboratories of the tetanus toxin, the diphtheria toxin, and the "Service des Anaérobies" from 1956 to 1968. Among his works on toxins, he studied the antigenicity of certain snake venoms, such as those from *Vipera aspis.* He also investigated various methods of preserving antivenomous sera for optimal therapeutic use, despite drastic field constraints such as exposure to heat [42].

4. Period of 1972–2004: Optimization of Serotherapy, Characterization of Biological Activity of Venom Components, and Natural Inhibitors from Snake Blood

Cassian Bon (1944–2008) (Figure 5) was born in Vietnam and completed his studies at the Ecole Normale Supérieure in Paris. In 1972, he joined the Institut Pasteur in Paris and worked on scorpion venoms with F. Tazieff-Depierre. This experience proved decisive, setting the course of his career on the study of venoms, primarily snake venoms, and antivenom sera across his dual scientific roles at the CNRS (Centre National de la Recherche Scientifique) and the Institut Pasteur [43]. C. Bon defended his PhD in 1979 on the major neurotoxins in snake venoms, ceruleotoxin and crototoxin. He later became a research scientist at the CNRS and led a venom laboratory at the Institut Pasteur. This laboratory was part of the cellular pharmacology unit directed by Bernardo-Boris Vargaftig (1937-), a Brazilian-born scientist who completed his medical studies in Sao Paulo (1963) and his scientific studies in Paris (PhD 1972). The research unit of B. Vargaftig evolved into the unit of "Pharmacology of the Mediators of Inflammation and Thrombosis" (1985–1997), and then the unit of "Biology, Cellular and Molecular Pharmacology of Pulmonary Inflammation" (1998–2004). B. Vargaftig is well known for his research on the role of the platelet-activating factor, particularly in bronchoconstriction and platelet aggregation.

Figure 5. Cassian Bon (1944–2008) in his laboratory in 1995. Photo was provided by the Institut Pasteur.

In 1990, C. Bon was appointed as the director of the "Unité des Venins" at the Institut Pasteur, a role he held until 2004 [43]. His research focused on the mechanism of the presynaptic neurotoxicity of *Viperidae* toxins in snake venom, and he later developed studies on antivenom immunotherapy, both in experimental and human cases. This line of inquiry followed the Pasteur tradition initiated by Albert Calmette in 1894.

C. Bon, along with his collaborators, studied the functional chaperon role and synergistic action of the acidic subunit of crotoxin. Crotoxin is a heterodimeric β-neurotoxin from the venom of the South American rattlesnake *Crotalus durissus terrificus*. The venom was purchased from the Instituto Butantan (Sao-Paulo, SP, Brazil) and crotoxin was purified from the venom using low-pressure gel filtration and ion-exchange chromatography. It is a toxic protein formed by the non-covalent association of a basic PLA_2 subunit of low toxicity (CB) and a nontoxic acidic subunit (CA) devoid of catalytic activity that potentiates the toxic effect of CB [44]. Crotoxin binds to the presynaptic membrane, inducing the complete failure of neuromuscular transmission by impairing the release of acetylcholine at the neuromuscular junctions [45]. The CA subunit enhances the capacity of CB to reach its target at the neuromuscular junction, thereby increasing its lethal potency, but it reduces the enzymatic activity of CB [46–50].

Grazyna Faure, a Pasteurian collaborator of C. Bon, discovered several isoforms of crotoxin in individual venom samples of *Crotalus durissus terrificus* [51]. She purified 16 natural isoforms of crotoxin and compared their molecular structures and biological activities [52]. The work of G. Faure and C. Bon demonstrated that crotoxin variants result from the random association of four isoforms of the CA subunit (CA_{1-4}) and four isoforms of the CB subunit (CBa_2, CBb, CBc, and CBd). They described two classes of crotoxin, Class I and Class II, which differ in their pharmacological properties [52]. The comparison of crotoxin isoforms revealed that the stability of the complex plays a major role in its pharmacological action [53]. The origins of the isoforms were also identified. Multiple CA isoforms result from post-translational modifications occurring on a precursor pro-CA, while CB isoforms result from the expression of different messenger RNAs [54–56]. The presence of various PLA_2 isoforms with diverse pharmacological activities could be explained by the accelerated evolution of exon regions after the duplication of a common ancestral gene, indicating the rapid adaptation of snakes for defense and predation.

During work in the Unité des Venins, G. Faure and Igor Krizaj demonstrated that crotoxin binds with a high affinity to a protein receptor from the presynaptic membrane of neuromuscular junctions in the electric organ of *Torpedo marmorata* [57,58]. They purified a 48 kDa crotoxin acceptor protein from *Torpedo* (CAPT) and characterized its binding to the receptor by surface plasmon resonance (SPR) characterization [58,59]. This study showed the formation of a ternary complex, CA CB Receptor, and the dissociation of CA at equilibrium.

Furthermore, G. Faure and Jonas Perales in the Unité des Venins identified, purified, and characterized a natural PLA_2 inhibitor, the crotoxin inhibitor from *Crotalus* serum (CICS), from the blood of *Crotalus durissus terrificus* [60–62]. The CICS is an acidic 130 kDa oligomeric glycoprotein formed by the non-covalent association of 23–25 kDa subunits. This natural PLA_2 inhibitor protects the rattlesnake from its own venom. It neutralizes the lethal and PLA activities of crotoxin, CB, and other PLA_2s from the *Viperidae* family by binding to the CB subunit and preventing the association of CB with CA. Since the molecular mechanism underlying the interaction between crotoxin and CICS seems to be identical to that of crotoxin with its protein receptor CAPT, it was suggested that CICS acts physiologically as a false soluble crotoxin receptor, retaining the toxin in the vascular system of the snake and thereby preventing its toxic effects on the neuromuscular junction [60].

Valerie Choumet and colleagues from the C. Bon unit investigated the immunological aspects of crotoxin and other neurotoxic PLA_2s [63–69]. Notably, they discovered that the acidic subunit CA of crotoxin interacts with the basic, single-chain ammodytoxin or agkistrodotoxin from *Viperidae* venom, in agreement with the amino acid sequence similarities between CB and these single-chain toxins with PLA_2 activity. They also produced monoclonal antibodies (mAbs) against the crotoxin CA and CB subunits and determined their dissociation constants, cross-reactivity, and neutralization ability. This research led them to propose functional regions in the crotoxin components, the toxic and enzymatic sites on CB, and to suggest interacting regions on the two components. Further crystallographic studies by G. Faure's group established the CA-CB binding interface of crotoxin [70]. The knowledge of this interface could be useful in mapping the epitope of the neutralizing monoclonal antibody A56.36.

Subsequently, C. Bon and his colleagues at the Institut Pasteur explored the envenomation process and its treatment by serotherapy. They determined the toxicokinetics and toxicodynamics of snake and scorpion venom in experimental animal models such as rats and rabbits. The distribution of venom in the organism was analyzed after an injection by the intravenous or intramuscular route and in the absence or presence of mAb fragments, either able (Fab) or unable (Fab'2) to cross the renal route. Toxin levels were assessed using an enzyme-linked immunosorbent assay (ELISA) and radiolabeled proteins. The kinetic parameters were analyzed after the administration of varying toxin doses. After an intravenous injection, the toxin exhibited a biexponential decline corresponding to the distribution and subsequent elimination. An intramuscular injection allowed the toxin to reach the vascular compartment, after which it exhibited a monoexponential decrease. The severity of symptoms was correlated with the toxin levels in the plasma. An intravenous injection of antivenom serum was the most efficient route. Antivenom antibodies were able to neutralize the totality of the toxin in the vascular compartment. Fab fragments were less efficient than Fab'2 fragments, possibly due to their differing pharmacokinetics. It is likely that complexes of antibody fragments/venom constituents were eliminated by phagocytosis rather than the renal route due to their high molecular weight. Interestingly, treatment with Fab was found to induce oliguria, which might induce adverse effects of serotherapy in humans. When examining experimental scorpion envenomation in rats, it was found that a Fab'2 intravenous injection neutralized venom more rapidly than Fab. However, Fab was more effective when administrated intramuscularly. Furthermore, Fab was more efficient at preventing the early symptoms of envenomation than Fab'2. Both antibody fragments were equally effective at preventing late symptoms, irrespective of the administration route. Based on these results, it was recommended to intravenously inject a combination of Fab and Fab'2 for the treatment of scorpion envenomation [66,71–80]. This research allowed for the optimization of the envenomation treatment by serotherapy. C. Bon was a renowned international expert in this field and was regularly invited by the WHO to share his expertise on this topic.

C. Bon and his collaborators began exploring the involvement of snake venoms in blood coagulation, aiming to develop novel therapeutic molecules for treating thrombotic events. Phospholipids (PLs) play an important role in the coagulation process. Anionic PLs are exposed on the membrane of activated platelets and facilitate the formation of complexes containing PL and protein enzymes involved in the coagulation cascade. Certain mammalian-secreted PLA2s and some snake PLA2s exhibit an anticoagulant effect through PL hydrolysis and/or by competing with coagulation factors, thus preventing their assembly into complexes involved in the coagulation cascade. Anticoagulant PLA2s have been reported in venoms from *Viperidae, Crotalidae, Elapidae,* and *Hydrophidae*. An analysis of snake PLA$_2$ has helped to better define the mode of action of the anticoagulant PLA$_2$s. Conversely, snake venoms also contain metalloproteases that activate coagulation factors, inducing procoagulant activity. C. Bon and collaborators purified and characterized some of these proteases, such as the prothrombin activator from *Bothrops atrox*, a serine protease from *Bungarus fasciatus* that activates blood coagulation factor X, and a thrombin-like serine protease from *Bothrops lanceolatus* that activates fibrinogen into fibrin [81–92].

C. Bon was actively involved in teaching about venomous animals at the MNHN and about human envenomation and its treatment at various universities. He was active in the international network of the Institut Pasteur, which produced antivenomous sera, notably in Algeria, Tunisia, Morocco, and Iran. He was a member of numerous scientific societies, including the French Society of Biochemistry and Molecular Biology, the French Association of Pharmacologists, the Society of Neurosciences, the International Society of Toxinology, and the International Society of Thrombosis and Hemostasis. He also served on the scientific council of the Institut Pasteur of Paris (1995–1999) and the Institut Pasteur of Iran (1998–2004). C. Bon was a co-founder of the French Society for the Study of Toxins

(SFET, Société Française d'Étude des Toxines) in 1992 and served as president from 2000 to 2008. In 2004, the unit of venoms at the Institut Pasteur was closed, and C. Bon moved to the Laboratory of Chemistry of Natural Substances of MNHN. He was warmly welcomed by Max Goyffon, finding a familiar environment for teaching. During his last years, he was primarily focused on teaching, advising the WHO on envenomation, and organizing SFET meetings [43,93].

5. Period of 2004–2023: Novel Therapeutic Peptides Based on Snake Toxins and Structural Analysis of Their Binding Interface with Biological Targets

G. Faure, an expert research associate at the Institut Pasteur, continued the work on the structure–function relationships and the mechanisms of action of snake venom toxins with various protein targets, including the coagulation factor Xa, proton-gated ion channels, and the cystic fibrosis transmembrane regulator (CFTR) chloride channel. She led a group on PLA_2 toxins in the Structural Immunology Unit (2004–2011) directed by Graham Bentley, and later in the channel-receptors unit (2011–2023) directed by Pierre-Jean Corringer.

Collaborating with Frederick Saul in the Structural Immunology Unit, Faure's group determined the crystal structure of the heterodimeric crotoxin from *Crotalus durissus terrificus*, isoform CA_2CBb of Class I [70], two isoforms of the basic CB subunit of crotoxin (CBb and CBd) [70,94], and two isoforms of ammodytoxin (AtxA and AtxC) from *Vipera ammodytes ammodytes* [95]. The three-dimensional structure of crotoxin revealed the nature of the binding interface between the CA and CB subunits and allowed for the identification of the key amino acid residues responsible for significant differences in the stability, toxicity, and enzymatic activity of the two classes of crotoxin complexes [70,96].

Snake venom PLA_2s exhibit a wide range of toxic and pharmacological effects, including neurotoxic (pre-synaptic or post-synaptic), myotoxic, and cardiotoxic activities; anticoagulant effects; the inhibition of platelet aggregation; hemolytic activity; internal hemorrhage; anti-hemorrhage activity; convulsing and hypotensive activity; the induction of edema; organ necrosis or tissue damage; bactericidal, anti-humoral, anti-HIV (human immunodeficiency virus), anti-*Leishmania*, and anti-*Plasmodium* activity; and anti-viral activity against the dengue and yellow fever viruses [97]. Faure's group identified a number of *Viperidae* venom PLA_2s that inhibit blood coagulation factor Xa (FXa) via a non-catalytic PL-independent mechanism [98]. The interaction sites on PLA_2 and FXa were mapped using SPR protein–protein interaction measurements, mutagenesis studies, and molecular docking simulations [98,99]. Comparative structural studies of natural PLA_2 isoforms, which differ in their neurotoxicity and anticoagulant activity, contributed to a better understanding of their mode of binding to human FXa and calmodulin [95,96,100]. Faure's group also discovered that PLA_2 binding to FXa prevents the oligomerization of PLA_2 [101]. The identification of the anticoagulant sites on Atx and CB and an analysis of the spatial arrangement of the PLA_2-FXa interface led to a better understanding of the hemostatic process. These results are important for the elaboration of novel anticoagulant agents (non-competitive FXa inhibitors) [100].

Working in the channel-receptors unit at Pasteur, Faure's group along with other collaborators discovered that the CB subunit of crotoxin binds with a high affinity to novel protein targets such as the pentameric proton-gated channel GLIC [102] as well as the cystic fibrosis transmembrane regulator (CFTR) and its mutant ΔF508-CFTR, which is implicated in cystic fibrosis [103]. They demonstrated that CB is a negative allosteric modulator of GLIC [102] and a positive allosteric modulator of CFTR [103]. By a direct interaction with the nucleotide-binding domain NBD1 of CFTR, CB potentiates the chloride channel current and corrects the trafficking defect of misfolded ΔF508CFTR inside the cell [103]. Faure's group identified the CB-ΔF508CFTR interface by molecular docking and by HDX-MS (hydrogen–deuterium exchange–mass spectrometry) studies [103]. For the therapeutic development of new anti-cystic fibrosis agents, G. Faure and collaborators used a structure-based in silico approach and designed peptides mimicking this CBb-ΔF508NBD1 interface [104]. Using electrophysiological and biophysical methods, they identified several peptides that interact with the ΔF508NBD1 domain of CFTR and increase the chloride channel activity [104]. These significant results provide a new class of CFTR potentiators and describe a novel approach for developing therapeutic peptides for the treatment of cystic fibrosis. Thus, the biochemical and structural characterization of the functional and pharmacological sites of snake venom PLA_2 (Figure 6) or fragments of PLA_2 in complexes with their biological targets is essential for the structure-based design of novel therapeutic agents [94].

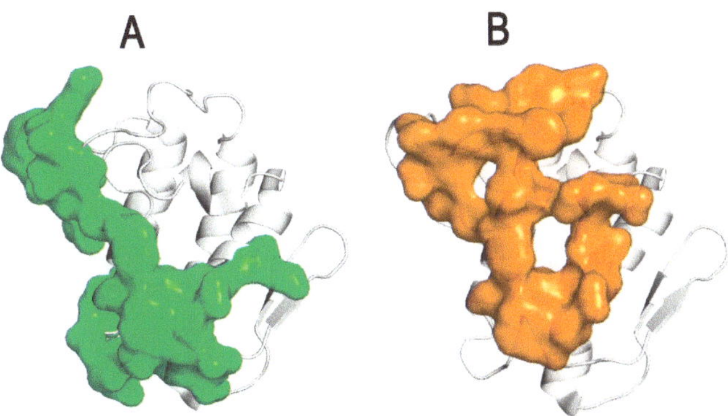

Figure 6. Identification of the pharmacological binding sites of CB, the PLA_2 subunit of crotoxin, which is important for the structure-based design of new anticoagulant and anti-cystic fibrosis agents. (**A**) FXa-binding site of CB (in green) (adapted from Nemecz et al., 2020) [94]. (**B**) F508CFTR-binding site of CB (in orange) (adapted from Nemecz et al., 2020) [94].

6. Concluding Remarks

Venomous animals pose a significant threat, and it has long been a concern to develop effective measures for the prevention and treatment of envenomation. Following in the tradition of the emerging field of microbiology during Pasteur's era, C. Phisalix and G. Bertrand from the MNHN and A. Calmette from the Institut Pasteur achieved a breakthrough in combating envenomation. They demonstrated that prevention can be acquired by immunization with attenuated venoms and that treatment can be implemented through serotherapy. Later, G. Ramon refined the process of venom attenuation using formalin treatment. These discoveries led to the mass production and distribution of antivenomous sera, particularly to high-risk areas such as Asia and Africa, largely through the international network of the Institut Pasteur. In addition to applications for human health, an interest in the basic knowledge of venoms emerged, including investigations into venom constituents and their biological activities. Pasteurians have contributed widely to venom knowledge, as have scientists from national and international institutions outside the scope of this review (Supplementary Table S1). Briefly, at the French level, one should mention the contributions of the team led by André Menez (CEA) and later by Denis Servent on the structure/function relationships of snake and scorpion venom toxins, and that of Jordi Molgo and Evelyne Benoit (CNRS, CEA) on marine toxins and, notably, on conotoxins [105–108], as well as numerous colleagues from the Universities of Aix-Marseille (Pierre Bougis, Marie-France Martin-Eauclaire, Pascale Marchot, Hervé Rochat, . . .), Anger (Christian Legros, Cesare Mattei . . .), Côte d'Azur (Sylvie Diochot, Pierre Escoubas, Michel Lazdunski . . .), Nantes (Michel de Waard, Michel Ronjat . . .), or Montpellier (Sébastien Dutertre . . .). Regarding the French contribution, the annual SFET meetings, called "Rencontres in Toxinology", offer an opportunity to present the most recent findings and updates in animal and bacterial toxins. The in-depth characterization of animal toxins allows for comparisons to be made with plant and bacterial toxins or, in contrast, allows for their unique properties to be highlighted. The extreme potency of toxins primarily results from their ability to target or hijack key physiological processes. Therefore, due to their high specificity for the host targets, toxins are very efficient tools for exploring cellular processes. The unique properties of certain toxins to reverse or inhibit specific pathological effects make them suitable therapeutic tools or enable the design of novel therapeutic approaches. Recent advances in animal toxins support their therapeutic application.

Supplementary Materials: The following supporting information can be downloaded at: https://www.mdpi.com/article/10.3390/toxins15070462/s1, Table S1: Institut Pasteur's collaborators working on animal toxins from 1986 to 2023.

Author Contributions: M.R.P. and G.F. wrote the manuscript. S.L. provided historical documentation and D.L. revised the manuscript. All authors have read and agreed to the published version of the manuscript.

Funding: This review received no external funding.

Institutional Review Board Statement: Not applicable.

Informed Consent Statement: Not applicable.

Data Availability Statement: Not applicable.

Acknowledgments: We thank Michaël Davy and Marie Martin, Photothèque Institut Pasteur, for providing the photos.

Conflicts of Interest: The authors declare no conflict of interest.

References

1. Charitos, I.A.; Gagliano-Candela, R.; Santacroce, L.; Bottalico, L. Venoms and Poisonings during the Centuries: A Narrative Review. *Endocr. Metab. Immune Disord. Drug Targets* **2022**, *22*, 558–570. [CrossRef] [PubMed]
2. Cilliers, L.; Retief, F.P. Poisons, poisoning and the drug trade in ancient Rome. *Akroterion* **2000**, *45*, 88–100. [CrossRef]
3. Cavaillon, J.M. From bacterial poisons to toxins: The early works of Pasteurians. *Toxins* **2022**, *14*, 759. [CrossRef] [PubMed]
4. Goyffon, M.; Chippaux, J.P. La découverte du sérum antivenimeux (10 février 1894). *Biofutur* **2008**, *292*, 32–35.
5. Bochner, R. Paths to the discovery of antivenom serotherapy in France. *J. Venom. Anim. Toxins Incl. Trop. Dis.* **2016**, *22*, 20. [CrossRef]
6. Calmette, A. The Treatment of Animals Poisoned with Snake Venom by the Injection of Antivenomous Serum. *Br. Med. J.* **1896**, *2*, 399–400. [CrossRef]
7. Hawgood, B.J. Doctor Albert Calmette 1863–1933: Founder of antivenomous serotherapy and of antituberculous BCG vaccination. *Toxicon* **1999**, *37*, 1241–1258. [CrossRef]
8. Brygoo, E.R. La découverte de la sérothérapie antivenimeuse en 1894: Physalix et Bertrand ou Calmette? *Bull. Ass. Anc. Elèves Inst. Pasteur* **1985**, *106*, 10–22.
9. Calmette, A. Contribution à l'étude du venin des serpents. Immunisation des animaux et traitement de l'envenimation. *Ann. Inst. Pasteur* **1894**, *8*, 275–291.
10. Popoff, M.R.; Legout, S. Anaerobes and Toxins, a Tradition of the Institut Pasteur. *Toxins* **2023**, *15*, 43. [CrossRef]
11. Gilbrin, E. Gaston Ramon (1886–1963). Le soixantième anniversaire des anatoxines. *Hist. Des Sci. Médicales* **1984**, *18*, 53–60.
12. Ramon, G.; Boquet, P.; Richou, R.; Delaunay-Ramon, R. Sur la production accélérée des sérums antivenimeux de différentes sortes au moyen des anavenins spécifiques et des substances stimulantes de l'immunité. *Rev. Immunol.* **1941**, *6*, 353–362.
13. Ramon, G. Procédé pour accroître la production des antitoxines. *Ann. Inst. Pasteur* **1926**, *40*, 1–10.
14. Boquet, P.; Détrait, J.; Farzanpay, R. Biochemical and immunological studies on snake venom. 3. Study of analogues of the *Naja nigricollis* venom alpha antigen. *Ann. Inst. Pasteur* **1969**, *116*, 522–542.
15. Boquet, P.; Izard, Y.; Detrait, J. Studies on the spreading factor of snake venoms. *C. R. Seances Soc. Biol. Fil.* **1958**, *152*, 1363–1365.
16. Cheymol, J.; Boquet, P.; Bourillet, F.; Detrait, J.; Roch-Arveiller, M. Comparison of the principal pharmacologic properties of different venoms of *Echis carinatus* (Viperides). *Arch. Int. Pharmacodyn. Ther.* **1973**, *205*, 293–304.
17. Detrait, J.; Boquet, P. Separation of the constituents of Naja naja venin by electrophoresis. *C. R. Hebd. Seances Acad. Sci.* **1958**, *246*, 1107–1109.
18. Detrait, J.; Izard, Y.; Boquet, P. Antigenic relations between a lethal factor in the venom of *Echis carinata* and neurotoxins in the venoms of Na ja na ja and Na ja nigricollis. *C. R. Seances Soc. Biol. Fil.* **1960**, *154*, 1163–1165.
19. Tazieff-Depierre, F.; Pierre, J. Curarizing action of alpha toxin of *Naja nigricollis*. *C. R. Acad. Hebd. Seances Acad. Sci. D* **1966**, *263*, 1785–1788.
20. Boquet, P.; Poilleux, G.; Dumarey, C.; Izard, Y.; Ronsseray, A.M. An attempt to classify the toxic proteins of *Elapidae* and *Hydrophiidae* venoms. *Toxicon* **1973**, *11*, 333–340. [CrossRef]
21. Boquet, P. Snake venoms. 2. The chemical constitution of snake venoms and immunity against venoms. *Toxicon* **1966**, *3*, 243–279. [CrossRef]

22. Boquet, P. Snake Venoms. I. Physiopathology of Envenomation and Biological Properties of Venoms. *Toxicon* **1964**, *15*, 5–41. [CrossRef]
23. Boquet, P.; Dumarey, C.; Joseph, D. The antigenicity of toxic proteins from the venoms of *Elapidae* and *Hydrophiidae*. *C. R. Acad. Hebd. Seances Acad. Sci. D* **1977**, *284*, 2439–2442. [PubMed]
24. Boquet, P.; Izard, Y.J.M.; Meaume, J. Biochemical and immunological research on snale venom. I. Trial separations of the antigens of *Naja nigricollis* venom by Sephadex filtration. *Ann. Inst. Pasteur* **1966**, *111*, 719–732.
25. Boquet, P.; Izard, Y.; Ronsseray, A.M. Attempt at classification of toxic proteins extracted from snake venoms. *C. R. Acad. Hebd. Seances Acad. Sci. D* **1970**, *271*, 1456–1459.
26. Boquet, P.; Izard, Y.; Ronsseray, A.M. An attempt to classify by serological techniques the toxic proteins of low molecular weight extracted from *Elapidae* and *Hydrophiidae* venoms. *Taiwan Yi Xue Hui Za Zhi* **1972**, *71*, 307–310. [PubMed]
27. Meunier, J.C.; Olsen, R.W.; Menez, A.; Fromageot, P.; Boquet, P.; Changeux, J.P. Some physical properties of the cholinergic receptor protein from *Electrophorus electricus* revealed by a tritiated alpha-toxin from *Naja nigricollis* venom. *Biochemistry* **1972**, *11*, 1200–1210. [CrossRef]
28. Tazieff-Depierre, F. Pharmacologic properties of toxins from scorpion venom (*Androctonus australis*). *C. R. Acad. Hebd. Seances Acad. Sci. D* **1968**, *267*, 240–243.
29. Lièvremont, M.; Czajka, M.; Tazieff-Depierre, F. Calcium cycle at the neuromuscular junction. *C. R. Acad. Hebd. Seances Acad. Sci. D* **1969**, *268*, 379–382.
30. Tazieff-Depierre, F.; Czajka, M.; Lowagie, C. Pharmacologic action of pure Naja nigricollis venom fractions and calcium liberation in striated muscles. *C. R. Acad. Hebd. Seances Acad. Sci. D* **1969**, *268*, 2511–2514.
31. Tazieff-Depierre, F. Tetrodotoxin, calcium and scorpion venom. *C. R. Acad. Hebd. Seances Acad. Sci. D* **1970**, *271*, 1655–1657. [PubMed]
32. Tazieff-Depierre, F. Scorpion venom, calcium and acetylcholine emission by nerve fibers in the ileum of guinea pigs. *C. R. Acad. Hebd. Seances Acad. Sci. D* **1972**, *275*, 3021–3024. [PubMed]
33. Tazieff-Depierre, F.; Goudou, D.; Metezeau, P. Desensitization of the guinea pig ileum to scorpion venom by various concentrations of ions in the external medium. *C. R. Acad. Hebd. Seances Acad. Sci. D* **1974**, *278*, 2703–2706.
34. Tazieff-Depierre, F.; Choucavy, M.; Métézeau, P. Pharmacologic properties of the toxins isolated from the sea anemone (*Anemonia sulcata*). *C. R. Acad. Hebd. Seances Acad. Sci. D* **1976**, *283*, 699–702.
35. Tazieff-Depierre, F.; Métezeau, P. Effect of scorpion venom (*Androctonus australis*) on neuromuscular transmission inhibited by botulinum toxin in the frog. *C. R. Acad. Hebd. Seances Acad. Sci. D* **1977**, *285*, 1335–1338. [PubMed]
36. Molgó, J.; Lemeignan, M.; Tazieff-Depierre, F. Enhancement by *Anemonia sulcata* toxin II of spontaneous quantal transmitter release from mammalian motor nerve terminals. *Toxicon* **1986**, *24*, 441–450. [CrossRef] [PubMed]
37. Métézeau, P.; Bourneau, R.; Mambrini, J.; Tazieff-Depierre, F. Effects of a neurotoxin isolated from the sea anemone, *Anemonia sulcata*, at frog neuromuscular junction (author's transl). *J. Physiol.* **1979**, *75*, 873–879.
38. Coraboeuf, E.; Deroubaix, E.; Tazieff-Depierre, T. Effect of toxin II isolated from scorpion venom on action potential and contraction of mammalian heart. *J. Mol. Cell. Cardiol.* **1975**, *7*, 643–653. [CrossRef]
39. Tazieff-Depierre, F. Cardiotoxicity in rats of purified gamma toxin isolated from venom of *Naja nigricollis* and of toxins extracted from scorpion venom. *C. R. Acad. Hebd. Seances Acad. Sci. D* **1975**, *280*, 1181–1184.
40. Tazieff-Depierre, F.; Trethevie, E.R. Action of calcium chloride in the cardiotoxicity in the rat of purified gamma toxin isolated from *Naja nigricollis* venom. *C. R. Acad. Hebd. Seances Acad. Sci. D* **1975**, *280*, 137–140.
41. Hallion, L.C. Delezenne. *Annls Physiol. PhysicoChimie Biol.* **1932**, *8*, 785–797.
42. Lamy, R.; Rouyer, M. Relation entre les valeurs antigéniques des venims blancs et jaunes de *Vipera aspis*. *Bull. Acad. Med.* **1947**, *131*, 114–116.
43. Goyffon, M.; Faure, G. Obituary: Cassian Bon—A scientist fascinated by toxins and their therapeutic potential. *Toxicon* **2008**, *52*, 400–403.
44. Hendon, R.A.; Fraenkel-Conrat, H. Biological roles of the two components of crotoxin. *Proc. Natl. Acad. Sci. USA* **1971**, *68*, 1560–1563. [CrossRef] [PubMed]
45. Bon, C.; Changeux, J.P.; Jeng, T.W.; Fraenkel-Conrat, H. Postsynaptic effects of crotoxin and of its isolated subunits. *Eur. J. Biochem.* **1979**, *99*, 471–481. [CrossRef] [PubMed]
46. Bon, C.; Bouchier, C.; Choumet, V.; Faure, G.; Jiang, M.S.; Lambezat, M.P.; Radvanyi, F.; Saliou, B. Crotoxin, half-century of investigations on a phospholipase A2 neurotoxin. *Acta Physiol. Pharmacol. Latinoam.* **1989**, *39*, 439–448. [PubMed]
47. Délot, E.; Bon, C. Differential effects of presynaptic phospholipase A2 neurotoxins on *Torpedo* synaptosomes. *J. Neurochem.* **1992**, *58*, 311–319. [CrossRef]
48. Délot, E.; Bon, C. Model for the interaction of crotoxin, a phospholipase A2 neurotoxin, with presynaptic membranes. *Biochemistry* **1993**, *32*, 10708–10713. [CrossRef]
49. Ollivier-Bousquet, M.; Radvanyi, F.; Bon, C. Crotoxin, a phospholipase A2 neurotoxin from snake venom, interacts with epithelial mammary cells, is internalized and induces secretion. *Mol. Cell. Endocrinol.* **1991**, *82*, 41–50. [CrossRef]
50. Radvanyi, F.R.; Bon, C. Catalytic activity and reactivity with p-bromophenacyl bromide of the phospholipase subunit of crotoxin. Influence of dimerization and association with the noncatalytic subunit. *J. Biol. Chem.* **1982**, *257*, 12616–12623. [CrossRef]

51. Faure, G.; Bon, C. Several isoforms of crotoxin are present in individual venoms from the South American rattlesnake *Crotalus durissus terrificus*. *Toxicon* **1987**, *25*, 229–234. [CrossRef] [PubMed]

52. Faure, G.; Bon, C. Crotoxin, a phospholipase A2 neurotoxin from the South American rattlesnake *Crotalus durissus terrificus*: Purification of several isoforms and comparison of their molecular structure and of their biological activities. *Biochemistry* **1988**, *27*, 730–738. [CrossRef]

53. Faure, G.; Harvey, A.L.; Thomson, E.; Saliou, B.; Radvanyi, F.; Bon, C. Comparison of crotoxin isoforms reveals that stability of the complex plays a major role in its pharmacological action. *Eur. J. Biochem.* **1993**, *214*, 491–496. [CrossRef] [PubMed]

54. Bouchier, C.; Boulain, J.C.; Bon, C.; Ménez, A. Analysis of cDNAs encoding the two subunits of crotoxin, a phospholipase A2 neurotoxin from rattlesnake venom: The acidic non enzymatic subunit derives from a phospholipase A2-like precursor. *Biochim. Biophys. Acta* **1991**, *1088*, 401–408. [CrossRef] [PubMed]

55. Faure, G.; Choumet, V.; Bouchier, C.; Camoin, L.; Guillaume, J.L.; Monegier, B.; Vuilhorgne, M.; Bon, C. The origin of the diversity of crotoxin isoforms in the venom of *Crotalus durissus terrificus*. *Eur. J. Biochem.* **1994**, *223*, 161–164. [CrossRef] [PubMed]

56. Faure, G.; Guillaume, J.L.; Camoin, L.; Saliou, B.; Bon, C. Multiplicity of acidic subunit isoforms of crotoxin, the phospholipase A2 neurotoxin from *Crotalus durissus terrificus* venom, results from posttranslational modifications. *Biochemistry* **1991**, *30*, 8074–8083. [CrossRef]

57. Krizaj, I.; Faure, G.; Gubensek, F.; Bon, C. Re-examination of crotoxin-membrane interactions. *Toxicon* **1996**, *34*, 1003–1009. [CrossRef]

58. Krizaj, I.; Faure, G.; Gubensek, F.; Bon, C. Neurotoxic phospholipases A2 ammodytoxin and crotoxin bind to distinct high-affinity protein acceptors in *Torpedo marmorata* electric organ. *Biochemistry* **1997**, *36*, 2779–2787. [CrossRef]

59. Faure, G.; Copic, A.; Le Porrier, S.; Gubensek, F.; Bon, C.; Krizaj, I. Crotoxin acceptor protein isolated from *Torpedo* electric organ: Binding properties to crotoxin by surface plasmon resonance. *Toxicon* **2003**, *41*, 509–517. [CrossRef]

60. Faure, G. Natural inhibitors of toxic phospholipases A(2). *Biochimie* **2000**, *82*, 833–840. [CrossRef]

61. Faure, G.; Villela, C.; Perales, J.; Bon, C. Interaction of the neurotoxic and nontoxic secretory phospholipases A2 with the crotoxin inhibitor from *Crotalus* serum. *Eur. J. Biochem.* **2000**, *267*, 4799–4808. [CrossRef] [PubMed]

62. Perales, J.; Villela, C.; Domont, G.B.; Choumet, V.; Saliou, B.; Moussatché, H.; Bon, C.; Faure, G. Molecular structure and mechanism of action of the crotoxin inhibitor from *Crotalus durissus terrificus* serum. *Eur. J. Biochem.* **1995**, *227*, 19–26. [CrossRef] [PubMed]

63. Bon, C.; Choumet, V.; Delot, E.; Faure, G.; Robbe-Vincent, A.; Saliou, B. Different evolution of phospholipase A2 neurotoxins (beta-neurotoxins) from *Elapidae* and *Viperidae* snakes. *Ann. N. Y. Acad. Sci.* **1994**, *710*, 142–148. [CrossRef] [PubMed]

64. Choumet, V.; Bouchier, C.; Délot, E.; Faure, G.; Saliou, B.; Bon, C. Structure and function relationship of crotoxin, a heterodimeric neurotoxic phospholipase A2 from the venom of a South-American rattlesnake. *Adv. Exp. Med. Biol.* **1996**, *391*, 197–202. [CrossRef] [PubMed]

65. Choumet, V.; Faure, G.; Robbe-Vincent, A.; Saliou, B.; Mazié, J.C.; Bon, C. Immunochemical analysis of a snake venom phospholipase A2 neurotoxin, crotoxin, with monoclonal antibodies. *Mol. Immunol.* **1992**, *29*, 871–882. [CrossRef]

66. Choumet, V.; Jiang, M.S.; Radvanyi, F.; Ownby, C.; Bon, C. Neutralization of lethal potency and inhibition of enzymatic activity of a phospholipase A2 neurotoxin, crotoxin, by non-precipitating antibodies (Fab). *FEBS Lett.* **1989**, *244*, 167–173. [CrossRef]

67. Choumet, V.; Lafaye, P.; Demangel, C.; Bon, C.; Mazié, J.C. Molecular mimicry between a monoclonal antibody and one subunit of crotoxin, a heterodimeric phospholipase A2 neurotoxin. *Biol. Chem.* **1999**, *380*, 561–568. [CrossRef]

68. Choumet, V.; Lafaye, P.; Mazié, J.C.; Bon, C. A monoclonal antibody directed against the non-toxic subunit of a dimeric phospholipase A2 neurotoxin, crotoxin, neutralizes its toxicity. *Biol. Chem.* **1998**, *379*, 899–906. [CrossRef]

69. Choumet, V.; Saliou, B.; Fideler, L.; Chen, Y.C.; Gubensek, F.; Bon, C.; Delot, E. Snake-venom phospholipase A2 neurotoxins. Potentiation of a single-chain neurotoxin by the chaperon subunit of a two-component neurotoxin. *Eur. J. Biochem.* **1993**, *211*, 57–62. [CrossRef]

70. Faure, G.; Xu, H.; Saul, F.A. Crystal structure of crotoxin reveals key residues involved in the stability and toxicity of this potent heterodimeric β-neurotoxin. *J. Mol. Biol.* **2011**, *412*, 176–191. [CrossRef]

71. Dekhil, H.; Wisner, A.; Marrakchi, N.; El Ayeb, M.; Bon, C.; Karoui, H. Molecular cloning and expression of a functional snake venom serine proteinase, with platelet aggregating activity, from the *Cerastes cerastes* viper. *Biochemistry* **2003**, *42*, 10609–10618. [CrossRef] [PubMed]

72. Jandrot-Perrus, M.; Lagrue, A.H.; Leduc, M.; Okuma, M.; Bon, C. Convulxin-induced platelet adhesion and aggregation: Involvement of glycoproteins VI and IaIIa. *Platelets* **1998**, *9*, 207–211. [CrossRef] [PubMed]

73. Krifi, M.N.; Savin, S.; Debray, M.; Bon, C.; El Ayeb, M.; Choumet, V. Pharmacokinetic studies of scorpion venom before and after antivenom immunotherapy. *Toxicon* **2005**, *45*, 187–198. [CrossRef] [PubMed]

74. Parry, M.A.; Jacob, U.; Huber, R.; Wisner, A.; Bon, C.; Bode, W. The crystal structure of the novel snake venom plasminogen activator TSV-PA: A prototype structure for snake venom serine proteinases. *Structure* **1998**, *6*, 1195–1206. [CrossRef] [PubMed]

75. Hammoudi-Triki, D.; Lefort, J.; Rougeot, C.; Robbe-Vincent, A.; Bon, C.; Laraba-Djebari, F.; Choumet, V. Toxicokinetic and toxicodynamic analyses of *Androctonus australis hector* venom in rats: Optimization of antivenom therapy. *Toxicol. Appl. Pharmacol.* **2007**, *218*, 205–214. [CrossRef]

76. Audebert, F.; Urtizberea, M.; Sabouraud, A.; Scherrmann, J.M.; Bon, C. Pharmacokinetics of *Vipera aspis* venom after experimental envenomation in rabbits. *J. Pharmacol. Exp. Ther.* **1994**, *268*, 1512–1517.

77. Choumet, V.; Audebert, F.; Rivière, G.; Sorkine, M.; Urtizberea, M.; Sabouraud, A.; Scherrmann, J.M.; Bon, C. New approaches in antivenom therapy. *Adv. Exp. Med. Biol.* **1996**, *391*, 515–520. [CrossRef]

78. Rivière, G.; Choumet, V.; Audebert, F.; Sabouraud, A.; Debray, M.; Scherrmann, J.M.; Bon, C. Effect of antivenom on venom pharmacokinetics in experimentally envenomed rabbits: Toward an optimization of antivenom therapy. *J. Pharmacol. Exp. Ther.* **1997**, *281*, 1–8.

79. Rivière, G.; Choumet, V.; Saliou, B.; Debray, M.; Bon, C. Absorption and elimination of viper venom after antivenom administration. *J. Pharmacol. Exp. Ther.* **1998**, *285*, 490–495.

80. Calderón-Aranda, E.S.; Rivière, G.; Choumet, V.; Possani, L.D.; Bon, C. Pharmacokinetics of the toxic fraction of *Centruroides limpidus limpidus* venom in experimentally envenomed rabbits and effects of immunotherapy with specific F(ab')2. *Toxicon* **1999**, *37*, 771–782. [CrossRef]

81. Stroka, A.; Donato, J.L.; Bon, C.; Hyslop, S.; de Araújo, A.L. Purification and characterization of a hemorrhagic metalloproteinase from *Bothrops lanceolatus* (Fer-de-lance) snake venom. *Toxicon* **2005**, *45*, 411–420. [CrossRef] [PubMed]

82. Lôbo de Araújo, A.; Kamiguti, A.; Bon, C. Coagulant and anticoagulant activities of *Bothrops lanceolatus* (Fer de lance) venom. *Toxicon* **2001**, *39*, 371–375. [CrossRef]

83. Wisner, A.; Braud, S.; Bon, C. Snake venom proteinases as tools in hemostasis studies: Structure-function relationship of a plasminogen activator purified from *Trimeresurus stejnegeri* venom. *Haemostasis* **2001**, *31*, 133–140. [CrossRef] [PubMed]

84. Braud, S.; Bon, C.; Wisner, A. Snake venom proteins acting on hemostasis. *Biochimie* **2000**, *82*, 851–859. [CrossRef] [PubMed]

85. Mounier, C.M.; Bon, C.; Kini, R.M. Anticoagulant venom and mammalian secreted phospholipases A(2): Protein-versus phospholipid-dependent mechanism of action. *Haemostasis* **2001**, *31*, 279–287. [CrossRef]

86. Mounier, C.; Franken, P.A.; Verheij, H.M.; Bon, C. The anticoagulant effect of the human secretory phospholipase A2 on blood plasma and on a cell-free system is due to a phospholipid-independent mechanism of action involving the inhibition of factor Va. *Eur. J. Biochem.* **1996**, *237*, 778–785. [CrossRef]

87. Hofmann, H.; Bon, C. Blood coagulation induced by the venom of *Bothrops atrox*. 2. Identification, purification, and properties of two factor X activators. *Biochemistry* **1987**, *26*, 780–787. [CrossRef]

88. Zhang, Y.; Wisner, A.; Xiong, Y.; Bon, C. A novel plasminogen activator from snake venom. Purification, characterization, and molecular cloning. *J. Biol. Chem.* **1995**, *270*, 10246–10255. [CrossRef]

89. Lôbo de Araújo, A.; Donato, J.L.; Bon, C. Purification from *Bothrops lanceolatus* (fer de lance) venom of a fibrino(geno)lytic enzyme with esterolytic activity. *Toxicon* **1998**, *36*, 745–758. [CrossRef]

90. Braud, S.; Parry, M.A.; Maroun, R.; Bon, C.; Wisner, A. The contribution of residues 192 and 193 to the specificity of snake venom serine proteinases. *J. Biol. Chem.* **2000**, *275*, 1823–1828. [CrossRef]

91. Zhang, Y.; Xiong, Y.L.; Bon, C. An activator of blood coagulation factor X from the venom of *Bungarus fasciatus*. *Toxicon* **1995**, *33*, 1277–1288. [CrossRef] [PubMed]

92. Hofmann, H.; Bon, C. Blood coagulation induced by the venom of *Bothrops atrox*. 1. Identification, purification, and properties of a prothrombin activator. *Biochemistry* **1987**, *26*, 772–780. [CrossRef] [PubMed]

93. Goyffon, M. Cassian Bon, 1944–2008. *Rencontres Toxinologie* **2008**, *RT16*, 9–10.

94. Nemecz, D.; Ostrowski, M.; Ravatin, M.; Saul, F.; Faure, G. Crystal Structure of Isoform CBd of the Basic Phospholipase A(2) Subunit of Crotoxin: Description of the Structural Framework of CB for Interaction with Protein Targets. *Molecules* **2020**, *25*, 5290. [CrossRef] [PubMed]

95. Saul, F.A.; Prijatelj-Znidarsic, P.; Vulliez-le Normand, B.; Villette, B.; Raynal, B.; Pungercar, J.; Krizaj, I.; Faure, G. Comparative structural studies of two natural isoforms of ammodytoxin, phospholipases A2 from *Vipera ammodytes* ammodytes which differ in neurotoxicity and anticoagulant activity. *J. Struct. Biol.* **2010**, *169*, 360–369. [CrossRef]

96. Faure, G.; Saul, F. Crystallographic characterization of functional sites of crotoxin and ammodytoxin, potent β-neurotoxins from *Viperidae* venom. *Toxicon.* **2012**, *60*, 531–538. [CrossRef]

97. Faure, G.; Porowinska, D.; Saul, F. Crototoxin from *Crotalus durissus terrificus* and Crotoxin-Related Proteins: Structure and Function Relationship. In *Toxins and Drug Discovery*; Gopalakrishnakone, P., Cruz, L.J., Luo, S., Eds.; Springer: Dordrecht, The Netherlands, 2017; pp. 3–20.

98. Faure, G.; Gowda, V.T.; Maroun, R.C. Characterization of a human coagulation factor Xa-binding site on *Viperidae* snake venom phospholipases A2 by affinity binding studies and molecular bioinformatics. *BMC Struct. Biol.* **2007**, *7*, 82. [CrossRef]

99. Faure, G.; Xu, H.; Saul, F. Anticoagulant Phospholipase A2 Which Bind to the Specific Soluble Receptor Coagulation Factor Xa. In *Toxins and Hemostasis: From Bench to Bedside*; Kini, R.M., Clemetson, K.J., Merkland, F.S., McClane, M.A., Moritta, T., Eds.; Springer: Dordrecht, The Netherlands, 2010; pp. 201–217.

100. Faure, G.; Saul, F. Structural and Functional Characterization of Anticoagulant, FXa-binding *Viperidae* Snake Venom Phospholipases A2. *Acta Chim. Slov.* **2011**, *58*, 671–677.

101. Ostrovski, M.; Žnidaršič, P.P.; Raynal, B.; Saul, F.; Faure, G. Human coagulation factor Xa prevents oligomerization of anticoagulant phospholipases A2. *Toxin Rev.* **2014**, *33*, 42–47. [CrossRef]

102. Ostrowski, M.; Porowinska, D.; Prochnicki, T.; Prevost, M.; Raynal, B.; Baron, B.; Sauguet, L.; Corringer, P.J.; Faure, G. Neurotoxic phospholipase A2 from rattlesnake as a new ligand and new regulator of prokaryotic receptor GLIC (proton-gated ion channel from *G. violaceus*). *Toxicon* **2016**, *116*, 63–71. [CrossRef]

103. Faure, G.; Bakouh, N.; Lourdel, S.; Odolczyk, N.; Premchandar, A.; Servel, N.; Hatton, A.; Ostrowski, M.K.; Xu, H.; Saul, F.A.; et al. Rattlesnake Phospholipase A2 Increases CFTR-Chloride Channel Current and Corrects ΔF508CFTR Dysfunction: Impact in Cystic Fibrosis. *J. Mol. Biol.* **2016**, *428*, 2898–2915. [CrossRef] [PubMed]
104. Ravatin, M.; Odolczyk, N.; Servel, N.; Guijarro, J.I.; Tagat, E.; Chevalier, B.; Baatallah, N.; Corringer, P.J.; Lukács, G.L.; Edelman, A.; et al. Design of Crotoxin-Based Peptides with Potentiator Activity Targeting the ΔF508NBD1 Cystic Fibrosis Transmembrane Conductance Regulator. *J. Mol. Biol.* **2023**, *435*, 167929. [CrossRef] [PubMed]
105. Guenneugues, M.; Ménez, A. [Structures and functions of animal toxins]. *C. R. Seances Soc. Biol. Fil.* **1997**, *191*, 329–344. [PubMed]
106. Ohno, M.; Ménez, R.; Ogawa, T.; Danse, J.M.; Shimohigashi, Y.; Fromen, C.; Ducancel, F.; Zinn-Justin, S.; Le Du, M.H.; Boulain, J.C.; et al. Molecular evolution of snake toxins: Is the functional diversity of snake toxins associated with a mechanism of accelerated evolution? *Prog. Nucleic Acid. Res. Mol. Biol.* **1998**, *59*, 307–364. [CrossRef] [PubMed]
107. Favreau, P.; Le Gall, F.; Benoit, E.; Molgó, J. A review on conotoxins targeting ion channels and acetylcholine receptors of the vertebrate neuromuscular junction. *Acta Physiol. Pharmacol. Ther. Latinoam.* **1999**, *49*, 257–267.
108. Molgó, J.; Aráoz, R.; Benoit, E.; Iorga, B.I. Physical and virtual screening methods for marine toxins and drug discovery targeting nicotinic acetylcholine receptors. *Expert. Opin. Drug Discov.* **2013**, *8*, 1203–1223. [CrossRef]

MDPI AG
Grosspeteranlage 5
4052 Basel
Switzerland
Tel.: +41 61 683 77 34

Toxins Editorial Office
E-mail: toxins@mdpi.com
www.mdpi.com/journal/toxins